2nd International Conference on Machine Learning, Security and Cloud Computing (ICMLSC 2020)

IOP Conference Series: Materials Science and Engineering
Volume 1042

Hyderabad, India
18 - 19 December 2020

Editors:

M.A. Jabbar
Rajanikanth Aluvalu
J.V.R. Ravindra

ISBN: 978-1-7138-3331-4
ISSN: 1757-8981

Printed from e-media with permission by:

Curran Associates, Inc.
57 Morehouse Lane
Red Hook, NY 12571

proceedings.com

Some format issues inherent in the e-media version may also appear in this print version.

Institute of Physics
Dirac House, Temple Back
Bristol BS1 6BE UK

Phone: 44 1 17 929 7481
Fax: 44 1 17 920 0979

techtracking@iop.org

Additional copies of this publication are available from:

Curran Associates, Inc.
57 Morehouse Lane
Red Hook, NY 12571 USA
Phone: 845-758-0400
Fax: 845-758-2633
Email: curran@proceedings.com
Web: www.proceedings.com

TABLE OF CONTENTS

IOP Conf. Series: Materials Science and Engineering 1042 (2021) 011001 doi:10.1088/1757-899X/1042/1/011001

On behalf of the Organizing Committee of ICMLSC-2020, we are delighted to welcome you to the 2nd International Conference on Machine Learning, Security and Cloud Computing. Not only does this conference set out to bring together some of leading academic scientists, researchers and research scholars to exchange and share their experiences and research results on all aspects of Machine Learning, Security and Cloud Computing but also it sets the tone for further research.

What this conference is intended to do is to provide a premier interdisciplinary platform for researchers, practitioners and educators to present and discuss the most recent innovations, trends, and concerns as well as practical challenges encountered and solutions adopted in the fields of Machine Learning, Security and Cloud Computing.

The primary focus of ICMLSC-2020 is to blend both the theory and the applications Machine Learning, Security and cloud computing. This conference is bound to bring together the academia, researchers, developers and practitioners from scientific organizations and industry to share and disseminate recent research findings in the fields of Machine Learning, Security and cloud computing and its applications. ICMLSC-2020 is an outstanding platform to discuss the key findings, exchanging novel ideas, listening to the world-class leaders and sharing experiences with peer groups.

ICMLSC-2020 is organized by Department of Computer Science and Engineering and Department of AI & ML, Vardhaman College of Engineering, Hyderabad, Telangana, India

A total of 128 research papers have been received and 33 papers have been accepted and registered.

We would like to express our sincere gratitude to our Keynote speakers, all the authors, and reviewers. We appreciate the contributions of all the student volunteers who have given their time to help make the conference successful.

Editors:

Dr. M. A. Jabbar, Professor & HOD, CSE (AI & ML), Vardhaman College of Engineering, Hyderabad

Dr. Rajanikanth Aluvalu, Professor & HOD, CSE, Vardhaman College of Engineering, Hyderabad

Dr. JVR Ravindra, Professor, Dept of ECE, Vardhaman College of Engineering, Hyderabad

IOP Conf. Series: Materials Science and Engineering 1042 (2021) 011002 doi:10.1088/1757-899X/1042/1/011002

Peer review declaration

All papers published in this volume of *IOP Conference Series: Materials Science and Engineering* have been peer reviewed through processes administered by the Editors. Reviews were conducted by expert referees to the professional and scientific standards expected of a proceedings journal published by IOP Publishing.

Type of peer review: Double blind

- **Conference submission management system: Microsoft Conference Management system**
- **Number of submissions received: 128**
- **Number of submissions sent for review: 127**
- **Number of submissions accepted:** 37
- **Acceptance Rate (Number of Submissions Accepted / Number of Submissions Received X 100): 28%**
- **Average number of reviews per paper: 3 (Three)**
- **Total number of reviewers involved: 28**
- **Contact person for queries:**

Dr MA.Jabbar
Professor and Head of the Department
Department of AI&ML
Vardhaman College of Engineering
Shamshabad,Hyderabad-501218
T.S,India
Mail id : akhiljabbar@ieee.org

Convolution Neural Network based Rain Noise Removal for Real Time Application

[1]A.Sowjanya,
[2]Sandeep Kumar,
[3] K. Sonali Swaroop

[1]Assistant Professor, Sreyas Institute of Engineering and Technology, India
[2]Professor, Sreyas Institute of Engineering and Technology, India
[3]Assistant Professor Sreyas Institute of Engineering and Technology, India

sowjanya.a@sreyas.ac.in,
er.sandeepsahratia@gmail.com,
sonali.k@sreyas.ac.in

Abstract:

Image quality is affected by rain streaks that have an adverse effect on the applications which involves the ideas of surveillance, automatic path navigation. Hence it is important to process the information present in the images affected by water droplets caused by the rain or the mist. So this forms the basic need to identify the problem which does not have any mathematical correlation about the same. This research tries to work on the methodology which will denoise the criteria based on Convolutional neural network (CNN). CNN modelling helps in finding out the connections between the noise input image and rain streak. The proposed method uses synthetic database available publicly.

Keywords: Rain streaks, Convolutional neural network, Denoising.

1. INTRODUCTION

Bad weather distortion in still images conditions can negatively affect the eye performance of several pictorial structures. Rain belongs to one of inclusion situations that caused substantial distinctions in pictures [1-3]. A study by Garg [4] shows this inclusion situations, drizzle or mist subject to variable climate. A variable climate incorporates components of generally huge dimensions gathered of cameras. Consistent climate like haze elements, are lighter parts and can be minimally recorded. Thus, these atmospheric conditions contributes to difficult variable intensities and tarnishes them evidence that is carried in the image or series of images [5-6]. The existing methods denoise the images using background and foreground images. Notwithstanding, these strategies may neglect to manage certifiable downpour inferable from the unpredictability of rain for example, the overlapping between downpour streaks and complex surface [7-10]. As of late, because of the amazing elemental analysis of convolutional neural network (CNN),deep learning based deraining strategies [11-12] have been proposed to eliminate downpour streaks in a solitary picture and accomplished promising execution [13-15]. They put forth an attempt to learn portrayals of various rain layers and afterward deduct these layers from the input picture.

Rest of the paper summarized as: section II represents the literature work of existing algorithms. Section III represents the proposed methodology. Section IV presents the result of the proposed methodology. Finally Section V represents the conclusion of overall framework

2. LITERATURE SURVEY

This section describes the related work on raindrop removal on still images. In the past few years, significant enhancement was observed in rain drop removal. Our work mainly focuses on rain drop removal on still images as it is very challenging using deep learning technique.

Table 1: Literature work of existing methodology based on rain noise removal

Year	Method	Pros/Cons	Dataset	Parameters Calculated
2018 [5]	1. Attempt to separate rain streaks from background using wavelet Transform. 2. Removal of haze veil using dark channel.	1. Better results are obtained. 2. Achieves better rain and haze veil removal.	NYU Depth V2	PSNR, SSIM, NIQE
2017 [6]	Rain removal is performed using deep learning networks.	Removes rain streaks even in extreme cases.	Rain12, Rain 100H, Rain 100L	PSNR, SSIM
2019 [7]	Supervised and Unsupervised Learning	Doesn't remove rain streaks in extreme conditions	Google images	PSNR
2019 [8]	Attempt to separate rain streaks from background	Better results obtained.	Own dataset	SSIM
2018 [1]	Convolutional Neural Network	It is effective and robust.	Rain12, BSD300	SSIM
2019 [2]	De-raining unit and dense connections are used	Better results obtained.	Rain100H, Rain100L, Rian1200	PSNR

2019 [9]	Multi-task deep network model	Removes rain streaks even in extreme cases.	Rain12, Rain100L, Rain100H, Rain800	PSNR, SSIM
2019[10]	3-stage pyramid: Heavy rain removal, Medium rain removal and low rain removal are employed to remove rain streaks	Achieves superior deraining performance	Own dataset	PSNR, SSIM
2020[3]	Coarse-fusion model, fine-fusion model, reconstruction model are used for rain streak removal	Better results obtained.	Rain800, Rain100H, Rain100L, Rain1200 Rain12	PSNR, SSIM, FSIM

A wide range of research has been carried out to denoise the images affected by the rain drops or mist. Most of the previous research was carried out using supervised learning and deep learning techniques, coarse fusion model etc. All the work carried out has minimal results under extremes conditions and hence we propose the CNN modelling to obtain the better results.

3. PROPOSED METHODOLOGY

A novel CNN algorithm is proposed for rain drop removal which also improves the performance. The input image is captured from the standard database as an RGB image, the basic stages for rain removal using image processing are as shown in Fig. 1.

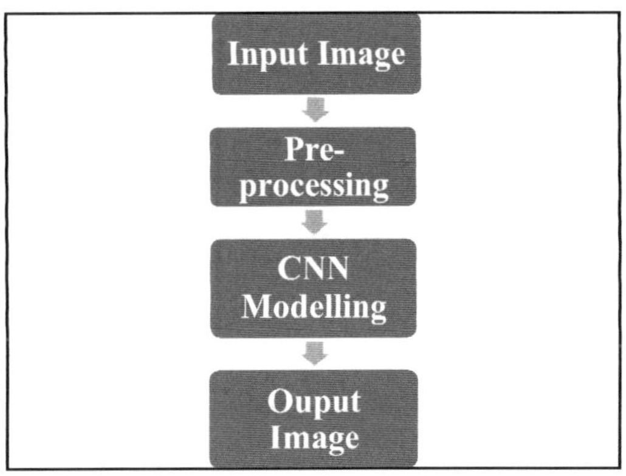

Fig. 1: Flow chart of the proposed model

3.1 Pre-processing:

To enhance or remove noise from the input image different pre-processing methods are used. Adaptive median filtering is a non-linear digital filtering technique employed to eliminate noise. It is very effective method as it preserves the edges and smoothens non-impulsive noise.

Fig. 2 (a) Input Image (b) Segmented Image

Algorithm: ADM

1. Initially check if intensity of the pixel is equal to the max of min values.
2. Find the distance of the pixel from the origin if it is a valid pixel and if it lies between the noise probabilities.
3. Let the distance be r and window size is now assumed as 2r+1.
4. The median is computed after the window size is found.

The main aim of Adaptive mean filter is to remove noise from the image using standard mean filter. The main difference between these filters lies in variable dimensions of the mask. Based on the average pixels in the current mask this distinction depends. The adaptive mean filter deals the corrupted image filtering process with a potential noise impulse higher than 0.2.

3.2 CNN Modelling:

CNN is a deep learning model which assigns importance to objects in input image and distinguishes it from other. CNN performs image identification and image recognition. CNN contains four layers: convolutional layer, pooling layer, fully connected layer and activation layer.

At first each information picture will go through a progression of convolution layers with channels (Kernals), Pooling, Fully associated layers (FC).Finally,softmax work, which is an activation function is applied to order an item with probabilistic qualities somewhere in the range of 0 and 1. The below figure Fig. 3 is a progression of CNN to handle a picture and arranges the articles dependent on values.

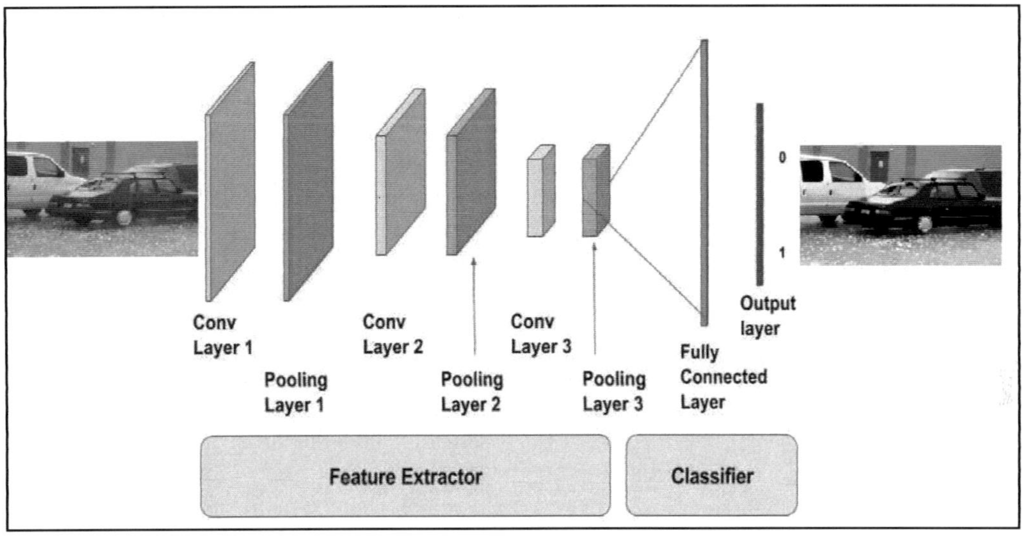

Fig. 3: Architecture of the Proposed Methodology

- *Convolutional layer*: A "kernal" is rolled over the image, taking a few pixels at a time and creating a vector that identifies the set to which every property belongs.
- *Pooling layer (down sampling):* It diminishes the amount of data in each component acquired in the convolutional layer and jam the most prominent data
- *Fully connected input layer (flatten):* It takes the yield of the past layers, "flattens" them and transforms them into a solitary vector that can be a contribution for the following stage
- *The first fully connected layer:* It takes the contributions from the element investigation and applies loads to foresee the right mark.
- *Fully connected output layer*: gives the finishing probabilities for every label.

3.3 Softmax function: It is a normalized exponential function. It is used as an activation function to normalize the output of system to a probabilistic distribution over a predicted output division.

The softmax function for the input vector x of k real numbers is defined as:

$$\sigma(x_i) = \frac{e^{x^i}}{\sum_{j=1}^{k} e^{x^j}} \quad \text{for i=....k and x= } (x_1,\ldots x_k) \in R^k$$

4. RESULTS

4.1 Evaluation Parameters:

4.1.1 Peak signal to noise ratio (PSNR) is ratio between the maximum signal power to the power of noise signal. It is generally used to calculate the quality of reconstructed image that have been compressed. PSNR is most easily defined via the mean squared error (MSE). The MSE for a given image I with pxq rows and columns is defined mathematically as:

$$MSE = \frac{1}{pq}\sum_{r=0}^{a-1}\sum_{s=0}^{b-1}[I(r,s) - K(r,s)]^2$$

$$PSNR = \log_{10}\frac{MAX_I}{\sqrt{MSE}}$$

Where MAXI is maximum possible pixel value of image.

Structural similarity index measure (SSIM) is a metric which has three components which are luminance(l), contrast(c), structure(s) which are defined mathematically as follows.

$$\mathrm{l(a,b)} = \frac{(2\mathrm{m_a}\ \mathrm{m_b} + \mathrm{d_1})}{(\mathrm{m_a^2} + \mathrm{m_b^2} + \mathrm{d_1})}$$

$$\mathrm{c(a,b)} = \frac{(2\mathrm{v_a v_b} + \mathrm{d_2})}{\mathrm{v_a^2} + \mathrm{v_b^2} + \mathrm{d_2}}$$

$$\mathrm{s(a,b)} = \frac{\mathrm{v_{ab}} + \mathrm{d_3}}{\mathrm{v_a v_b} + \mathrm{d_3}}$$

m_a is mean value of a

m_b is mean value of b

v_a^2 is variance of a

v_b^2 is variance of a

$v_a v_b$ is covariance of a & b

d_1=(n1,T)2

d_2= (n2, T)2

d_3= $d_2/2$

n1=0.01 and n2=0.03 by default

T is the dynamic range of pixel values

Now, SSIM is weighted combination of these measures:

$$\mathrm{SSIM(a,b)} = [\mathrm{l(a,b)}^{\delta}\mathrm{c(a,b)}^{\vartheta}\mathrm{s(a,b)}^{\theta}]$$

Whereδ,ϑ,θ is 1.

4.2 Datasets:

In the proposed work synthetic dataset used which is publically available. This dataset contains more than 4k images. Sample images of synthetic datasets are shown in Fig. 4.

Fig. 4: Sample datasets of Rain (Synthetic Datasets)

4.3 Results:

The experimentation was carried in MATLAB 2015 software. The system specifications on which the images are processed includes 4 GB RAM with NVDIA graphic card.

Table 2: Comparison of Proposed work with PSNR & SSIM parameters with existing methodologies

S. No.	**Author Name**	**Parameters Evaluated**	
		PSNR(db)	**SSIM**
1.	Chen et al. [1], 2018	-	0.94
2.	Liang et al. [5], 2018	30.19	0.95
3.	Yupei et al. [10], 2019	28.3	0.91
4.	Yang et al. [6],2017	36.05	0.96
5.	Kui et al.[3], 2020	22.48	0.904
6.	Wei et al. [7], 2019	26.98	-
7.	Proposed Work	21.81	0.971

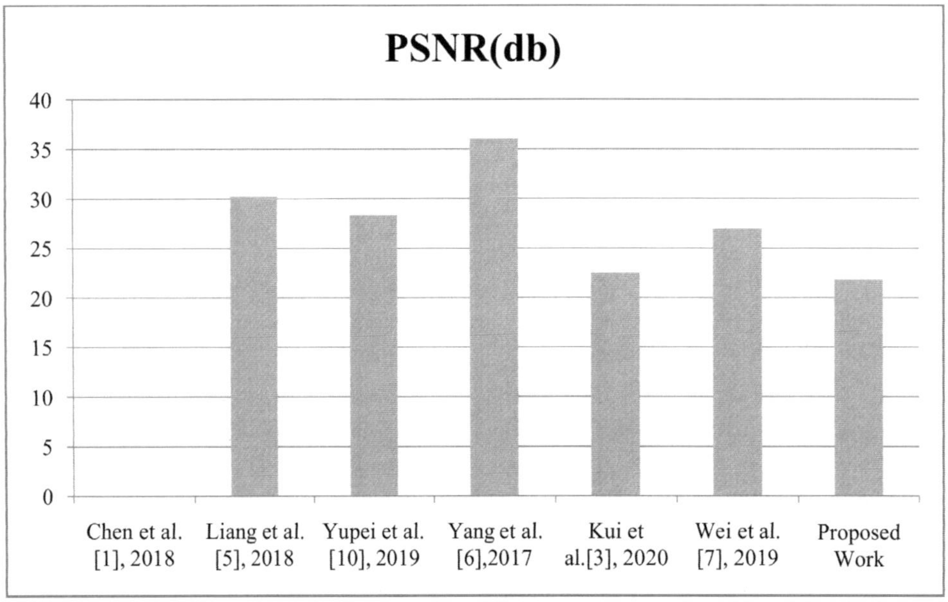

Fig. 5: Comparison of Proposed work with PSNR parameters with existing methodologies

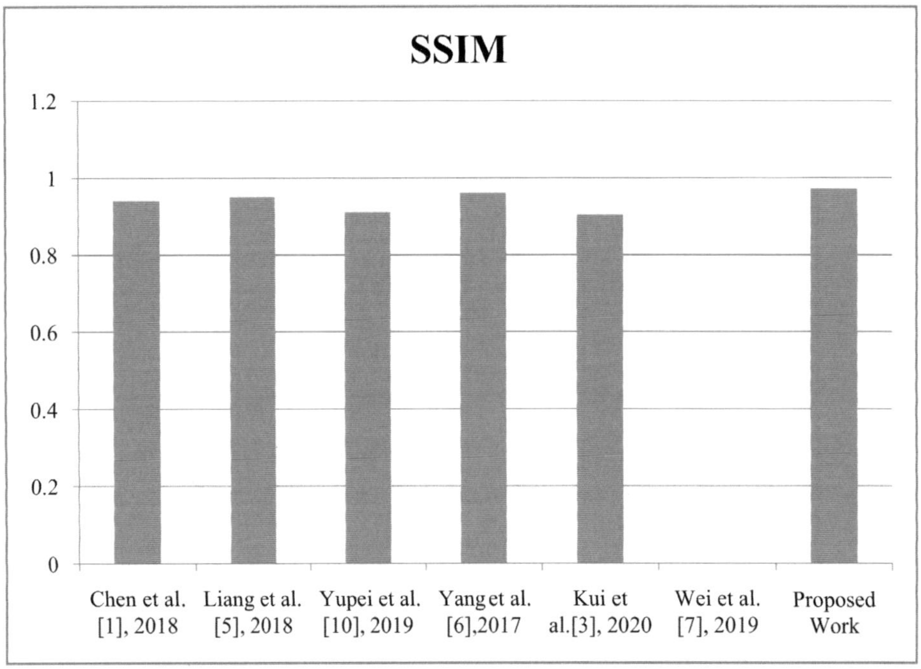

Fig. 6: Comparison of Proposed work with SSIM parameters with existing methodologies

5.CONCLUSION

The paper mainly focused on the effective noise removal caused by the rain streaks in the still images using convolution neural networks. The results are vivid and conclusive that the proposed methodology has a very good yield. The similarity index has increased to 0.971 when compared to the previous work. The PSNR values obtained is 21.81dB which is in the acceptable range comparatively. The results can be more effectively used to improve the quality of the images using hybrid models.

References

[1]Chen, T., & Fu, C, "Single-image-based Rain Detection and Removal via CNN", Journal of Physics: Conference Series, 1004, 012007, 2018.

[2] Cong Wang, Man Zhang, Jinshan Pan, ZhixunSu, "Single image rain removal via densely connected contextual and semantic correlation net", Journal of Electronic Imaging 28 (3), 033018, 2019.

[3] KuiJiang,Zhongyuan Wang, Peng Yi, Chen Chen, Baojin Huang, Yimin Luo, Jiayi Ma, Junjun Jiang,"Multi-Scale Progressive Fusion Network for Single Image Deraining", 28 Mar 2020.

[4]K. Garg and S. K. Nayar, "Detection and removal of rain from videos," Proceedings of the 2004 IEEE Computer Society Conference on Computer Vision and Pattern Recognition, 2004. CVPR 2004., Washington, DC, USA, 2004, pp. I-I.

[5]Liang Shen ,Zihan Yue , Quan Chen, Fan Feng and Jie Ma "Deep joint rain and haze removal from a single image" IEEE 2018 24th International Conference on Pattern Recognition (ICPR) Beijing, China, August 20-24, 2018.

[6]W. Yang, R Tan, J. Feng, J. Liu, Z. Guo and S. Yan, "Deep joint rain detection and removal from a single image", IEEE Conference on Computer Vision and Pattern Recognition (CVPR-2017), pp. 1685-1694, Honolulu, HI, USA, July 21-26, 2017.

[7]Wei Wei,Deyu Meng, Qian Zhao , Zongben Xu , Ying Wu, "Semi-supervised Transfer Learning for Image Rain Removal", IEEE/CVF Conference on Computer

Vision and Pattern Recognition (CVPR-2019). Pp. 3872-3881,USA,June15-20, 2019.

[8]Y. Himabindu, R. Manjusha, Latha Parameswaran, "Detection and Removal of RainDrop from Images Using DeepLearning", ICCVBIC 2019, pp. 1355-1362.

[9] Wenhan Yang, Robby T Tan, Jiashi Feng, Zongming Guo, Shuicheng Yan, Jiaying Liu, "Joint rain detection and removal from a single image with contextualized deep networks", IEEE transactions on pattern analysis and machine intelligence 42 (6), 1377-1393, 2019.

[10] Yupei Zheng, Xin Yu,Miaomiao,Shunli Zhang, "Residual Multiscale Based Single Image Deraining", bmvc2019.

[11] Xueyang Fu, Jiabin Huang, Xinghao Ding, Yinghao Liao, and John Paisley. Clearing the skies: A deep network architecture for single-image rain removal. IEEE Transactions on Image Processing, 26(6):2944–2956, 2017.

[12] Yu Li, Robby T Tan, Xiaojie Guo, Jiangbo Lu, and Michael S Brown. Rain streak removal using layer priors. In Proceedings of the IEEE conference on computer vision and pattern recognition, pages 2736–2744, 2016.

[13] Ponnapalli, Venkata A. Sankar, VY Jayasree Pappu, and B. Srinivasulu. "Design of thinned rhombic fractal array antenna using GA and PSO optimization techniques for space and advanced wireless applications." In Microelectronics, Electromagnetics and Telecommunications, pp. 719-727. Springer, Singapore, 2018.

[14] Subbarao, D. "The Influence Of Electronic Communication On Machine Learning." International Journal of Advanced Research in Computer Science 2.3 (2011).

[15] Prasad, Kantipudi MVV, and H. N. Suresh. "Spectral Estimation Using Improved Recursive Least Square (RLS) Algorithm: An Investigational Study." Emerging Research in Computing, Information, Communication and Applications. Springer, New Delhi, 2015. 363- 376.

Artery-Vein Detection using Neural Network in Retina Images

[1] K. Sonali Swaroop,
[2]Sandeep Kumar,
[3] A.Sowjanya
[1]Assistant Professor, Sreyas Institute of Engineering and Technology, India
[2]Professor, Sreyas Institute of Engineering and Technology, India
[3]Assistant Professor Sreyas Institute of Engineering and Technology, India
sonali.k@sreyas.ac.in,
er.sandeepsahratia@gmail.com
sowjanya.a@sreyas.ac.in

ABSTRACT:

Retinal veins segmentation play a vital role in the life because of entire world moves toward the digital words. The paper presents an approach to identify the blood vessels from the optically affected areas of an eye. This method is proposed as the manual extraction of vessels from the fundus image is a very tedious task also it takes higher amount of time. Therefore, the paper mainly focuses on the developing an approach for vessel extraction. The methodology reportedly excels in terms of accuracy also being simple it finds many applications in image diagnosis for predicting various pathological diseases opathalmologically. It can create a new perspective in disease detection in a more accurate way. The input images are initially preprocessed; next the features are extracted from the images. The extracted images form the training set and the testing process is performed using the proposed methodology. Finally, the proposed work achieved 96.8% accuracy on benchmark datasets.

Keywords: Fundus images, vessel extraction, features, ophthalmologic diseases, segmentation.

1. INTRODUCTION

Day by day demands are increasing in medical sectors because of huge populations in the world [1-3]. In medical field, Retinal veins segmentation play a vital role in the life because of entire world moves toward the digital words [4]. For the most part project a minor variance outward circulation and show up as a wire work like pattern or nodes-like pattern [5-7], [17]. Their variant highlights, for example, length and width, is critical in the early location and treatment of various angiocardiopathy and visual infections, for example, stroke, vein impediments, diabetes and arteriosclerosis [4], [9-10], [20]. The investigation of morphological highlights of retinal veins is helpful for distinguishing and rewarding an ailment in time when it is still in its beginning phase. Since angiocardiopathy and visual illnesses seriously affect human's life, the investigation of retinal veins is of incredible hugeness in numerous clinical applications to uncover significant data of fundamental infections and bolster conclusion and treatment [11-13]. Accordingly, the necessity of vessel examination framework develops quickly, in which vessel division is the first and one of the most significant advances [14-15].

Vessel division is a first process needed for the numerical investigation of optically affected retinal pictures [10], [16-17]. The divided vascular tree can be utilized to extricate the qualities of veins, for example, length, width, stretch and edges. Besides, the vascular pattern has been received in multimodal retinal picture enlistment [18-20] and retinal mosaic [11] as the steadiest element in the pictures. In the improvement of a PC helped demonstrative framework for ophthalmic disarranges, programmed division of retinal vessels has been acknowledged as an essential and testing step [7], [21-22]. The size and shape of retinal vessels can fluctuate colossally in various neighborhoods. The width of veins frequently runs from 1 to 20 pixels, contingent upon both the anatomical width of the vessel and the picture goals. The presence of vessel intersection, fanning and centerline reflex makes it troublesome [23-25] to section the vessels precisely utilizing misleadingly structured highlights. Pathologies as sores and exudates can additionally muddle the programmed division. In the last decades, various researchers proposed different techniques and they can be partitioned into two classifications: unaided and managed strategies. The main objective of this research is to identify the artery/vein in the retinal images such that the veins can further be processed to identify the pathologies in the fundus images. These pathologies can be used for early detection of various impediments of the human body.

This paper summarizes the following information: section II represents the literature work of existing algorithms. Section III represents the proposed methodology. Section IV describes the result of the proposed methodology. Finally Section V represents the conclusion of overall framework.

2. LITERATURE REVIEW

In everyday life, medical sector play an important role because of our hectic schedule in the real life. Hence no of researcher and scientist are working in the field of retinal vain detection. Various authors done good work in the field of retinal detection and few of description are shown in Table 1.

Table 1: Literature work of Retinal Artery/Vein Separation techniques

S.No	Name of author	Year	Methodology	Accuracy
1	Rolando Estrada et al. [13]	2014	Vessel topology estimation	92%
2	Vinayak Joshi Joseph et al. [19]	2014	Graph search	91%
3	Qing shao, et al. [24]	2014	Spectral domain – optical coherence tomography (SD-OCT)	86.4%
4	Nima Hatami, et al. [23]	2015	Local binary pattern (LBP)	75%
5	Shishir Maheshwari et al. [15]	2016	Empirical wavelet transform and correntropy	96.6%
6	Mei zhou et al. [25]	2017	Luminosity and contrast adjustment	90%
7	Fan Huang et al. [3]	2018	genetic-search feature selection	90.2%
8	Sehrish qummar et al. [14]	2019	5 CNN models ensemble (resnet50, inception v3, xception , dense121, dense169)	85%
9	Dong ming li et al. [8]	2019	dual-tree complex wavelet transform and morphology	85%
10	Ahmed Soomro et al. [16]	2019	Fuzzy c mean , CNN	95%
11	Pan xiuquin et al.	2019	Residual module and u-net network combination	96%

12	Yuhao luo et al. [21]	2019	Cascade refined u net	97%
13	Temitope mapayi et al. [18]	2015	Global thresholding technique	95%
14	Daniel jiminez et al. [5]	2017	Graph cut methodology , computed tomography	85%
15	Fan huang et al. [3]	2017	Linear discriminate analysis, reflection feature in fundus image	86.9%

According to the Table 1, it can be concluded that research has been performed on the retinal data which can be used to detect various pathologies. Various researchers has followed methods like vessel search topology estimation, graph search method, LBP, Genetic search, Empirical wavelet transform and core entropy etc. out of which wavelet transformation has given the maximum accuracy till now but the system complexity is more in core entropy[9].

3. SYSTEM MODEL

A novel CNN algorithm is proposed, in order to eliminate the complexity as well as improve the accuracy of Retinal Artery model. Input image has been taken from the standard benchmark database as an RGB image. The features maps are obtained through CNN Algorithm as shown in the Figure 1.

3.1 Pre-processing:

The input image is fed to the processing unit. Since the inserted image will be a complex image, it must be converted to a grayscale image. This process is called pre- processing. The next step is to remove noise in the input image using an adaptive median filter.

Figure 1: Flow chart of the proposed model

3.1.1 Adaptive Median Filtering:

In real time applications, the input images are generally associated with the noise components which cause some complexity in the analysis of the system and hence the noise has to be removed. The process has to be done such that the mean square error is minimal. The most traditional way of removing noise is by using median filter which generally uses bubble sort. The number of computations used here increases as the dimension of the frame changes, also the dimension of for every application is fixed. For instance, if a window of size 3*3 is used then it requires 30 computations to calculate the median of the sorted array.

The adaptive median filter is made to eliminate errands with a standard median filter. The variable dimensions of the mask are the main distinction between the two filters. This difference is based on the average pixels in the current mask. If the average value is shot noise, the dimension of the window will expand. Otherwise, the image performs additional processing within the specifications of the current window.

Accordingly, the Adaptive Media Filter eliminates the image point noise and reducing image degradation. It can handle the damaged image filtering process with a potential noise impulse higher than 0.2. In particular, if the pixel p satisfies $I(s,t) = I_{max}$ or $I(s,t) = I_{min}$, hence point p is considered as impulse point, for by and large the value of noise is neighborhood greatest or nearby least. If $I(s\pm m,t\pm n)$ point is valid pixels, $I(s\pm m,t\pm n)<a$ and $I(s\pm m,t\pm n)> b$ assign its position (m,n) and find the distance from the origin I(s,t) denoted by r, and it can be represented by

$$r=\begin{cases} |s-m|, |s-m| \geq |t-n| \\ |t-n|, |s-m| \geq |t-n| \end{cases}$$

where a and b are the probabilities of noise and m,n are the coordinates of the pixel values .

Once the r value is calculated for one origin pixel, the same process is continued for all the window sizes and the most repeated value is takes as the magnitude. And the window size is 2r+1. After the calculation of the window size, the median value is now calculated for the sorted sequence.

Algorithm: AMF

1. Check if intensity of the pixel is equal to the max of min values
2. Compute the distance of the pixel from the origin if it is a valid pixel and if it lies between the noise probabilities
3. Let the distance be r
4. The window size is now assumed to be 2r+1
5. The median is computed after the window size is computed.

3.2 CNN Modeling:

CNN includes the knowledge of biology, mathematical modeling and computer sciences attached to it.CNN comes under the class of deep learning or deep neural networks that is most commonly applied to access medical images.

CNN handles the tasks of image recognition and image classification. CNN classifications used an input image, analyses it and distinguish it under certain categories. A computer sees an input image as array of pixels and it depends on the image resolution. CNN working procedure can be classified as 2 basic steps. They are:

- Feature learning

- Connected layers

There are four main layers in CNN. They are conversion convolution layer, pooling layer, activation functional layer and fully connected layer as shown in Figure 2.

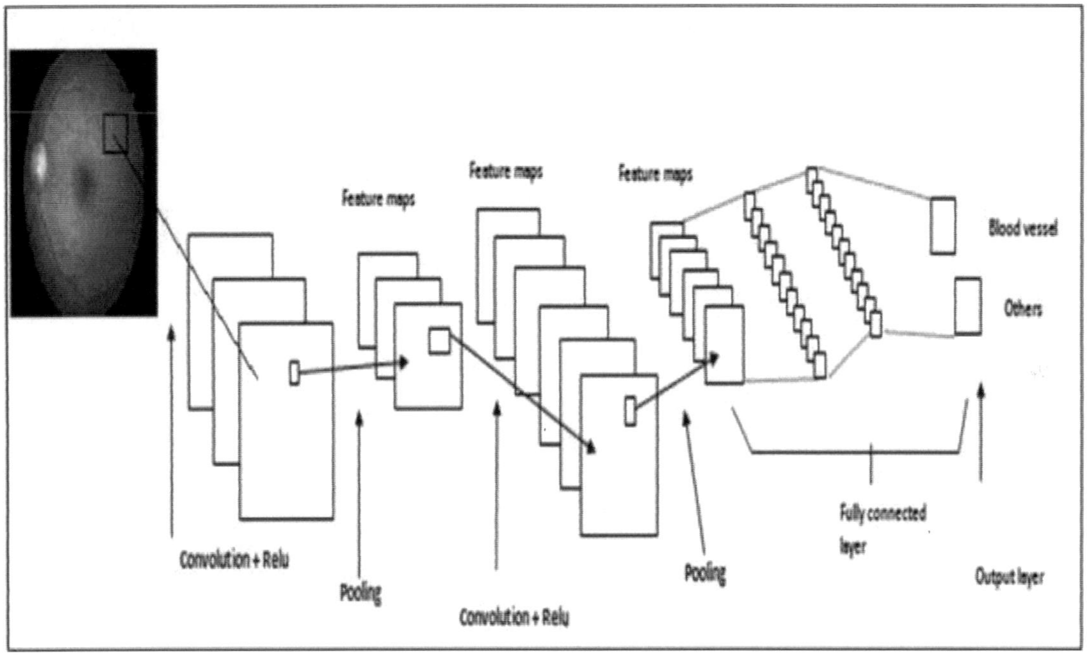

Figure 2: Architecture of the Proposed Methodology

3.2.1 Convolution + Relu

Convolution is the primary process to calculate the properties of an retinal portion of the eye. Convolution captures the information between pixels by stored picture properties utilizing minimum masks of data. It is a quantitative analysis which uses two information sources, for instance, image matrix and a filter or kernel. The matrix multiplication operation can be done between input (p) and filter (k) of the image. The deep convolution filter should match with the depth of the image. Higher the number of different filters higher the feature maps. All this conversation operations text at a place known as receptive field. The processor performs image classification by considering low level features such as edges and curves, and then creating more detailing concepts by a series of convolution layers. That is, by change of filter matrix we can obtain the edges, blur and sharpening etc.. and if a filter does not match

IOP Conf. Series: Materials Science and Engineering 1042 (2021) 012002 doi:10.1088/1757-899X/1042/1/012002

perfectly padding can be used ReLu can be modeled as noisy ReLUs by the following mathematical model

$$f(i)=max(0,i+N) \text{ with } N\sim Noise((0,\sigma(x)))$$

3.2.2 POOLING:

According to Max-Pooling layers, it takes the biggest value in the window size (m,n) for the further procedure as shown in Figure 3 and it also performs noise suppressant. While down sampling, it discards the noise value through activations function. Hence, this process also covers dimensionality reduction. This convolution and pooling is repeated for several times until the required level is reached which is already pictured in the structure.

(a) 4x4 Input Image (b) 2x2 Max. Pool Image

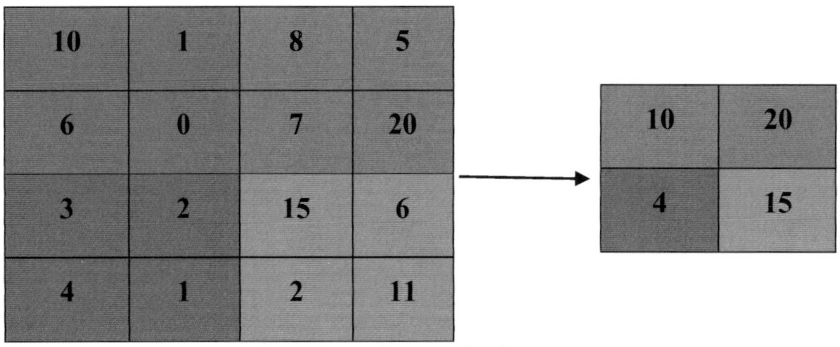

Figure 3: Pooling mechanism

3.2.3 Activation Functional Layer & Fully Connected Layer

It gives a curvilinear connection among input and output layers. This layer influences the system execution. Non-direct learning of the system happens through activation function. There are various kinds of enactment capacities, for example, Linear, Sigmoid, tangent, however the nonlinear ReLU (Rectified Linear Unit) actuation work is normally utilized in CNN. At last, fully connected layers required for classification of the retinal veins.

Assume a CNN with Y layers where the final 3 layers are fully connected layers. Likewise assume αl mean the pace for y^{th} layer in the system. It calibrates just the updated part of the system for $\alpha l = 0$ and $y \neq Y$. The degree changes relates to preparing a linear classifier with the features created in layer Y-1 . Similarly, the last 2 layers of the system can be calibrated by setting αl =0 for y≠Y , Y-1 . This degree of adjusting relates to preparing a counterfeit neural system with 1 concealed layer, which can be seen as preparing a nonlinear classifier utilizing the highlights created in layer Y-2 . Also, adjusting layers Y, Y-1, and Y-2 are basically comparable to preparing a artificial neural system with 2 hidden layers. Counting the past convolution layers in the update procedure further adjusts the pre-prepared CNN to the current application however may require increasingly marked preparing information to abstain from over fitting. By and large, the early layers of a CNN learn low level picture features, which are relevant to most vision assignments, yet the late layers learn significant level features, which are explicit to the current application.

To close, with recovery of CNNs attributable to the improvement of incredible processing strategies, the clinical imaging writing has seen another age of computer- aided identification frameworks that show improved execution. Examples: polyp discovery for colonoscopy recordings, PC helped location of aspiratory embolism (PE) in CT datasets, programmed recognition of mitotic cells in histopathology pictures, PC supported identification of lymph hubs in CT pictures, and PC supported life systems location in CT volumes. Utilizations of CNNs in clinical picture examination are not restricted to just computer-aided detection systems, in any case and thus is utilized in arrangement of retinal veins

3.2.4 Softmax function:

It is a normalized exponential function. It is used as an activation function to normalize the output in the form of probability distribution. The softmax function σ for the input vector x of k real numbers is defined as:

$$\sigma(x_i) = \frac{e^{x^i}}{\sum_{j=1}^{k} e^{x^j}} \quad \text{for i=....k and x= } (x_1,\dots x_k) \in R^k$$

3.3 Post Filtration:

At last the post-filtration operations applied on the processed image so that the output is much presentable and is easy to be understood by the end user. This includes the background exclusion also.

4. NUMERICAL RESULTS

4.1 Data Sets & Evaluation Parameter:

The data sets used for the experimentation are DRIVE and STARE. The images used in this data set are of JPEG standards. The images were provided by the screening program in Netherlands. The program constituted of the data from the people belonging to the age group of 25-40 years. The images are 8 bit images with 768*584 dimensions. The Stare database was provided by university of California. Blood vessel segmentation datasets consists of images from 30 people. Out of the total samples present in the database 70% of the images were the training samples and 30% were used as test samples.

Evaluation parameters used is

$$\text{ACCURACY} = \frac{TP + TN}{TP + TN + FP + FN}$$

where TP, FN, FP and TN represent the number of true positives, false negatives, false positives and true negatives, respectively.

4.2 Comparison of Results:

The entire experiment is performed on a PC setup having compatibility of 8GB RAM and NVIDA graphic card interface. The images were processed in MATLAB 2015 software.

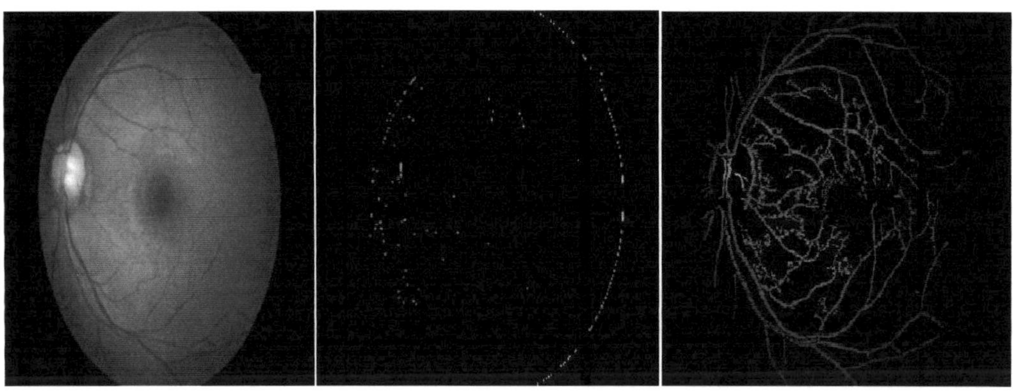

Figure 4: (a) Input image (b) Pre-processed image (c) Segmentation Results

Table 2: Comparison of Proposed work with existing methodology

S.No	Name of author	Methodology	Accuracy
1	Nima Hatami, et al. [23], 2015	Local binary pattern (LBP)	75%
2	Shishir Maheshwari et al. [15], 2016	Empirical wavelet transform and correntropy	96.6%
3	Mei zhou et al. [25], 2017	Luminosity and contrast adjustment	90%
4	Fan Huang et al. [3], 2018.	genetic-search feature selection	90.2%
5	Dong ming li et al. [8], 2019	dual-tree complex wavelet transform and morphology	85%
6	Ahmed Soomro et al. [16], 2019	Fuzzy C Mean	95%
7	**Proposed Work**	**CNN**	**96.8%**

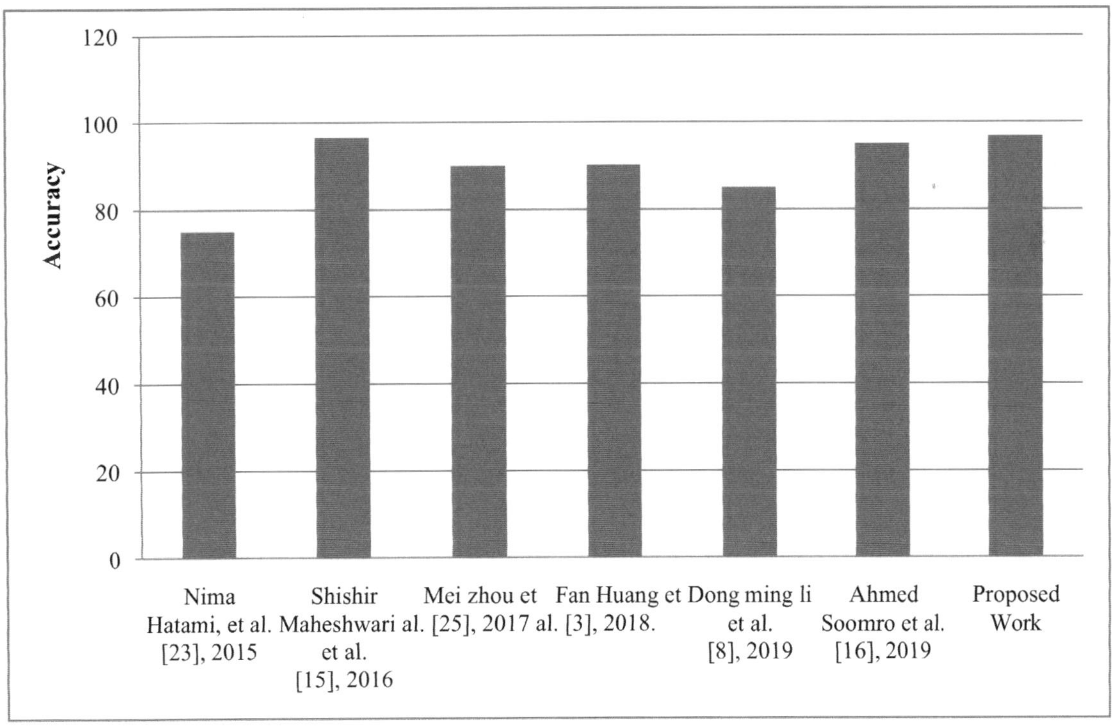

Figure 5: Segmentation Results

5. CONCLUSION

In this paper, the proposed work used AMF & convolution based network to improve the accuracy of vein detection. The proposed method was estimated on publically available DRIVE and STARE benchmark databases. Finally, the proposed work gives better results on vein detection as compare to existing methods. Overall, the proposed work achieved 96.8% accuracy on benchmark datasets. In future, we want to improve the algorithm with deep learning process so that retinal based module will be more efficient & precise in the real life.

REFERENCES

[1]. C. A. Lupas, D. Tegolo, and E. Trucco, ―FABC: Retinal vessel segmentation using AdaBoost,‖ *IEEE Trans. Inf. Technol. Biomed.*, vol. 14, no. 5, pp. 1267–1274, Sep. 2010.

[2]. D. Marín, A. Aquino, M. E. Gegúndez-Arias, and J. M. Bravo, ―A new supervised method for blood vessel segmentation in retinal images by using gray-level and moment invariants-based features,‖ *IEEE Trans.Med. Imag.*, vol. 30, no. 1, pp. 146– 158, Jan. 2011.

[3]. Huang, F., Dashtbozorg, B., Tan, T. and ter Haar Romeny,B.M., 2018. Retinal artery/vein classification using genetic-search feature selection. Computer methods and programs in biomedicine, 161, pp.197-207.

[4]. J. J. Kanski and B. Bowling, *Clinical Ophthalmology: A Systematic Approach*. London, U.K.: Elsevier Health Sciences, 2011.

[5]. J. Staal, M. D. Abramoff, M. Niemeijer, M. A. Viergever, and B. van Ginneken, ―Ridge-based vessel segmentation in color images of the retina,‖ *IEEE Trans. Med. Imag.*, vol. 23, no. 4, pp. 501–509, Apr. 2004.

[6]. Jimenez-Carretero, D., Bermejo-Peláez, D., Nardelli, P.,Fraga, P., Fraile, E., Estépar, R.S.J. and Ledesma-Carbayo, M.J., 2019. A graph-cut approach for pulmonary artery- vein segmentation in noncontrast CT images. Medical image analysis, 52, pp.144-159.

[7]. J. V. B. Soares, J. J. G. Leandro, R. M. Cesar, H. F. Jelinek, and M. J. Cree, ―Retinal vessel segmentation using the 2-D Gabor wavelet and supervised classification,‖ *IEEE Trans. Med. Imag.*, vol. 25, no. 9, pp. 1214–1222, Sep. 2006.

[8]. Li, D., Zhang, L., Sun, C., Yin, T., Liu, C. and Yang, J., 2019.Robust retinal image enhancement via dual-tree complex wavelet transform and morphology-based method. IEEE Access, 7, pp.47303-47316.

[9]. Maheshwari S., Pachori R.B., Kanhangad V., Bhandary .V., AcharyaU.R. ‒Iterative variational mode decomposition based automateddetection of glaucoma using fundus images‖ Computers in Biology and medicine, 88 , pp. 142-149.2016.

[10]. M. W. K. Law and A. C. S. Chung, ‒Segmentation of intracranial vessels and aneurysms in phase contrast magnetic resonance angiography using multirange filters and local variances,‖ *IEEE Trans. Image Process.*, vol. 22, no. 3, pp. 845–859, Mar. 2013.

[11]. M. M. Fraz *et al.*, ‒Blood vessel segmentation methodologies in retinal images—A survey,‖ *Comput. Methods Programs Biomed.*, vol. 108, no. 1, pp. 407–433, 2012.

[12]. Qummar, S., Khan, F.G., Shah, S., Khan, A., Shamshirband,S., Rehman, Z.U., Khan, I.A. and Jadoon, W., 2019. A Deep Learning Ensemble Approach for Diabetic Retinopathy Detection. IEEE Access, 7, pp.150530-150539.

[13]. Rolando Estrada, Priyatham mettu, Sina farsiu, "Vesseltopology estimation", 2014.

[14]. Sehrish qummar, Sajid shah, Ahmad khan, "5 CNNmodels ensemble (resnet50,inception v3, xception,dense121, dense169)", 2019.

[15]. Shishir Maheshwari, Ram bilas, Rajendra acharya,"Empirical wavelet transform and correntropy", 2016.

[16]. Soomro, T.A., Afifi, A.J., Shah, A.A., Soomro, S., Baloch, G.A., Zheng, L., Yin, M. and Gao, J., 2019. Impact of Image Enhancement Technique on CNN Model for Retinal Blood Vessels Segmentation. IEEE Access, 7, pp.158183-158197.

[17]. S. Abbasi-Sureshjani, M. Favali, G. Citti, A. Sarti, and B. M. ter Haar Romeny, ‒Curvature integration in a 5D kernel for extracting vessel connections in retinal images,‖ *IEEE Trans. Image Process.*, vol. 27, no. 2, pp. 606–621, Feb. 2018.

[18]. Temitope mapayi, Serestina veriri, Raymond tapamo,"Global thresholding technique", 2015.

[19]. Vinayak joshi, Joseph, Mona K Garvin, "Graphsearch", 2014.12. Xiuqin, P., Zhang, Q., Zhang, H. and Li, S., 2019. A fundus retinal vessels segmentation scheme based on the improved deep learning U-Net model. IEEE Access, 7,pp.122634-122643.

[20]. Y. Cheng, X. Hu, J. Wang, Y. Wang, and S. Tamura, ‒Accurate vessel segmentation with constrained B-snake,‖ *IEEE Trans. Image Process.*, vol. 24, no. 8, pp. 2440–2455, Aug. 2015.

[21]. Yuhao luo, Jian bo mao, Rui zheng, "Cascade refined unet", 2019.

[22]. Zhang, S., Zheng, R., Luo, Y., Wang, X., Mao, J.,Roberts, C.J. and Sun, M., 2019. Simultaneous arteriole and venule segmentation of dual-modal fundus images using a multi-task cascade network. IEEE Access, 7, pp.57561-57573.

[23]. Hatami N, Goldbaum M. Automatic identification of retinal arteries and veins in fundus images using local binary patterns. arXiv preprint arXiv:1605.00763. 2016 May 3.

[24]. Schuman JS. Spectral domain optical coherence tomography for glaucoma (an AOS thesis). Transactions of the American Ophthalmological Society. 2008 Dec;106:426.

[25]. Zhou M, Jin K, Wang S, Ye J, Qian D. Color retinal image enhancement based on luminosity and contrast adjustment. IEEE Transactions on Biomedical engineering. 2017 May 3;65(3):521-7.

[26]. Ponnapalli, Venkata A. Sankar, VY Jayasree Pappu, and B. Srinivasulu. "Design of thinned rhombic fractal array antenna using GA and PSO optimization techniques for space and advanced wireless applications." In Microelectronics, Electromagnetics and Telecommunications, pp. 719-727. Springer, Singapore, 2018.

[27]. Subbarao, D. "The Influence Of Electronic Communication On Machine Learning." International Journal of Advanced Research in Computer Science 2.3 (2011).

[28]. Prasad, Kantipudi MVV, and H. N. Suresh. "Spectral Estimation Using Improved Recursive Least Square (RLS) Algorithm: An Investigational Study." Emerging Research in Computing, Information, Communication and Applications. Springer, New Delhi, 2015. 363-376.

IOP Conf. Series: Materials Science and Engineering 1042 (2021) 012003 doi:10.1088/1757-899X/1042/1/012003

Computational Effective Multimodal Medical Image Fusion in NSCT Domain

Dr. G.A.E.Sathish Kumar[1*], Dr. H. Devanna[2]

[1]Department of ECE, Vardhaman College of Engineering, Hyderabad, Telangana
[2]Department of ECE, St.Johns College of Engineering, Yerrakota ,Yemmiganur

*gaesathi@vardhaman.org

Abstract: Multimodal Medical Image Fusion provides comprehensive data in recent medical image diagnosis applications. This paper presents a novel Multimodal Medical Image fusion framework based on the Angular Consistency (AC) and Sub Band Adaptive Filtering (SAF). In the proposed model, the Non-subsampled Contourlet Transform (NSSCT) decomposes the source image pair into high frequency (HF) and low frequency (LF) sub bands. The LF bands are integrated with Angular consistency rule that can fuse the finer details which are very much helpful in medical image diagnosis. Further the HF bands are processed through sub band adaptive filtering such that the redundant information is filtered out and only significant information is preserved. At last the fused sub bands are processed for Inverse Non-subsampled Contourlet Transform to produce a fused image. Experimental evaluation is accomplished through a real time image data set and the performance is measured through Edge Based Similarity index, Local Quality Index and Computational time. The obtained results show that the proposed system achieves a satisfactory results compared to the conventional methods.

1. Introduction

In current years, medical imaging has gained considerable interest in studies owing to the growing need for diagnosis of disease and clinical research. Each image modal can demonstrate prior information about the human body due to different medical image modalities, but restricted to particular purposes only. For example, in the computed tomography (CT) image, bone and hard tissue structures are better visualized, while the detailed soft tissue structure is only described through MRI. Similarly, the image model MR-T1 depicts the information of anatomical structures, while the usual tissues and pathological tissues are portrayed by MR-T2. In this regard every image modal has its unique significance in delivering the required information about various issues related to medical images. Thus there is a need to use various image modalities to provide physicians with additional data for better diagnosis. In addition, the evaluation of the all imaging model creates an additional burden as well as also produces an extra time complexity. Hence there is a need to add the multi modal medical images to produce a fused image [1].

In Multimodal Medical Image Fusion (MMIF) one single image is produced by combine the complementary data from multimodal medical images [2]. The process of MMIF involves registering and consolidating various imaging models to improve the imaging quality, providing more accurate and comprehensive information. The signals from multi-modal images deal a greater diversity in terms of features which are helpful for applications of medical analysis, e.g., brain tissue mapping and volume identification, breast cancer detection and surgery, prostate localization and motion modeling etc. The integration of different models provides more information than one can derive from each of the single modality image alone. In the treatment planning, the fused imagery is best suited for helping the doctor [3].

Another advantage with MMIF is storage cost reduction. In medical diagnosis, only high-resolution images can deliver more prompt information for doctors. Hence the images captured should be high resolution in nature and these are needed to be stored for further utilization in the future. Thus, the required storage cost is more to store the images acquired in every stage, by which the additional storage requirements are necessary to provide sufficient storage for all images. This problem can be overcome with fusion techniques. This process reduces the storage cost.

A vast research is already carried out over image fusion in various orientations. According to the domain the fusion is accomplished [4], [5]. Since the spatial domain approaches considers the image pixels as it is, the computational complexity is less whereas the quality is fused image observed as less. Unlike the spatial domain approaches, the transform domain approaches initially transforms the source images and then fuses through some predefined fusion rules. Though the quality of fused image is high in these methods, the computational complexity is observed to be high. Hence to achieve an effective fused image with less complexity, this paper proposed a fusion scheme using NSSCT and SAF.

The paper is structured as follows; Section II discusses the literature survey. Section III explains the proposed fusion methodology. Section IV discusses the details of experimental outcomes and Section V gives the concluding remarks.

2. Literature Survey

The earlier fusion techniques are categorized into spatial-based and transform based. In spatial approach execute average on pixel-by-pixel, but this approach reduced contrast. Different methods are proposed by researchers. V. Aslantas et.al proposed a new technique for multi-focus image fusion (MFIF) using differential evolution algorithm [6]. Ishita De et.al presents block-based algorithms quad-tree structure is used to obtain an optimal subdivision of blocks for MFIF [7]. S. Li, J. T. Kwok et.al describes a method for pixel level MFIF by combining artificial neural networks and image blocks [8]. J. Tian et.al, suggested a new bilateral sharpness technique to determine the local content (sharp) information of the input images [9]. Shutao Li et.al, presented a new method for MFIF in dynamic scenes [10]. Yipeng Liu et.al propose a novel region level based MFIF method to locate the boundary of the focus region exactly [11].

3. Proposed Methodology

In this section, a new framework is proposed for fusion of medical images to resolve the issues with conventional methods. The NSCT is considered here to decompose source image into two subbands i.e low and high-frequency. Furthermore, the LF subband images are fused using AC and the HF subband images with their energies. The major benefit of the angular consistency is the provision of invariant brightness at LF subband images. Further the advantage with sub band adaptive filtering is reduction of computational cost and computational time. A simple schematic of NSCT based image decomposition is show in Fig.1.

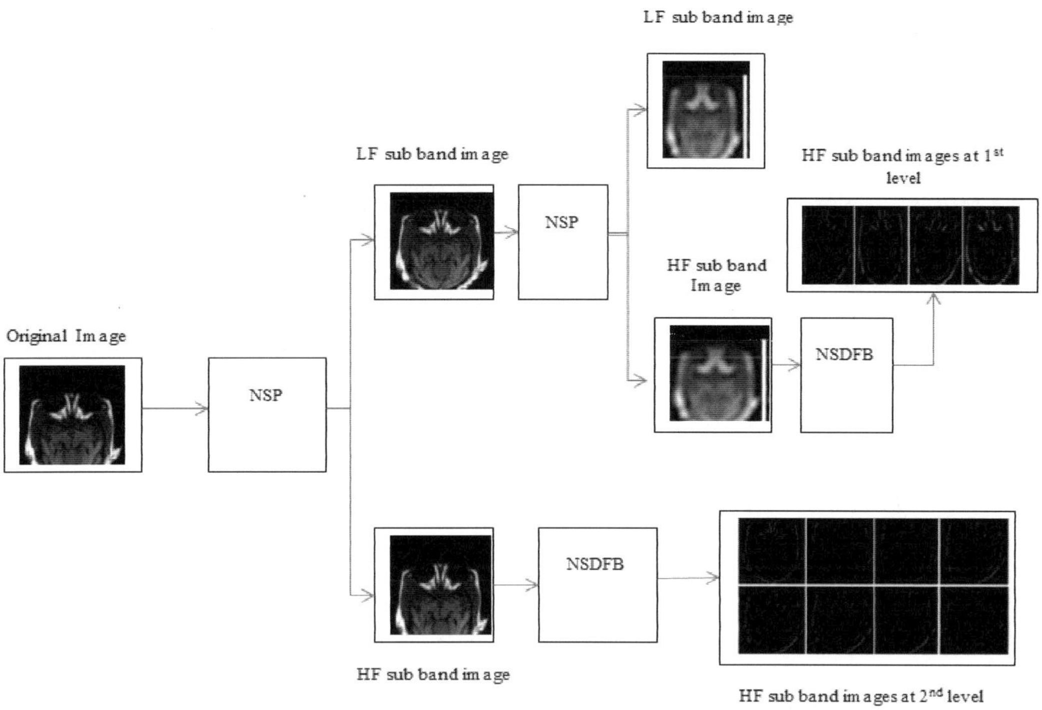

Fig.1. MRI image decomposition through NSCT

3.1. Angular Consistency

Angular consistency is a way of perceiving the features invariant to changes in illumination and contrast. This method is based on the concept of Local Energy which can evaluate the most important pixels at the point where the greatest consistency in the Fourier elements occurs. For identification of features, angular consistency may be used. This model offers feature localization and also the compensation of noise. At one point the angular consistency (x , y) is defined as:

$$AC(x,y) = \frac{\sum_n A_n(x,y)(|cos(\varphi_n(x,y))| - |sin(\varphi_n(x,y))|)}{\sum_n A_n(x,y)} \quad (1)$$

3.2. Adaptive Sub-Band Filtering (ASF)

The ASF method is based on the adaptive filter typical of the LMS. The convergence is focused on optimizing this LMS function, which uses weight features to optimize the mean error. A Normalized ASF (NASF) is suggested in to attain a faster convergence with less cost function.
The desired band *d(n)* in a ASF system can be defined by,

$$d(n) = u(n)W^o + v(n) \quad (2)$$

Further the obtained only informative sub-bands from both LF sub-bands and HF sub-bands are accomplished for fusion. The following section demonstrates the further fusion process.

3.3. Fusion Framework

Fig.2 describes the block diagram for the proposed fusion method. Because of the beneficial characteristics of NSCT, it is selected to decompose the source images and then to reduce the redundant data n sub-bands, ASF is accomplished. For every source image, he NSCT decomposes it into two sub-bands. Next, Gabor filter is utilized to process these bands in order to get the dominant features in every orientation. Here totally the Gabor filter is applied in eight orientations, the difference between every filter range is 45^0. Based on this range, totally we will get eight sub bands at 45^0, 90^0, 135^0, 180^0, 225^0, 270^0, 315^0, and 360^0. Fig.4.2 depicts the Gabor filter and feature extraction accomplishment over every sub band.

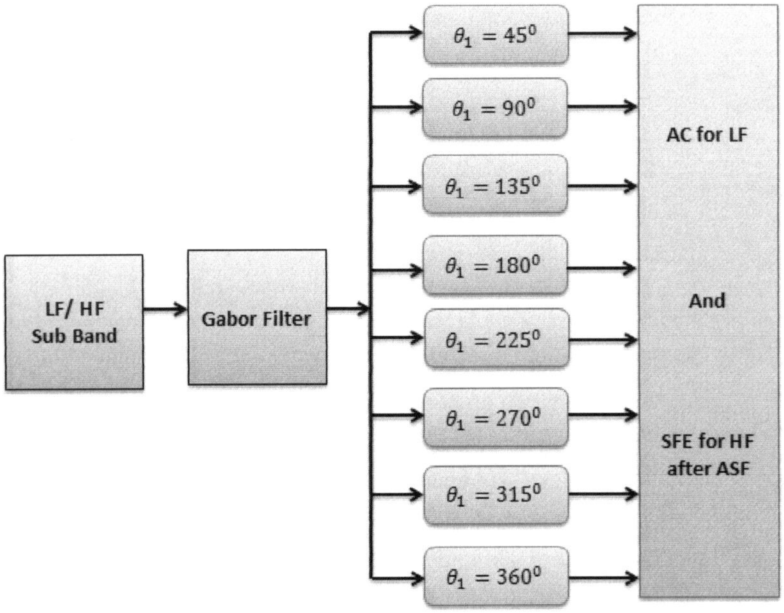

Fig.2 Gabor Filter and feature extraction accomplishment

In the fig. 2, the term AC denotes the angular consistency and SFE denotes the Spatio-frequency Orientational energy. Here every sub band, both LF and HF are represented through Orientational sub bands. Though this accomplishment increases the fusion performance, the additional complexity will rise due to the extra bands. Moreover, the eight Orientational sub bands consist mostly the redundant information by which there is no much use. Every Orientational sub band has only few dominant features and extraction of those features will improve the fusion performance. Further this important feature extraction also removes the redundant information which results in a reduced computational time.

Furthermore, fusion rules are applied to orientation sub-bands with only informative features. In the fusion of the medical image, the rules of fusion perform a significant role. Therefore, it is also essential to choose fusion rules which depend on the sub-bands acquired. To obtain good performance, AC and SFE are used to fuse the sub bands.

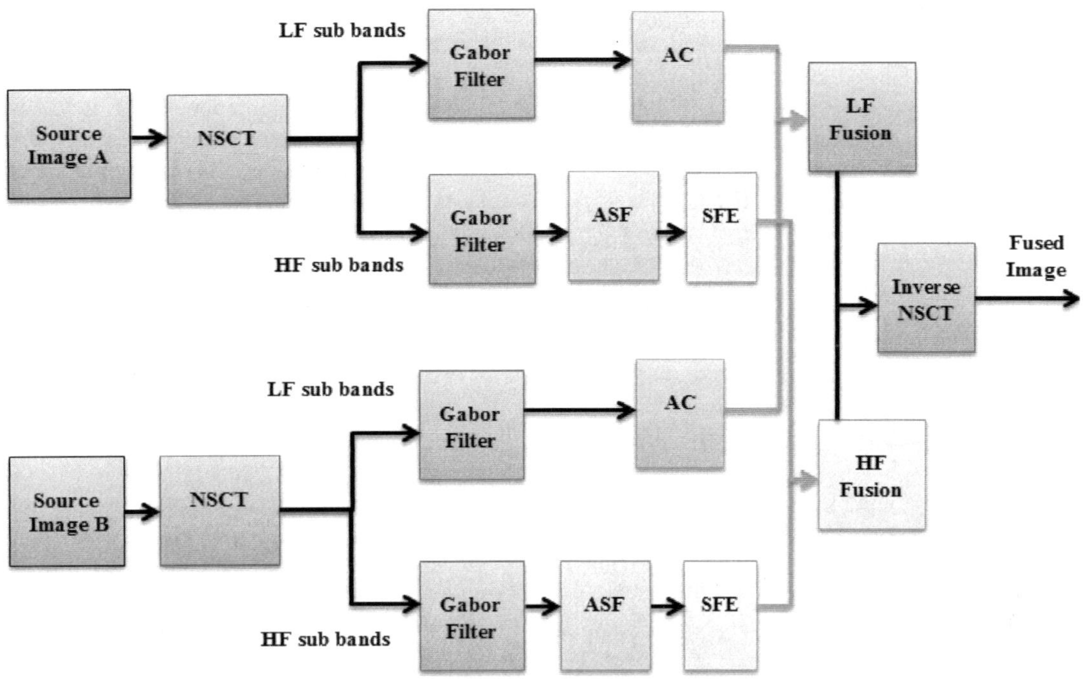

Fig..3. schematic of proposed image fusion framework

Step 1: Decompose the images A and B using NSCT into LF and t HF bands for one position l. Consider $LF^B_{k,l}$and $LF^A_{k,l}$are LF bands of B and A images, and also $\{HF^A_{k,l}\}$ and $\{HF^B_{k,l}\}$ are the HF bands of A and B images.

Step 2: For getting low frequency fused sub image the low frequency coefficients of images A and B are fused by using the proposed fusion rule. The fusion for the LF sub bands is given as

$$LF^F(x,y) = \begin{cases} LF^A(x,y) & if\ A_c^{A\theta}(x,y) > A_c^{B\theta}(x,y) \\ LF^B(x,y) & if\ A_c^{A\theta}(x,y) < A_c^{B\theta}(x,y) \\ \frac{1}{2}\left(LF^A(x,y) + LF^B(x,y)\right) & if\ A_c^{A\theta}(x,y) = A_c^{B\theta}(x,y) \end{cases} \quad (3)$$

Where $A_c^{B\theta}(x,y), A_c^{A\theta}(x,y)$, phase congruency is evaluated for the LF sub band images of the B and A images.

The criterion based on SFE is accomplished to fuse HF coefficients. Further there are large numbers of sub bands are there which are obtained after the accomplishment of Gabor filter over high frequency sub bands. The accomplishment of ASF over all these Orientational sub bands reduced the computational time for fusion process.

The HF sub bands fusion can be formulated as

$$HF^F(x,y) = \begin{cases} HF^A(x,y) & if\ E^A(x,y) \geq E^B(x,y) \\ HF^B(x,y) & if\ E^A(x,y) < E^B(x,y) \end{cases} \quad (4)$$

Where $E^A(x,y)$ and $E^B(x,y)$are Log Gabor energies of HF sub-bad images of A and B images.

Step 3: Obtain the fused image with INSCT.

4. Simulation Results

Under this section, this work undergoes for the accomplishment of proposed contributions over a real time image set acquired from a hospital. The medical images are acquired from **Gowri Gopal Hospital, Kurnool, and Andhra Pradesh, India**. The images acquired are of two different models, CT and MRI of a brain. These two images are belongs to a patient admitted for the Diagnosis of brain dysfunction. As an initial formality, the patient is undergone for CT scanning and MRI scanning and the obtained images are considered here for validation. The images acquired are in '.PNG' format and the size of each images is 300×375. Over these real time images, the proposed contributions are applied and obtained a fused image. Further to observe the conventional approaches performance, ten basic methodologies of conventional approaches is also applied and the obtained results are shown below;

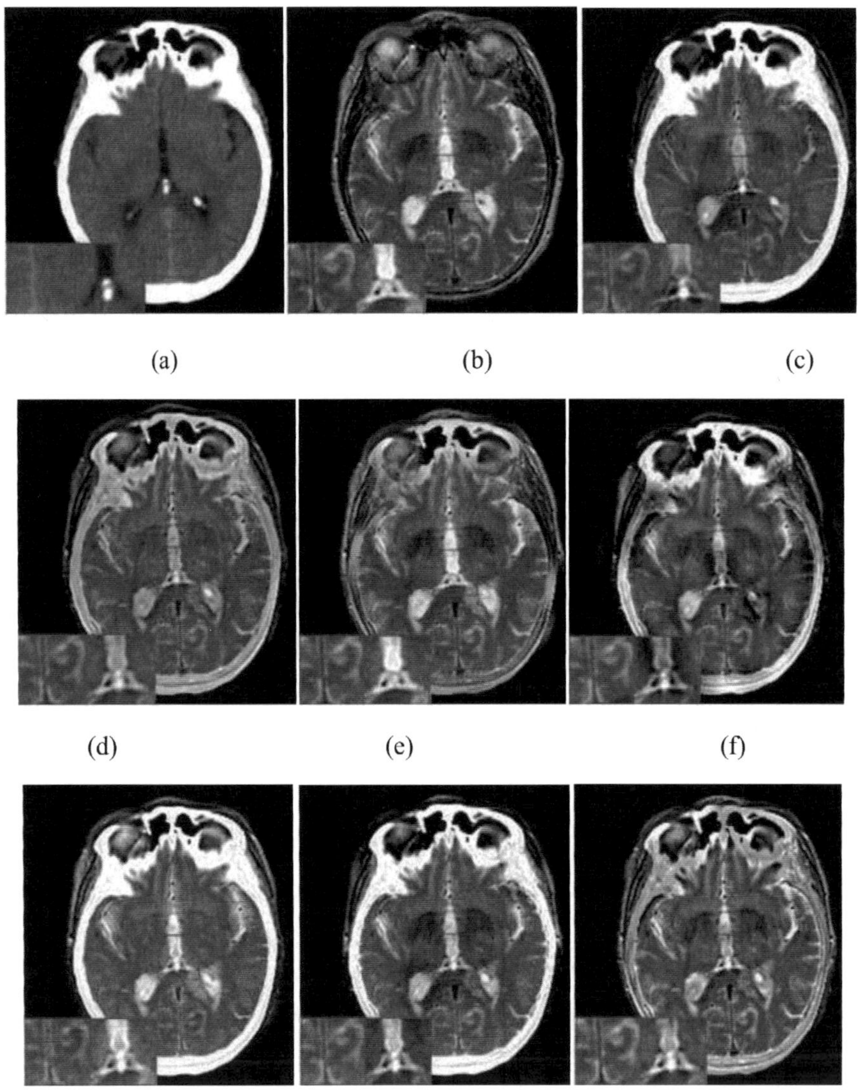

(a) (b) (c)

(d) (e) (f)

(g) (h) (i)

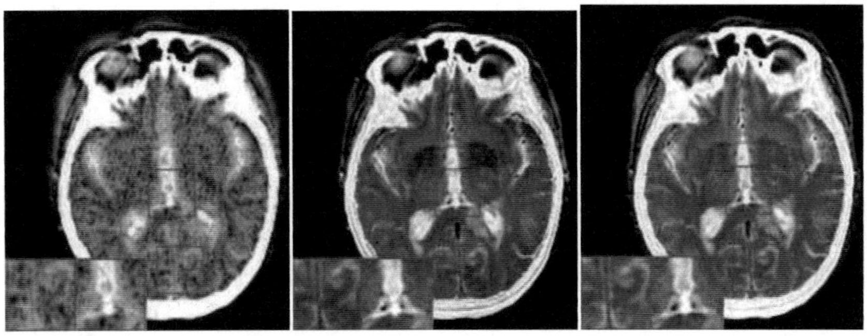

(j) (k) (l)

Fig..4 Reuslts if Validation (a) CT image (b) MRI image, fused image obatined through (c) DWT [37], (d) CT [20], (e) NSCT-1 [38], (f) NSCT-2 [23], (g) NSCT-3 [24], (h) NSCT-4 [32], (i) NSCT-5 [39], (j) NSCT-6 [6], (k) NSCT-SFE, and (l) NSCT-ASF

Fig..4 shows the results obtained through the proposed and conventional approaches. Initially to highlight the important regions in both CT and MRI images, a red colored squared is drawn which shows a major difference in CT and MRI images. Though the both images are acquired for diagnosis purpose, the CT image cannot captres all the data from brain. In Fig..4(a) the highlighetd region signifies bones structure whereas the highlighted region in Fig..4(b) signifies the soft tissues region in brain. Upon the fusion of these two different models, the obatined results are shown from Fig..4(c)-(l). It can be shown that in several regions the methods CT[12], NSCT-1[16], NSCT-2[13] and NSCT-5[17] decrease in intensity and contrast (see regions in Fig..4 (d)-(f) and (i)). But this problem is solved in the fused image obtained through NSCT-SFE (see Fig..4 (k)). The NSCT-3[14] method can conserve image strength, but fails to extract the MR image structural information (see Fig..4 (g)). The NSCT-4 [15] performs better than the NSCT-3 method, but some details are still not successfully extracted (see Fig..4 (h)). The NSCT-6[6] fused images experience severe noise like objects (see the soft-tissue regions in Fig..4 (j)). The final proposed method, NSCT-ASF performs well on both the energy and details preservation (see Fig..4 (l)). Further the performance is analyzed through the performance metrics and they are depicted in the following table.1.

Table.1 Performance Metrics for Real time image set

Method	**Metric**					
	$Q^{AB/F}$	Q_s	**MI**	Q_0	Q_W	Q_E
DWT	0.6573	0.7336	1.5242	0.6897	0.7458	0.5558
CT	0.6604	0.7458	1.6875	0.6912	0.7559	0.5689
NSCT-1	0.6855	0.7684	1.9222	0.7124	0.7769	0.5937
NSCT-2	0.6869	0.7702	1.9679	0.7155	0.7793	0.6047
NSCT-3	0.6874	0.7749	2.0217	0.7208	0.7806	0.6099
NSCT-4	0.6943	0.7841	2.1417	0.7236	0.7859	0.6112
NSCT-5	0.6967	0.7888	2.1989	0.7299	0.7901	0.6184
NSCT-6	0.6998	0.7914	2.2078	0.7314	0.7915	0.6233
NSCT-SFE	**0.7047**	**0.8096**	**2.3298**	**0.7449**	**0.8012**	**0.6469**
Proposed	**0.7089**	**0.8186**	**2.5589**	**0.7549**	**0.8058**	**0.6530**

As shown in the above table, the proposed two contributions show a greater improvement all performance metrics. Compared to the idle images which are captured under disturbance free and noise free environment, the image images acquired in real time somewhat distorted and hence the obtained metrics shows a smaller less values. From this validation, it can be declared that developed approach can achieve an effective performance for all types of images. On an average the developed work achieved an improvement of 0.0233 (2.33%) of $Q^{AB/F}$, 0.0444 (4.44%) of Q_s, 0.09854 (9.854%) of MI, 0.0356 (3.56%) of Q_0, 0.0278 (2.78%) of Q_W, and 0.0517 (5.17%) of Q_E from conventional approaches.

5. Conclusion

This paper proposed a new technique for multimodal medical image fusion based on Adaptive Sub band Filtering for reducing redundant information. Next the low-frequency sub-band images are fused by the AC and the HF sub-band images are fused by the SFE rule. Under simulation results, the proposed approach is subjected to extensive simulations over a real time multimodal medical images and Performance assessment is carried out through different quality metrics, including a $Q^{AB/F}$, Q_S which measures the structural similarity, MI, Q_E, Q_0, Q_W, and Computer Time (Sec). The obtained performance values show an superior performance in achieving the optimal quality in the fused image by transferring all the significant information likes edges, textures, boundaries etc., from sources images.

References

[1]. B. Solaiman, R. Debon, F. Pipelier, J. M. Cauvin, and C. Roux, "Information fusion: Application to data and model fusion for ultrasound image segmentation", *IEEE TBME*, vol. 46, no. 10, pp. 1171–1175, 1999.

[2]. B. V. Dasarathy, Editorial: Information fusion in the realm of medical applications-a bibliographic glimpse at its growing appeal, Information Fusion 13 (1) (2012) 1–9.

[3]. F. E. Ali, I. M. El-Dokany, A. A. Saad, and F. E. Abd El-Samie, "A curvelet transform approach for the fusion of MR and CT images", *Journal of Modern Optics*, vol. 57, no. 4, pp. 273–286, 2010.

[4]. Zhijun Wang, Djemel Ziou, Costas Armenakis, Deren Li, and Qingquan Li, "A Comparative Analysis of Image Fusion Methods", IEEE Transactions on Geoscience And Remote Sensing, Vol. 43, No. 6, June 2005.

[5]. H.Devanna, G.A.E.Satish Kumar, M.N.Giri Prasad, "A Survey on Multimodal Medical Image Fusion", *IOSR Journal of Computer Engineering (IOSR-JCE),* Volume 19, Issue 1, 2017.

[6]. V. Aslantas and R. Kurban, "Fusion of multi-focus images using differential evolution algorithm," *Expert Syst. Appl.*, Vol. 37, No. 12, pp. 8861–8870, Dec. 2010.

[7]. I. De and B. Chanda, "Multi-focus image fusion using a morphology based focus measure in a quad-tree structure," *Inf. Fusion*, Vol. 14, Vo. 2, pp. 136–146, Apr. 2013.

[8]. S. Li, J. T. Kwok, and Y. Wang, "Multi-focus image fusion using artificial neural networks," *Pattern Recognit. Lett.*, Vol. 23, No. 8, pp. 985–997, Jun. 2002.

[9]. J. Tian, L. Chen, L. Ma, and W. Yu, "Multi-focus image fusion using a bilateral gradient-based sharpness criterion," *Opt. Commun.*, Vol. 284, No. 1, pp. 80–87, Jan. 2011.

[10]. S. Li, X. Kang, J. Hu, and B. Yang, "Image matting for fusion of multi-focus images in dynamic scenes," *Inf. Fusion*, Vol. 14, No. 2, pp. 147–162, Apr. 2013.

[11]. Y. Liu, J. Jin, Q. Wang, Y. Shen, and X. Dong, "Region level based multi-focus image fusion using quaternion wavelet and normalized cut," *Signal Process.*, Vol. 97, pp. 9–30, Apr. 2014.

[12]. Hui Huang, Xi'an Feng, and Jionghui Jiang, "Medical Image Fusion Algorithm Based on Nonlinear Approximation of Contourlet Transform and Regional Features", *Journal of Electrical and Computer Engineering,* Vol.2017, pp.1-9, August 2017.

[13]. G. Bhatnagar, Q. M. J. Wu, and Z. Liu, "A new contrast based multimodal medical image fusion framework", *Neurocomputing*, Vol. 157, No.1, pp. 143–152, 2015.

[14]. Z. Y. Jin and Y. J.Wang, "Multi-modality medical image fusion method based on non-subsampled contourlet transform", *Chinese Journal of Medical Physics*, Vol. 33, No. 5, pp. 445–450, 2016.
[15]. Yong Yang, *Member, IEEE*, Song Tong, Shuying Huang, and Pan Lin, "Multi-focus Image Fusion Based on NSCT and Focused Area Detection", *IEEE Sensors Journal*, Vol. 15, No. 5, pp.2824-2838, May 2015.
[16]. S. Das and M. K. Kundu, "NSCT-based multimodal medical image fusion using pulse-coupled neural network and modified spatial frequency," *Medical and Biological Engineering and Computing*, vol. 50, no. 10, pp. 1105–1114, 2012.
[17]. Y. Yang, S.Tong, S. Huang, andP.Lin, "Log-Gabor energy based multimodal medical image fusion in NSCT domain," *Computational and Mathematical Methods in Medicine*, Vol. 2014, pp.1-12, August 2014.

A Novel Error Free Dimensional Projection Coding

[1] **Shaik Mohammed Rafi,** [2] **G D Basavaraj,** [3] **Dr. Shaik Fairooz.**
[1,2]Assistant Professor, Sreyas Institute of Engineering and Technology, India
[3]Associate Professor, Sreyas Institute of Engineering and Technology, India
mohammed.rafi@sreyas.ac.in, basavaraj.g@sreyas.ac.in, fairooz.sk@sreyas.ac.in

Abstract: In this paper, dimensional coding of projection has been developed with the hardware design for RS algorithms. The result of simulation of Memory Interface Unit for basic and dimensional projection coding with scale and Implemented Video Interface module using VHDL and Target on to the FPGA implemented (XCV300-bg432-6). It provided operating speed of 732.5 MHz, power consumption of 1249 BEL and 128.3 mW in FPGA.

Keywords: FPGA, BEL's, VHDL, HD, Video Codec, Projectional Coding.

1. Introduction:

Unlike most other information technologies, whose exceptional growth has increased in recent decades, the resolution of the screen has largely stopped being reduced to less resolution digital video screens[1] and now offers high-resolution screen scaling , high-resolution digital video scaling [4] open opportunity , they have proposed a general framework for video presentation and flexibility[5-7]. As the size of the video increases, pixels, which are part of the video, begin to appear faster, make the video visible. On the contrary, it is obvious to reduce a video which leads to better and sharpness. It has a codec and a numerical video screen, which allows the user to convert the low-resolution [2] content to the signal in the form of HD content in that controller. Based on the quality of the projector located above the launch, the yield feature of the output video shown can be improved.

2. Traditional System Design Issues:

High resolution video problems when evaluating a high-resolution (HR) [3,8] version that is equivalent to low-resolution equivalent (LR) video stream. First, the properties of this recommended method to reproduce more selective human resources between high quality and frames, locally required models can be proposed.

3. Proposed Architecture:

Architectures designed for RS algorithms have been modified according to the upper level block diagram, shown in Figure 1.

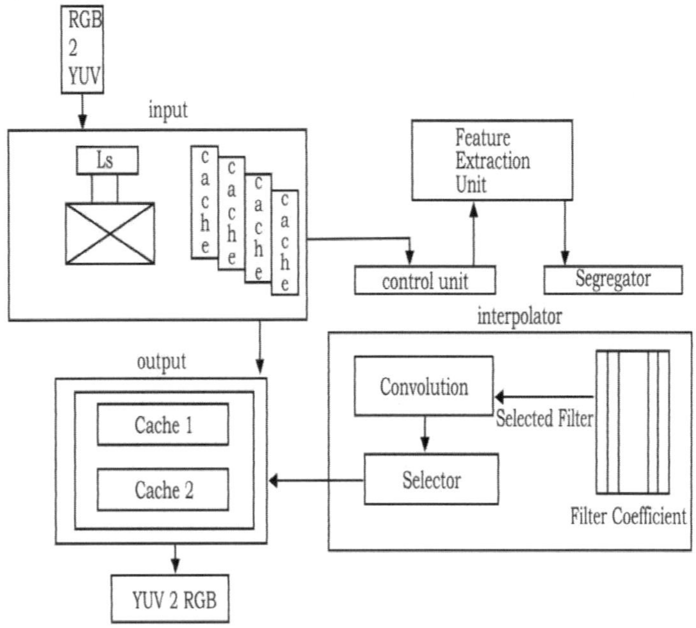

Figure 1: Top Level Block Diagram

3.1 (a) Functional Description

The structure shown above is made up of different functional blocks; the RS adjustment algorithm is used in color-coded signals by luminance path (Y) and pixel replication. On each pixel input, the controller reads the column cache of 4 x 1 pixels and provides 5 x 5 pixel windows and 3x3 extraction and classification. The function is to eliminate the vector and then there is another dimension other than 8 x1, which again selects a predefined model, the division of vector characteristics and the appropriate filter for the pixel input environment. Fixed transaction multiplier was used to adjust and the output result of the pixel was placed in the cache memory.

1. Output - With HD Pixel format, find the window analogous to the 5x5 precision in this neighborhood and remove, compile and interpolate features.

2. Input - For each SD pixel format, select the analogous coordinates of 4 HD pixels.

End the final pixel interpolation output if the LV or LH (decreasing ratio is high) [9] is the wrong value. Two key factors affecting the efficiency of implementation is a fast data entry / exit and a destination that is defined as below:

$$D_r = \frac{N_e}{N_p} \quad (1)$$

Ne represents the number of elements for feature vector

N_p represents the number of processing paths for feature extraction

To achieve the essential performance levels with dissimilar data and resource sharing, the core clock frequency, Fclk_core, the frequency input of the FClk_in pixel clock must be associated with the given formula below:

$$F_{clk\_core} = [D_r]F_{clk\_in} \quad (2)$$

3.1(b) Control Unit

Control unit (CU) provides information access to the data path blocks in a timely manner and format of control unit.

1. Feature Extraction uses a sliding window , memory conversion pixel window input column 5 x 1 by 5 x 5 pixels.
2. Control unit produce memory address and control signals for input memory and output memory blocks
3. It produces exact synchronization signal in video standards
4. It produces X_H and X_V values mentioned in the subsequent equation.

$$X_{Vs} = [Q_{vs}(Y_{Vs}/L_{Vs} - R_{Vs}) + €s\ ] \quad (3)$$

$$X_{Hs}=[Q_{Hs}(Y_{Hs}/L_{Hs} - R_{Hs}) + €s\ ] \quad (4)$$

Where,

X_V = Projection plane Vertical.

Y_H = Projection plane Horizontal.

Y_V = Represents the high resolution pixel using vertical coordinates.

Y_H = Represents the high resolution pixel using horizontal coordinates.

€ = Constant (10^{-6}.)

R_V = Low resolution pixel using vertical coordinates.

R_H = High resolution pixel using horizontal coordinates.

5. For projection ratios less than 2, the pixels are given as,

$$R_{Vs} = [Y_{Vs}/L_{Vs}] \quad (5)$$

$$R_{Hs} = [Y_{Hs}/L_{Hs}]$$

3.1(c) Input Memory

Input memory module (IM) input clock frequency Fclk_in runs in blocks of 4 cache memory stores, 5x1 pixel columns for the CU is provided. The extent of the store line is the horizontal size of the video input.

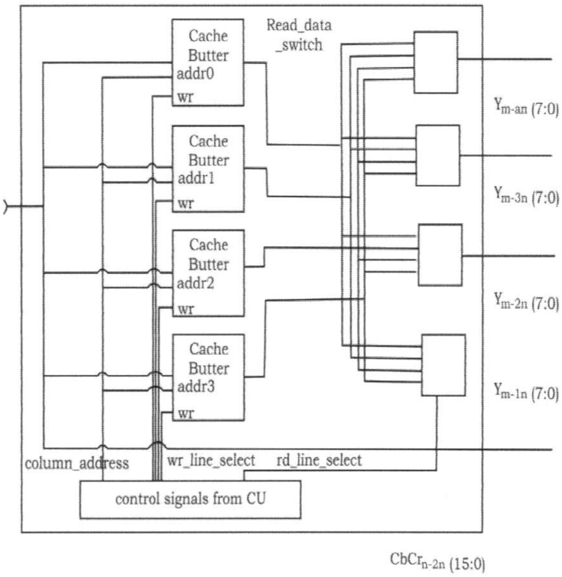

Figure 2: Input interface Unit

3.1(d) Segregator Unit

The function works with the central clock of the CPU and produces 4 high-resolution pixels using specified Fclk_core. The number of pipeline registers is shown in Figure 3 with a slight increase in reference classification and pixel extraction unit.

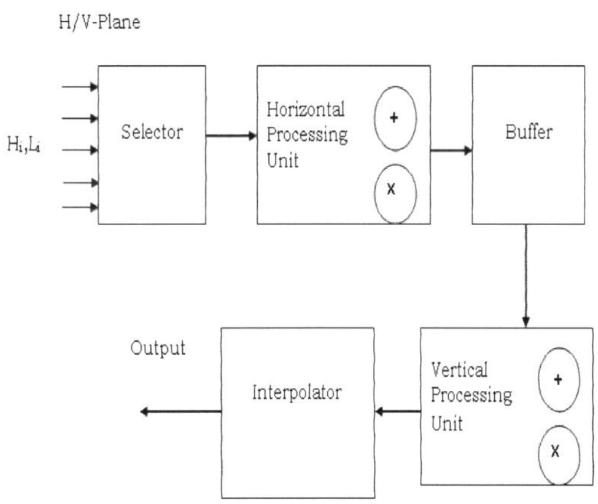

Figure 3: Context Classification and Pixel Extraction.

3.1(e) Interpolation Unit

The interpolation unit (IN) is used as an adjustment function, which produces 4 high definition pixels every cycle in Fclk.

3.1(f) Output Unit

The output memory module (OM) works by the original clock frequency. Basically, select the appropriate pixel between the high-resolution pixels generated by these units and adjust the scanning scans according to the mapping relationship.

4. Simulation Results:

Simulation result of memory interface unit for original and dimensional projection coding with scale ratio shown below.

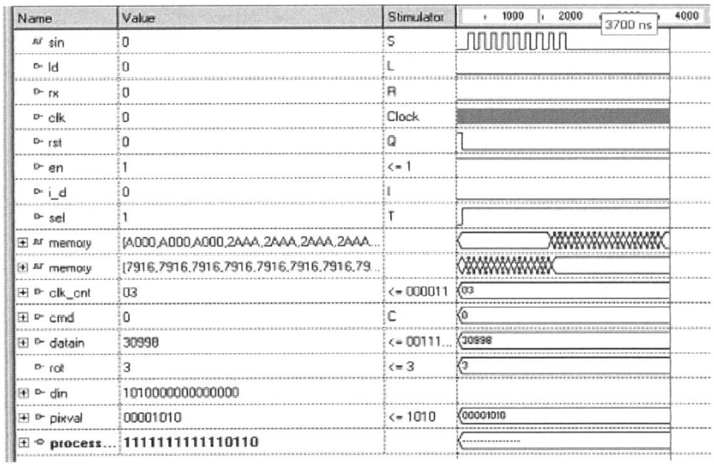

Figure 4: Simulation plot for memory interface unit

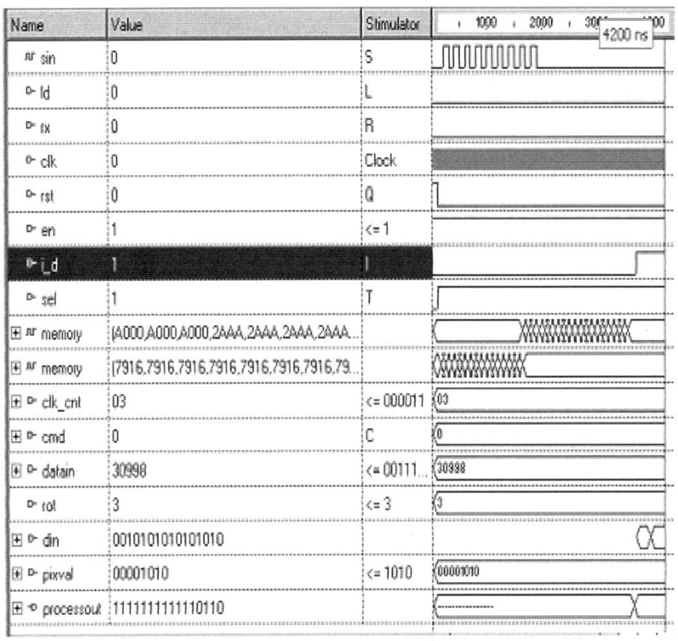

Figure 5: Simulation plot for memory interface unit.

The total width of the 8-bit bus has been provided so that the signal can provide a rotation of 256 pixels at a time. The data stored in the buffer is stored in the memory unit, where the input is done through Signal and the processed data is distributed as the output of the process. Depending on the control signal RD and the signal line SL, the writing and reading task is selected.

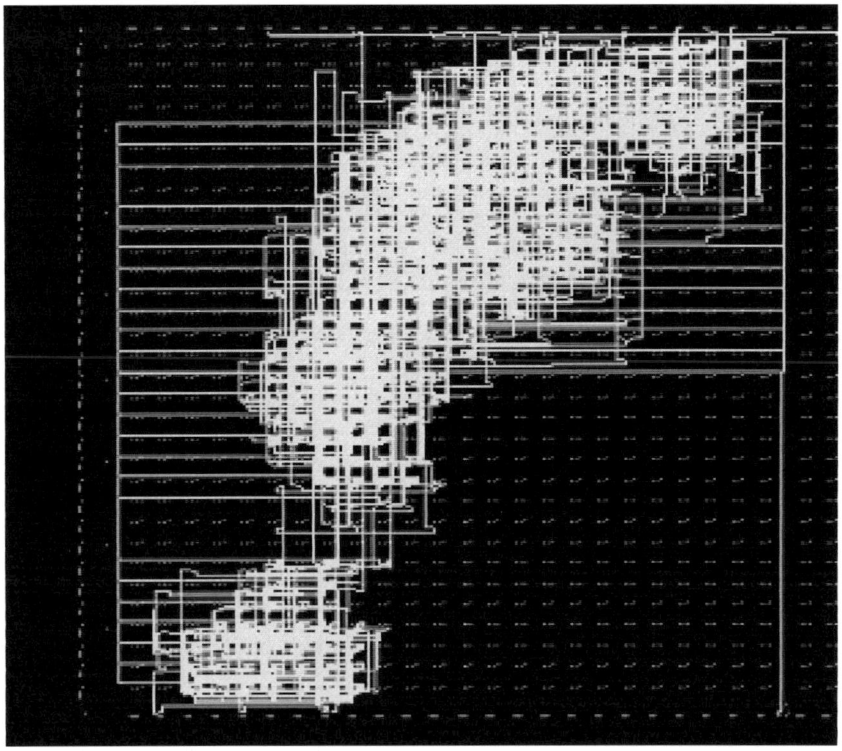

Figure 6: Routing of the implemented core into the targeted FPGA

Figure 6 shows the implementation of FPGA (XCV-300-BG 432-6) in the FPGA, through implementation in basic routing and advanced VHDL video interface module. To monitor the collection of results from several frames, it is executed with the Interface and Multisim Tool with the testbench. The processing of five frames are passed to a system and the results obtained are shown. in the Figure 7, Figure 8 with the ratios of 1:2:5.

Figure 7: Sequence of input video

Figure 8: Sequence of video scaled at 1:2.5 ratios

An illustration shows clearly the accuracy of the recovery in terms of visual quality compared to traditional technology based on Fourier coding, as shown in Figure 10. Comparison time is better compared to comparison to B-bar cubic interpolation, as shown in Figure 10, Fourier interpolation techniques.

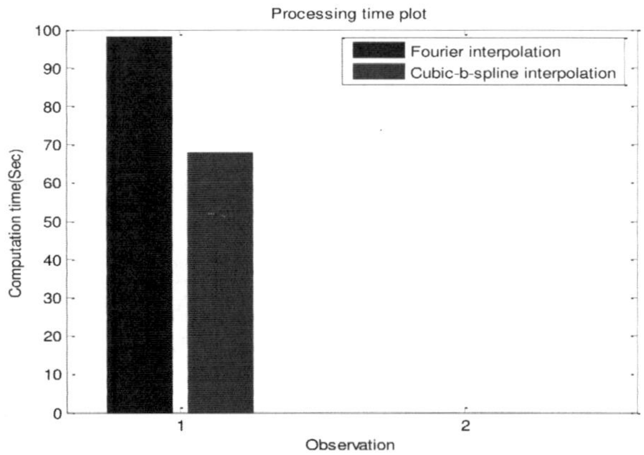

Figure 9: Computation time taken for the two methods

While computing the sequence of tables and calculating the time used in the project, the system has been developed to see if it has been completed.

5. Conclusion:

In the recommended approach, data for pixel-encoded frame coding passes through decimal methods at lower resolutions. At the reception, it has been suggested that the interpolation logic uses horizontal / vertical projection to achieve high resolution screens. This proposed interpolation argument shows better operation in each encoding coding using B-band projection. The objective is 732.5 MHz operating speed, 1249 BEL power consumption and FPGA provides 128.3 mW.

References

[1] Dzung Nguyen, Minh Dao, Trac Tran, "Video Error Concealment Using Sparse Recovery And Local Dictionaries", IEEE International Conference On Acoustics, Speech, And Signal Processing, 2011.

[2] Yui-Lam Chan, Hoi-Kin Cheung, and Wan-Chi Siu, "Compressed-Domain Techniques for Error-Resilient Video Transcoding Using RPS", IEEE Transactions on Image Processing, Vol. 18, No. 2, 2009.

[3] Muhammad Shoaib, "Key Picture Error Concealment Using Residual Motion-Copy in Scalable Video Coding", World Congress on Computer Science and Information Engineering , 2009.

[4] M. Tun, K. K. Loo and J. Cosmas, "Error-Resilient Performance of Dirac Video Codec Over Packet-Erasure Channel" ,IEEE Transactions on Broadcasting, Vol. 53, No. 3, 2007.

[5] Zhenyu Wu and Jill M. Boyce, "Adaptive Error Resilient Video Coding Based On Redundant Slices of

H.264/AVC", IEEE International Conference on Multimedia and Expo, 2007.

[6] Daniel Persson, Thomas Eriksson, and Per Hedelin, "Packet Video Error Concealment with Gaussian Mixture Models", IEEE Transactions on Image Processing, Vol. 17, No. 2, 2008.

[7] Yi J. Liang, John G. Apostol opoulos and Bernd Girod, "Analysis of Packet Loss for Compressed Video: Effect of Burst Losses and Correlation between Error Frames", IEEE Transactions on Circuits and Systems for Video Technology, Vol. 18, No. 7, 2008.

[8] Antonios Argyriou, "Error-Resilient Video Encoding and Transmission in Multirate Wireless LANs " ,IEEE Transactions On Multimedia, Vol. 10, No. 5, 2008.

[9] Dieu Thanh Nguyen, Michio Hayashi and Joern Ostermann, "Adaptive Error Protection for Scalable Video Coding Extension of H.264/AVC", IEEE International Conference on Multimedia and Expo, 2008.

IOP Conf. Series: Materials Science and Engineering 1042 (2021) 012005 doi:10.1088/1757-899X/1042/1/012005

Design of Deterministic Sparse Fractal Array with Haferman Counter-Diagonal Fractal Tapering Technique for Modern Wireless Systems

V. A. Sankar Ponnapalli *, Praveena A and Vinod Babu P

Department of Electronics and Communication Engineering, Sreyas Institute of Engineering and Technology, Hyderabad, India.

* *vadityasankar3@gmail.com*

Abstract. Deterministic fractal structure is a repetitive geometry based structure having applications in various regions of science and engineering. Fractal antennas and array antennas are artistic structures and they are finding much attention in recent years owing to their feasible properties. A few deterministic fractal arrays owing to the fractal dimensional design methodology are thinned arrays, but the great concern for these types of array antennas is the great number of antennas present at higher level iteration and as well as expansion levels. The present research work discusses the feasibility of reducing antenna elements at higher expansion and iteration levels of fractal arrays with a new type of tapering technique named as Haferman counter-diagonal fractal tapering. Different types of fractal arrays created using a sub array of concentric elliptical ring design methodology have been considered for the application proposed tapering technique. Owing to the proposed technique, nearly 40% to 50% of the antenna elements thinned while maintaining fully populated array factor properties. The proposed technique can simplify the design complexity of fractal arrays and reduce the designing cost.

1. Introduction

When a complex wireless based communication system like a celestial system demands better gain, a more directive pattern with low side lobe (SdLb) ratio, multibeam characteristics, or any other attribute that a solitary antenna element cannot perform, an array antenna made up of group of distinct antenna elements may proffer an answer to this type of problems [1-6]. Ultra wide band and multibeam performance of an antenna and the antenna arrays are instrumental in celestial and other modern wireless based communication system [7, 8]. Fractal array antennas (FAA's) are the better solution for this type of antenna performance. FAA's are repetitive geometry based thinned and multibeam arrays having applications in satellite and celestial communications [9-12]. Geometrical construction of these arrays is basically three types like conventional array antennas. Linear FAA's are one dimensional arrays having fractal geometric nature. Even and odd numbered Cantor linear array antennas are the best example for this type of arrays [13, 14]. Generally, deterministic FAA's are generated by geometric design procedures. Concentric circular and a sub-array of elliptical ring designing technique and concentric sphere generators are ideal for 2D and 3D fractal geometric methodologies. Triangular, square, pentagonal, hexagonal, heptagonal, octagonal, and Sierpinski carpet arrays are ideal for planar FAA's [15-18]. The array factor performance of these arrays is dependent on the number of iteration levels (p) and the scaling or expansion component (S). These two components are dependent on the fractal geometric design procedures of FAA. Fractal organization pattern of array elements generates a thinned FAA in turn accomplish desired radiation performance. Generally tapering techniques can be helpful in the process of thinning antenna elements with reduction in side lobe(SdLb) ratio. Array-tapering means eliminating a few antenna elements over an array surface according to array factor requirements [19, 20].

This report investigated deterministic rhombic FAA with Haferman anti-digonal tapering technique (HADTT). Fractal numerical sequences are obeying repetitive nature as similar to fractal geometric shapes. Technical findings reported on fractal tapering techniques applied to FAA's are quite less. A few findings reported various sequences such as Morse-Thue, atomic functional, and Fibonacci as the current -amplitudes in order to decrease SdLb ratio in conventional and also fractal arrays [20-25]. Haferman counter-diagonal fractal tapering, related formula, and expressions have discussed in the section 2 of this paper. Design equations of the considered fractal array presented in the section 3, corresponding array factor details and graphs presented in the results and discussion session of this paper. Finally need of the proposed paper and future ways of this research work presented in the conclusions session.

2. Haferman's Counter0diagonal fractal Tapering Technique

Hagerman's counter-diagonal sequence pattern is ideal for partly fractal sequence. This type of sequences never follows the fractal nature up to nth iteration level. Haferman's matrix is the modified variant of Sierpinski matrix. Actually in the Sierpinski matrix, the value of 'M_{22}' is zero and remaining all values is one. But in this case both diagonal and counter-diagonal matrices are zero. This tapering technique for the design of sparse deterministic rhombic FAA is obtained from the Haferman's matrix. In the implementation procedure of tapering technique, the count of antenna elements which are in 'ON' condition for every single iteration can be determined by equation (1)

$$\frac{1}{14}\left[(-1)^{p}5^{p+1}+9^{p+1}\right] \quad (1)$$

Generally, this type of sequence can have infinite iterations. In the present work, successive five iterations are taken to achieve thinning of array antenna. Due to the expanding nature of the taken sequence, all the five iterations have reported nearly fifty percentage of thinning. The thinning percentage is maintained constantly throughout all the iterations. The complexness involved in designing of deterministic rhombic FAA can be lowered with the simple tapering behavior. The conventional Haferman matrix with corresponding counter-diagonal bit pattern till nth iteration is given below;

Iteration -I (p=1) of Hagerman's matrix,

$$\begin{bmatrix} 010 \\ 101 \\ 010 \end{bmatrix}$$

Counter-Diagonal Sequence of above matrix is,

[011000110]

Iteration two (p=2) of Haferman's matrix,

$$\begin{bmatrix} 111010111 \\ 111101111 \\ 111010111 \\ 010111010 \\ 101111101 \\ 010111010 \\ 111010111 \\ 111101111 \\ 111010111 \end{bmatrix}$$

Counter-Diagonal Sequence of above matrix is,

[1111110110111110000001111111110110111111111111101101111110000001111101101111111]

n^{th} iteration (p=n) of Haferman's carpet matrix,

$$\begin{bmatrix} 111010------n \\ 111101------n \\ 111010------n \\ ----------- \\ -----------n \end{bmatrix}$$

Counter-Diagonal Sequence of above matrix is,

[1111111111011011111000000011-------111000000-----n]

3. Design Equations of Deterministic Rhombic FAA

The minimum number of elements required to generate a rhombic array antenna is four. And this can be reiterated up till infinite times. This works reports till fourth iteration. The antenna elements in the array maintained uniform distance for expansion behavior (S) of one and two. Owing to recursive character of the rhombic FAA near to ternary of the total populated antenna elements set to OFF condition in every single iteration and thus led to multibeam characteristics. The expansion behavior of the FAA determines the actual number of elements required and their respective positions. This section handout the two different expansion factors which are taken into consideration for rhombic fractal array. The Fig. 1 depicts the antenna array pattern when expansion factor is one. The corresponding array factor is determined by equation-(2).

$$A.F_p(\theta,\phi) = \prod_{p=1}^{P} [\sum_{m=1}^{M} \sum_{n=1}^{N} I_{mn} e^{jk(1)^{p-1}\Psi_{mn}}] \quad (2)$$

Where, 'p' represents iteration value or the iteration level, 'm' gives count of concentric rings, 'n' is gives the number of antenna elements, I_{mn} represents the current-amplitudes, k gives wave count, 'S' is the representation of expansion factor, where S = 1. Equation "(3)" determines the antenna Array factor when expansion factor is two. The behavior of array is changed based on its expansion factor. Dimensional structure of rhombic fractal array when expanding factor is two is show in Fig 2. Because of the expansion nature, no elements in the array are thinned for the given configuration.

$$A.F_P(\theta,\phi)=\prod_{p=1}^{P}[\sum_{m=1}^{M}\sum_{n=1}^{N} I_{mn}e^{jk(2)^{p-1}\Psi_{mn}}] \quad (3)$$

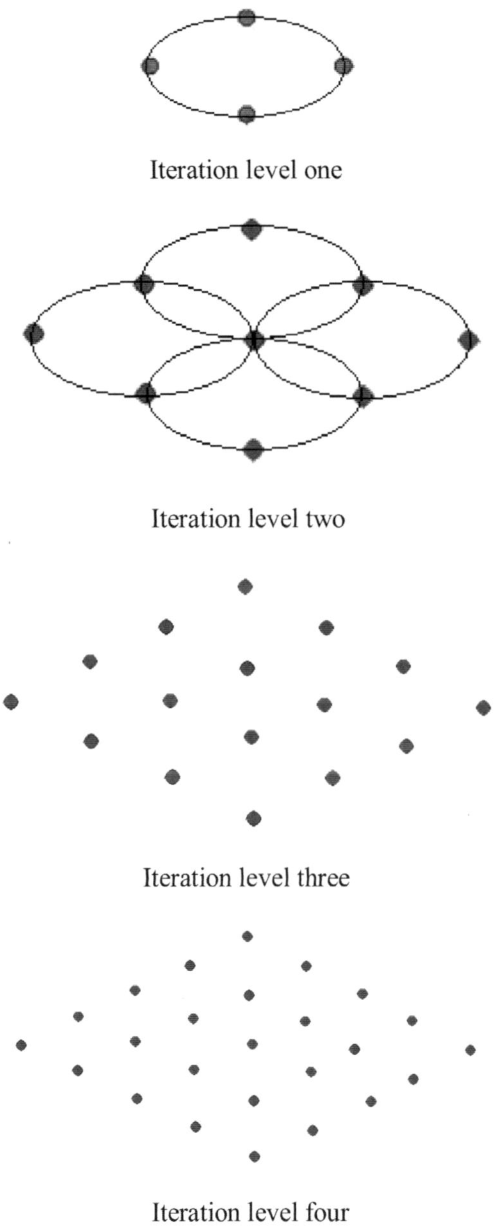

Figure 1. Successive four iterations levels of rhombus fractal array antenna where expansion factor is taken as one.

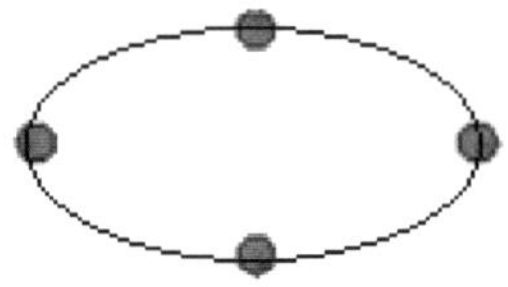

Iteration level one

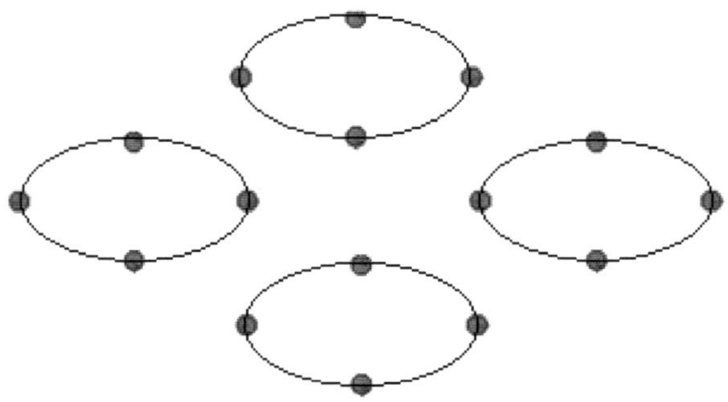

Iteration level two

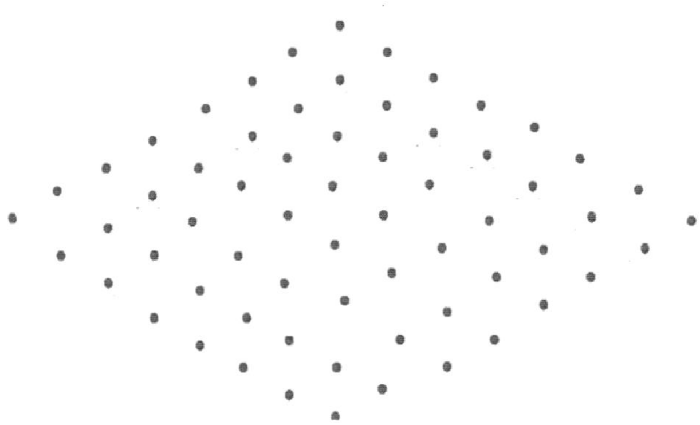

Third level Iteration

Figure 2. Successive three iterations levels of rhombus fractal array for where expansion factor is taken as two.

4. Results and Discussion

A wide amount of thinning is achieved in the Haferman counter-diagonal tapered rhombic array antennas for both expansion levels (S= 1 and 2). Nearly 50% of thinning has obtained in all iterations of deterministic rhombic FAA. Array factor properties are shown in Fig 3 (a) and 3 (b). Percentage of thinning and the parameters are given in table 1. The proposed tapered arrays are providing superior performance in terms of thinning percentage and the array factor behavior than an array which consists of all the actual number of elements. Haferman's counter-diagonal tapering method has achieved high and uniform thinning percentage in comparison with Sierpinski counter-diagonal tapering for all levels of iterations. First iteration of FAA has achieved -4.6 SdLb level, 42° beamwidth, and 87° side level angle with 50% of thinning. Second iteration of FAA has achieved -13 SdLb level, 25° beamwidth, and 89° side lobe level angle with 55% of thinning. Third and forth iterations achieved -22 and -31.3 SdLb levels respectively with good thinning. It is clear from the considered sequence that if the iterations are increasing, the thinning percentage is also increasing. It is a good sign for the fractal array antenna thinning especially at higher orders.

The tapering techniques are feasible for practical design considerations due to its low profile, higher tapering efficiency, and less amplitude weights. These tapering methods can be suggested for satellite, celestial, and other modern wireless systems, where antenna aperture plays a vital role.

Table 1. Array properties of rhombic fractal antenna array with Haferman counter-diagonal binary semi-fractal distribution function of four successive level of iterations (P=1, 2, 3, 4) where expansion factors is taken as S=1 and 2.

S	*Expansion Factor (S) = 1*				
Ite. (P)	**Total No. of elements**	**SdLb level (dB)**	**HPBW (Degree)**	**SdLb Level Angle (Degree)**	**No. of off eleme nts**
1	04	-4.6	42	87	02
2	09	-13	25	89	05
3	16	-22	19.4	89	06
4	25	-31.3	16	89	08
S	*Expansion Factor (S) = 2*				
Ite. (P)	**Total No. of elements**	**SdLb level (dB)**	**HPBW (Degree)**	**SdLb Level Angle (Degree)**	**No. of off eleme nts**
1	04	-4.6	42	87	02
2	16	-15.4	15.4	34.8	08
3	64	-22.3	7.4	40	32
4	256	-22.4	3.6	21	128

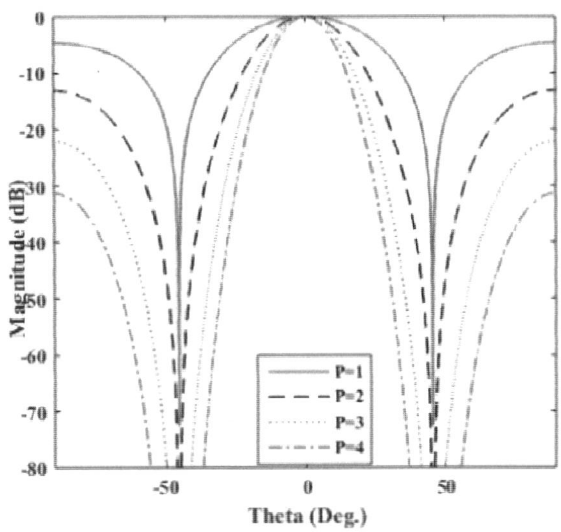

(a)

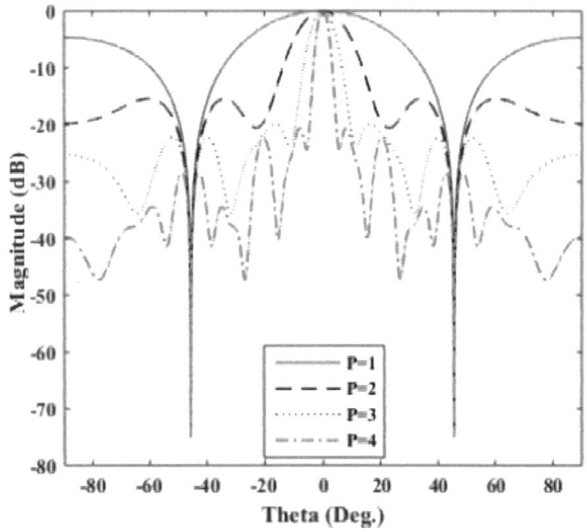

(b)

Figure 3. Array factor (P=1, 2, 3, 4) of Haferman counter-diagonal tapered rhombic fractal antenna array where(a) S=1, (b) S=2.

5. Conclusion

Haferman counter-diagonal fractal tapering technique to achieve thinning of antenna elements in deterministic rhombic FAA for two different expansion factors has been presented. In both cases, nearly 50% of elements become OFF and SLL lowered in most of the iterations compare to fully-populated FAA. The comparative analysis gives remarkable improvement of SLL with notable decrement of number total elements required by the proposed technique, which led to reduction of complexity and cost-effective.

References

[1] A Robert J. Mailloux, Phased Array Antenna Handbook. *Massachusetts, USA: Artech House, 2005.*

[2] R. C. Hansen, Phased Array Antennas. New Jersey, *USA: John Wiley & Sons, 2009.*

[3] M. T. Ma, Theory and Application of Antenna Arrays, New York, USA: John Wiley & Sons, *1974.*

[4] C. A. Balanis, Antenna Theory: Analysis and Design, New York, *USA: John Wiley & Sons, 1997.*

[5] Kraus, J. D. and R. J. Marhefka, Antennas: For All Application, *New York, USA: McGraw-Hill, 1997.*

[6] Praveena A, Ponnapalli VS. "A Review on Design Aspects of Fractal Antenna Arrays*" IEEE International Conference on Computer Communication and Informatics (ICCCI , pp. 1-3. 23 Jan 2019.*

[7] Kashyap, S. Sreenath, Kantipudi MVV Prasad, and Vipul M. Dabhi. "Novel Microstrip Band Pass Filter for C-Band Wireless Applications." *International Journal of Engineering & Technology 7.4.6 (2018): 227-229.*

[8] D. S. S. Satyanarayana and K. M. V. V. Prasad, "Multilayered antenna design for smart city applications," *2nd Smart Cities Symposium (SCS 2019), Bahrain, Bahrain, 2019, pp. 1-7.*

[9] Steven. R. Best, Small and Fractal Antennas (chapter10), Modern Antenna Handbook, *USA: John Wiley & Sons, Inc., publication.*

[10] D.H.Werner and R.Mittra, Frontiers in Electromagnetics, *USA: IEEE Press, 2000.*

[11] V. A. Sankar Ponnapalli and P. V. Y. Jayasree, "Design of multi-beam rhombus fractal array antenna using new geometric design methodology," *Progress In Electromagnetics Research C,*

vol. 64, pp. 151-158, June 2016.
[12] Praveena A, V.A. Sankar Ponnapalli, Design of Thinned Triangular Fractal Antenna Array using Genetic Optimization Algorithm, *International Journal of Algorithms Design and Analysis, Vol. 6, No.1, pp.1-5, 2020.*
[13] R. T. Hussein and F. J. Jibrael, "Comparison of the radiation pattern of fractal and conventional linear array antenna," *Progress in Electromagnetics Research Letters, vol. 4, pp. 183–190, 2008.*
[14] V. Srinivasa Rao and V.A. Sankar Ponnapalli, "Study and analysis of fractal linear antenna arrays", *IOSR-JECE, vol.5, pp. 23–27, March 2013.*
[15] Douglas H. Werner, R.L.Haupt and P.L.Werner, "Fractal antenna engineering: The theory and design of antenna arrays," *IEEE Trans. Antennas Propag. Mag., vol. 41, pp. 37-59, Oct.1999.*
[16] V.A.Sankar Ponnapalli and P.V.Y.Jayasree, "Heptagonal fractal antenna array for wireless communications," *Lecture Notes in Electrical Engineering Series, vol.372, pp. 387-394, Jan. 2016.*
[17] Waroth Kuhirun, A new design methodology for modular broadband arrays based on fractal tilings, *Ph.D. Thesis, Department of Electrical Engineering, Pennsylvania State University, USA, 2003.*
[18] V.A.SankarPonnapalli and P.V.Y.Jayasree, "Design and analysis of novel fractal linear and planar array antennas for satellite applications," *Advanced Elctromagnetics (AEM) Journal, vol. 5, pp. 56-62, 2016.*
[19] Skolnik.M et al, ''Statistically Designed Density-Tapered Arrays,'' *IEEE Trans. on Antennas and Propagation, vol. 12, pp. 408–417, July 1964.*
[20] Will P. M. N. Keizer, "Synthesis of Thinned Planar Circular and Square Arrays Using Density tapering", *IEEE Trans. on Antennas and Propagation, vol.62, pp.1555 –1563, April 2014.*
[21] *www.oeis.org.*
[22] Xu F.,Shmat'ko A.A., Panov M.I.,He.S., "Antenna array with a given fractal distribution of current," Ultra wideband and Ultra short Impulse Signals, *pp. 212-214, Sevastopol, Ukraine, 17-21 eptember, 2012.*
[23] Vladimir M. Masyuk, "Methods for mathematical modelingof fractal antenna arrays," *International Conference on Antenna Theory and Techniques, pp. 220-223, Semstopol, Ukraine, 9-12 September 2003.*
[24] V. F. Kravchenko and V. M. Masyuk, "Application of new weighting functions for design of Two-dimensional fractal antenna arrays," *MSMW'O4 Symposium Proceedings, pp. 171-176, Kharkov, Ukraine, 21-26 June 2004.*
[25] V. A. Sankar Ponnapalli and P. V. Y. Jayasree, "An Investigation of Fractal Antenna Arrays for Side Lobe Reduction with a Fractal Distribution of Current," *pp.1757-176, IEEE Int. Conf. Adv. In Comp., Comm., and Infor., Kochi, India, 10-13 Aug. 2015.*

Ontology Based Website for Job Posting and Searching

N. Anil Kumar[1*], Ramu K[2]

[1]Department of IT, Vardhaman College of Engineering, Shamshabad, Kacharam, Hyderabad, Telangana 501218, India. *Email: anil230@gmail.com

[2]Department of CSE, Vardhaman College of Engineering, Shamshabad, Kacharam, Hyderabad, Telangana 501218, India.

Abstract. Searching for the jobs remains a tedious work. Whenever we search for job we getting some results which are not very relevant. So, we have to manually search through those results and find some suitable job for which we can apply. Many times, we miss out some job postings because these posting do not appear in the results. The same problem exists for recruiters also. When recruiters search for some specialization there are not getting the same results because the keywords used for searching are different from the keywords used in their resumes. So, to solve this problem we use ontology to post job and to search the jobs.

1. Introduction

Unemployment rate in 2017 is 5.52% according to [1]. With world population of 7.5 billion, 428 million people are unemployed. There is another category called underemployment, which means that highly qualified people who are working for under paying jobs [2]. Another problem is Skill shortage or Unemployability which means qualified degree holder is not fit for employment because of he / she lack the skill the job requires (Eg. Problem solving skill, communication skill, advanced skill like data analytics etc) as stated by former President of India APJ Abdul Kalam [3]. For example, India has more than 300 dental colleges in India, in contrast US has around 66 dental colleges. So obviously dentists have no jobs. In this paper we try to give an outline of how to build a job search portal where people can easily search the jobs for a given skill. They know where there is skill gap in their respective field. They can search for the salary they will be getting for a particular skill. [8] in their book explains how the web should be semantic not just display the pages means we should process the data and give results for our queries.

2. Tradition Job Search

Most of the current job search sites provide features like search a job by keywords, search a job by salary, search a job by type, location, qualification etc for the users. They would not provide answers for queries like how many jobs are there for a particular qualification. They do not get answers for queries like what specializations are there in their domain which are offering more jobs and its comparison to other specializations in their domain

3. Web Ontology Language

Ontology attempts to represent different entities and their relationships. Web Ontology Language (OWL) [4] is a programming Language used for knowledge representation. It is used to represent entities and their relationships so that machine can process the knowledge and give conclusions.

OWL has features to define a Class. To define a class OWL follow the XML syntax. In OWL the concept of class is similar to the concept of class in object-oriented programming (although there are some differences).

We can define the class as

```
<owl:Class RDF:ID="name"/>
```

Here "name" is defined as a class.

OWL has features to define a sub Class of the Class, properties of the class, to define that two objects or properties are equivalent, etc.

4. Web Ontology Language Ontology For Posting The Job:

To create ontology for job posting [10] we have to define the vocabulary for the educational qualifications. We define the course Graduation and Post-Graduation as the class.

We define the Bachelor of Science and Bachelor of Technology to be the subclass of Graduation and Master of Science as the subclass of Post-Graduation. We define MBA to be subclass of Post-Graduation.

```
<owl:Class rdf:ID="B.Sc">
  <rdfs:subClassOf rdf:resource="#Graduation">
</owl:class>
<owl:Class rdf:ID="B.Tech">
  <rdfs:subClassOf rdf:resource="#Graduation">
</owl:class>
<owl:Class rdf:ID="M.Sc">
  <rdfs:subClassOf rdf:resource="#Post Graduation">
</owl:class>
<owl:Class rdf:ID="MBA">
  <rdfs:subClassOf rdf:resource="#Post Graduation">
</owl:class>
```

We define the Chemistry to be class and Inorganic Chemistry to be subclass of Chemistry.

```
<owl:Class rdf:ID="Chemistry"/>
<owl:Class rdf:ID="Inorganic Chemistry">
  <rdfs:subClassOf rdf:resource="#Chemistry">
</owl:class>
```

If the recruiter wants post a job on the job portal, he has to follow some predefined format. As an example, the recruiter wants to recruit a person with qualification M.Sc with specialization in Chemistry, having experience of two years, work place is Hyderabad, with cost to the company of one lakh rupees and the last date to apply for the job is 15 aug 18. Let us post the job with ID XYZ. Then the posting will look like this

```
<owl:Class rdf:ID="XYZ">
  <job:studyQualification> MSc</job:studyQualification>
  <job:studySpecialization>Chemistry</job:studySpecialization>
  <job:Company> companyId</job:Company>
 <job:Experience> 2 years</job:Experience>
 <job:Location> Hyderabad </job:Location>
 <job:CostToCompany> 1 lakhs
</job:CostToCompany>
  <job:LastDateToApply> 15 aug 18
</job:LastDateToApply>
<job:VacantPosition>10</job:VacantPosition>
</owl:Class>
```

Suppose there are some words which have same meaning, these words can be included in the OWL as follows

<owl:Thing rdf:ID="SCM">
<owl:sameAs rdf:resource="Software Configuration Management"/>
</owl:Thing>
<owl:Thing rdf:ID="SCM">
<owl:sameAs rdf:resource="DevOp"/>
</owl:Thing>

When queried with different keywords also based on the meaning it gives correct results. If the user wants to query for the jobs he has to use the SPARQL query language [5] . Suppose the user wants to find the jobs which require specialization in Chemistry, he uses the following query.

SELECT ?jobid, ?company
WHERE {
?jobid job:studySpecialization Organic Chemistry
?jobid job:Company ?company
}

Now this query lists not only the jobs for Chemistry specialization but also the specialization of its subclasses. The above query lists the jobs posted with id XYZ because we have specified in the OWL that Organic chemistry is the subclass of chemistry. So the user does not miss the job posting of Chemistry. We get more accurate results.

5. Working of the website

Before you begin to format your paper, first write and save the content as a separate text file. The job portal has three different kinds of users a) employer/ recruiter b) user or the person who is looking for the job c) the website administrator

5.1. The functionalities of the employer:

i) To post the job: The employer fills a form which has the fields like qualifications, Specialization, Experience, job Location, Cost to Company, last Date to apply, etc.

For selecting qualification there will be dropdown menu for selection qualification from basic degree to highest degree. The recruiter has the option to select any PG with specialization or a particular PG with no specialization. Similarly, he has the options to select a particular PG and working experience of 2 years. Now the recruiter has the option to specify the Boolean operators like and, or, etc., with specialization and experience.

ii) Give feedback on the recent interviews conducted: After completion of the interview the recruiter has to give feedback on the average knowledge of the candidates he has interview, what are the deficiencies he has observed in the candidates, whether he got the right candidate he is looked for.

iii) Pay packages offered: What are the promotions, allowance apart from the pay he will be getting etc.

iv) Search the resumes in the portals: Sometimes it happens that the recruiter does not want post his job, he will search the job portal for matching resumes and he tries to contact them directly. While searching for the resumes the job portal provides a summary of the information based on the chosen criteria. The recruiter can contact the candidates through the portal. The portal has facilities to group mail the short listed candidates.

5.2. The functionality of User:

i) To post his resume: The user who wants to post his resume will first create an account. He has to select his qualification, experience from the dropdown menu. So, there will be no ambiguity in his qualification and experience. He has to select his expertise in the areas by dropdown menus.

IOP Conf. Series: Materials Science and Engineering 1042 (2021) 012006 doi:10.1088/1757-899X/1042/1/012006

ii) Search for jobs: For searching the jobs the user does not need to create the login. The user can get the summary of the current job posting for a particular educational qualification which he can query, so that he can decide his carrier path. The user can query what are specialization available for his educational qualification and how many jobs are there for that particular specialization including salary, job location. User will search the jobs and can apply based on his qualification, experience, specialization, etc. For example

The following is one job posting for a ‘companyId1’ company

```
<owl:Class rdf:ID="XYZ1">
 <job:studyQualification>
    B.Tech
</job:studyQualification>
<job:studySpecialization>
   Computer Science
</job:studySpecialization>
 <job:Company> companyId1</job:Company>
 <job:Experience> 0 -2 years</job:Experience>
<job:jobSepecialization>
        SCM
</job:jobSepecialization>
 <job:Location> Hyderabad </job:Location>
 <job:CostToCompany>
       3 lakhs
</job:CostToCompany>
 <job:LastDateToApply>
      15 aug 18
</job:LastDateToApply>
<job:VacantPosition>10</job:VacantPosition>
</owl:Class>
```

The following is another job posting for a ‘companyId2’ company

```
<owl:Class rdf:ID="XYZ2">
 <job:studyQualification>
        B.Tech
</job:studyQualification>
<job:studySpecialization>
      Computer Science
</job:studySpecialization>
<job:Company> companyId1</job:Company>
 <job:Experience> 0 -2 years</job:Experience>
<job:jobSepecialization>
      Software Configuration Management
</job:jobSepecialization>
 <job:Location> Hyderabad </job:Location>
 <job:CostToCompany> 2 lakhs
</job:CostToCompany>
 <job:LastDateToApply>
      15 aug 18
</job:LastDateToApply>
<job:VacantPosition>11</job:VacantPosition>
</owl:Class>
```

The following is another job posting for a ‘companyId3” company

```
<owl:Class rdf:ID–"XYZ2">
```

```
<job:studyQualification>
    B.Tech
</job:studyQualification>
<job:Company> companyId1</job:Company>
 <job:Experience> 0 -2 years</job:Experience>
<job:jobSepecialization>
     digital marketing
</job:jobSepecialization>
 <job:Location> Hyderabad </job:Location>
 <job:CostToCompany> 2 lakhs
</job:CostToCompany>
 <job:LastDateToApply>
     15 aug 18
</job:LastDateToApply>
<job:VacantPosition>5</job:VacantPosition>
</owl:Class>
```

Now the user queries for what are specializations available for B.Tech Computer Science students

```
SELECT (count (?positionVacant) as ?Total),
    ?jobSpecialization  WHERE {
?jobid job:studySpecialization Computer Science
?jobid job:studyQualification B.Tech
?jobid job:jobSpecialization ?jobSpecialization
?jobid Job:VacantPosition ?positionVacant
}
```

Given the above two ontology entries the above query gives the following result
26 Software Configuration Management
5 digital Marketing

Because the ontology understands SCM and Software Configuration Management as same and B.Tech Computer Science is subclass of B.Tech.

iii) Report inappropriate results: If the user feels that he got results for his search queries which are not appropriate for the query, a feedback form will be provided to him to report the results.

iv) Report relevant missing results: If the user feels that by giving a particular query he has not got some result but by giving some other query he got the result with is reverent to former query also, he can report it to the administrator by giving both the query and the results.

5.3. The functionality of administrator:

i) Accepting the jobs posted: Generate ontologies for the posted jobs by the employer. Checks for anomalies and if detected they will be reported to the employer.

ii) Update Ontology: If the employer added new qualification check its validity and add it to the ontology database. Similarly, if the User adds new qualifications to his resume he checks the validity and add it to the ontology database

iii) Fine tuning the search results: If the user reports inappropriate results he checks for the error and if required he added the ontology to the database or does some intuitive work like changing the page ranking, etc. If the user reports missing results he checks for the error and if required he added the ontology to the database or does some intuitive work.

iv) Generating the reports: Gives the summary of the jobs per location, per qualification, per position, etc and post them every two months. These reports are in OWL format so that we can query these reports.

6. Conclusion

Unemployed people can find the job by searching the web site easily based on their qualification. They can know what are the specializations available in their domain that are currently offering jobs. The unemployed people can learn these specializations and get jobs. Youth can choose their career path by search the available jobs in this website for that particular study.

With these basic ideas we hope to construct a full-fledged web site to help people with jobs searches [9]. There are many opensource webservers that supports OWL and SPARQL some examples are Apache Jena [6] , ClioPatria [7]. We hope that Commerical web sites like Shine.com, naukri.com, freshersworld.com, etc. implement this OWL in their job search.

References

[1] A reference Unemployment, total (% of total labor force) (modeled ILO estimate) | Data. [online] Available at: https://data.worldbank.org/indicator/sl.uem.totl.zs [Accessed 5 Oct. 2018].

[2] Postgraduates, PhD holders applying for sweepers, peons' posts in UP: Varun - Times of India. Retrieved from http://timesofindia.indiatimes.com/articleshow/51185254.cms?utm_source=contentofinterest&utm_medium=text&utm_campaign=cppst

[3] 'Unemployability' is a bigger crisis than unemployment, says Kalam. Retrieved from https://www.thehindu.com/todays-paper/tp-national/tp-karnataka/lsquoUnemployabilityrsquo-is-a-bigger-crisis-than-unemployment-says-Kalam/article15147193.ece

[4] OWL - Semantic Web Standards. [online] Available at: https://www.w3.org/OWL/ [Accessed 10 Oct. 2018]

[5] SPARQL Query Language for RDF. Retrieved from https://www.w3.org/TR/rdf-sparql-query

[6] Apache Jena (2018, August 24). Retrieved from https://en.wikipedia.org/wiki/Apache_Jena

[7] Wielemaker, J., Beek, W., Hildebrand, M., & Ossenbruggen, J. V. (2016). ClioPatria: A SWI-Prolog infrastructure for the Semantic Web. Semantic Web, 7(5), 529-541. doi:10.3233/sw-150191

[8] Alesso, H. P., & Smith, C. F. (2006). Thinking on the Web: Berners-Lee, Godel, and Turing. Hoboken, NJ: Wiley-Interscience.

[9] McGrath, S. (1998). XML by example: Building e-commerce applications. Upper Saddle River: Prentice Hall PTR.

[10] Daconta, M. C., Obrst, L. J., & Smith, K. T. (2003). The semantic web: A guide to the future of XML, web services, and knowledge management. Indianapolis: Wiley Pub.

Indian agriculture GDP and non performing assets: A regression model

Bhoomika Jethwani[1], Devasya Dave[2], Tamanna Ali[3], Shraddha Phansalkar[4] and Swati Ahhirao[5]

[12345]Department of Computer Science and Engineering, Symbiosis Institute of Technology, Lavale, Tal Mulshi, Pune, 412115, India
bhoomika.jethwani@sitpune.edu.in, devasya.dave@sitpune.edu.in, ali.tamanna@sitpune.edu.in, shraddhap@sitpune.edu.in, swatia@sitpune.edu.in

Abstract. Agriculture is one of India's crucial sectors in terms of its contribution to employment and the country's (Gross Domestic Product) GDP. It has primarily emerged as an essential - growing sector in the global economy since independence. [21] However, the non-realization of the reasonable price for agricultural crop production leads to the introduction of loan waivers, which impact the credit culture and weaken the farming economy and growth. The presented work aims to perform exploratory data analytics on the GDP data in agriculture public domain by performing feature engineering on the factors affecting the agricultural GDP using the data for the period 1961 to 2019. It further builds a multi-linear prediction model to forecast the Agriculture Sector's economic performance in terms of GDP and NPAs generated by the Agricultural Sector using Machine Learning Techniques. Keywords: Multiple Linear Regression, Ordinary least Squared, Agricultural GDP, Non-Performing assets, Indian Economy.

1. Introduction

In today's day, the identification of nations of the world is made by their economy's strength. High wage jobs and improved quality of life are some components that are driven by economic development. We can portray the economy's growth using indicators that attempt to present an overall idea of any particular geography's financial health. Economic growth is an essential indicator because it indicates growth, whether measured by GDP (Gross Domestic Product), GVA (Gross Value Added).

GDP data helps to measure economic health and future inflation. High GDP data portrays the strength of an economy. Higher growth increase investors automatically increase the country's exports, which have a direct impact on the jobs and infrastructure. [6] However, a shrinking GDP is an early sign of recession, decreasing wages, and declining business revenues. Thus, the GDP information is crucial for the health of the economy and everything else that is remotely associated with it. A flourishing economy's skeleton is a strong banking sector, failure of which can cause adverse effects on other industries, as well. In India, banks face lots of concerns regarding India is Non-Performing Assets (NPA). The rise in NPAs implies a high possibility of increased credit defaults that, in turn, affect banks' profitability and net worth and plunge the asset's value. Thus, there is an indispensable need to control banks' issues regarding NPA's in the Agricultural Industry.

Banks must identify whether crediting loans into a particular sector is beneficial or not to avoid NPAs. This work intends to study the agricultural sector's performance in terms of GDP and bad loans (NPA).

Contributions of this work:

1. Recognizing key features affecting the agricultural GDP.
2. Analyzing the performance of the Agricultural sector in terms of GDP.
3. Proposing a model that predicts the percentage of NPA in terms of Agricultural sector GDP.

2. Related Work

The NPA is a short form of 'Non-Performing Asset.' The built-up of NPA has been the dominant factor in destroying the public sector banks' profitability in India. [2] Narasimham Committee (II) has emphasized the need to reduce all banks' average NPAs. It has recommended prudential norms on income recognition, asset classification & provisioning. When interest or other dues to a bank remains outstanding for longer than 90 days, the NPA rule affirms that the entire bank loan automatically turns into a non-performing asset.

The major problem for banks and financial institutions is the recovery of loans. The plan of action is firstly avoiding NPA, and if unavoidable, then identify and manage the factors responsible for it.

[2] Emulating international best practices and ensuring greater transparency, the 90 days' overdue criterion for recognizing NPAs has been selected, from the fiscal year 2005. Consequently, a non-performing asset (NPA) will be a loan or an advance where:

i. The principal's interest and installment remain overdue for more than 90 days (for term loan).
ii. Said account remains out of order for more than 90 days, in respect of an Overdraft/Cash Credit.
iii. The bill remains unpaid and overdue for more than 90 days in the case of purchased and discounted bills.
iv. Interest and installment of the principal continue to be unpaid and overdue for two harvest seasons.

The problem of Non Performing Assets has been a matter of discussion for commercial systems worldwide. [28] NPA's harm the banks and, consequently, the whole economy. The state of well-being of the production and trade is reflected by the level of NPA's in Indian banks. The prime duty of banking is the granting of credit. However, lending also carries a risk called credit risk, which arises from the borrower's failure. The non-recovery of loans causes a significant hurdle in the process of the credit cycle. These loans affect the bank's profitability. The problem of Non Performing Assets has been a matter of discussion for commercial systems worldwide. [28] NPA's harm the banks and, consequently, the whole economy. The state of well-being of the production and trade is reflected by the level of NPA's in Indian banks. The prime duty of banking is the granting of credit. However, lending also carries a risk called credit risk, which arises from the borrower's failure. The non-recovery of loans causes a significant hurdle in the process of the credit cycle. These loans affect the bank's profitability. NPA control is the most significant hurdle encountered by the Banking Industry in India. Non Performing Assets (NPA) are a loss to the economy. Although the causes of NPA are varied, one of them is poor credit decisions by bank management. These decisions are taken due to the unavailability of data, inability to forecast the growth in a particular sector, policies of direct credit, that is, the crude form of behest lending.

In Sergio's work, [24] in a study on non-performing loans in Italy, an increase in loan assets' riskiness is rooted in a bank's lending policy adducing relative unselective and inadequate evaluation of sectorial prospects.

In Gopalakrishnan's work, [26] the causes of NPAs can be classified into political, economic, social, and technological factors. He perceived that oversight of regular credit appraisal, absence of follow-up and surveillance, recessional stresses in the marketplace, a shift in government policies, and digression of funds are the major causes of NPAs.

[23]Challenges faced by the agricultural sector:

1. Fragmentation and uneconomical size of land holdings due to the *Zamindari* system and division of land amongst family members through numerous generations.
2. High wastage due to poor storage and supply chain infrastructure.
3. Significant dependency on monsoon and insufficiency of irrigation facilities.
4. Lack of extension duties in the sector, which include applying scientific research and new knowledge to agricultural practices.
5. Governments debt waiver schemes
6. Stress in the sector is inadequate to justify the growth in non-performing loans. According to the report of Standard Chartered Securities, *"The problem with farm loans is not slow growth but rising non-performing loans as farmers expect more debt waive"*.
7. Mandatory requirements by the government
8. An increase in NPA can be attributed mainly to existing policy prescriptions, substantially increasing agricultural credit.

However, no systems are analyzing the primary reasons contributing to high NPA ratios in agricultural finance and how the non-repayment of loans affects performance and the GDP.

3. Proposed Work

We propose a model that finds the key features relating to GPD and NPA, which can be tackled to solve the problem of NPA management in India. In this work, we focus on the agricultural sector, and the system is based on the following objectives:

i. Prediction of Agricultural GDP using the regression model.
ii. Prediction of NPA in the agricultural sector.
iii. Analyzing the performance of Agricultural GDP for the period 1961 - 2019.
iv. Investigating the States contributing to the maximum amount of NPA and the types of loans provided to farmers that often tend to go wrong.

3.1 Methodology

This work explores the capabilities of ML methods in the task of predicting the growth and presence of bad loans in the agricultural sector. We propose the method of supervised ML Regression algorithms in a two-phase process for this purpose. In the first phase, the dataset is split into the training and testing data. The second phase involves the application of the regression algorithms to build and identify the best model. This permits us to recognize the most relevant results and thereby increase accuracy.

3.1.1 Dataset Construction. The first phase of data analysis for predicting growth and presence of bad loans in the agricultural sector is collecting the data on essential features from various open sources, dataset repositories, official annual reports, and research papers. However, the collection and integration of data from such varied sources are challenging. A labeled dataset is required to develop an efficient prediction model. We have prepared such a dataset by manually labeling the features that will enable us to predict the relevant output.

Variables used in the dataset are as follows:

Area under agriculture (million hectares)
India GDP (billion USD)
Export Value (billion USD)
LokSabha Elections (Yes/No)
Outstanding credit amount (billion USD)
Agriculture Percentage in India GDP
Arable Land (million hectares)
Agriculture GDP (billion USD)
Fertiliser Consumption (kg/hectare)
Area under agriculture (million hectares)
Rainfall (mm)
Year (1960 - 2019)
Agricultural machinery (per 100 sq. km. of arable land)
Outstanding Credit as percentage of Agricultural GDP
Agriculture Credit Disbursed as amount of Agriculture GDP
Agriculture Credit Disbursed as percentage of Agriculture GDP

Total FDI amount (billion USD)
Percentage of Agricultural NPAs (Public Sector Banks)
Percentage of Agricultural NPAs (Private Sector Banks)
Rainfall deviation in Percentage from LPA

3.1.2 Exploratory Data Analysis. In this work, the raw dataset is transformed into standardized forms to develop efficient models. This standardization has taken place in multiple steps:

- ***Handling categorical data:*** The dataset consisted of only one categorical feature, namely 'LokSabha Elections'. To handle this feature, *LabelEncoder* has been used, where each column is converted to a number.
- ***Correlation amongst features:*** It helps in feature selection by showcasing the relationship between the features. Heat-maps have been used as they are a good visualization of the correlation. Observations have led to the inference that the features with high correlation with agricultural GDP are -Production, Yield, Export Value, Credit disbursed amount, Outstanding credit amount.
- ***Handling missing values:*** The features where the number of missing values were much higher than available values, such as '*Fertiliser Consumption*' and '*FDI Amount*', were dropped. For features with a few missing values, such as '*Machinery*', the data was filled using the mean of all available values.
- ***Feature Scaling:*** MinMax Scaler has been used to scale the features to normalise the data within a particular range and to speed up the calculations in the algorithm.
- ***Visualisation:*** The visualisation of data has led to the following inferences:
    1. Production, Yield, Export Value, Credit Disbursed, Outstanding Credit have a strong positive correlation with Agriculture GDP.
    2. Rural Population, Employment, Arable Land have a strong negative correlation with Agriculture GDP.
    3. All the variables that have high p-value (> 0.05), like Area, Rainfall, etc have very low correlation with Agriculture GDP, so they will be insignificant in our model building process.

3.1.3 Feature Selection. The first step in regression analysis is to determine the criterion variable. Following our objective and system, for our first model, the dependent variable is Agricultural GDP. For our second model, the dependent variable is Outstanding Credit as a percentage of Agricultural GDP. The independent variables for each of the model needs to be chosen after the dependent variable is found. For an efficient model, it is advised to have the minimum number of features that cause maximum variance in the dependent variable.

This work has incorporated multiple feature selection components such as RFE, VIF, and p-value.

- Recursive Feature Elimination (RFE): It removes attributes recursively and builds a model on the remaining attributes. [25] An accuracy metric is used to rank the features according to their importance. The model to be used and the number of required features are given as input to the RFE method. Then, the ranking of each variable (1st feature is most important) is returned. It also provides its support where 'True' is relevant, and 'False' is irrelevant.
- Variance Inflation Factor (VIF): The collinearity of two variables can be identified using this. A VIF of 1 indicates non-collinearity, i.e., no correlation. However, the higher the VIF, the higher is the collinearity.
- P-value: [19] A low p-value signifies a more significant effect on the output, and the result is of greater importance. Statistically, having a p-value < 0.05 is deemed vital. In this work, we have selected the features with a p-value of less than 0.05 for each model.

For both of the models in our system, feature selection has been done in two steps.

A) *Agricultural GDP Model*

i. Features with insignificant p-value such as 'percentage of NPAs in Public Sector Banks and Private Sector Banks', 'Arable land', 'Rainfall Deviation' were removed.
ii. Feature with non-coherent VIF such as 'Yield' has been removed.

B) *NPA Model*

i. Features with insignificant p-value such as 'Rural Population', 'Export Value' were removed.
b. Feature with non-coherent VIF such as 'Outstanding Credit Amount' has been removed.

4. Trends of Agricultural GDP

Depicting the correlation of features with the Agricultural GDP and analyzing the p-value helps us select features to build an accurate prediction model. The graphs of features having a linear relationship with GDP and p-value<0.05 are further analyzed along the period (1960 - 2019) below.

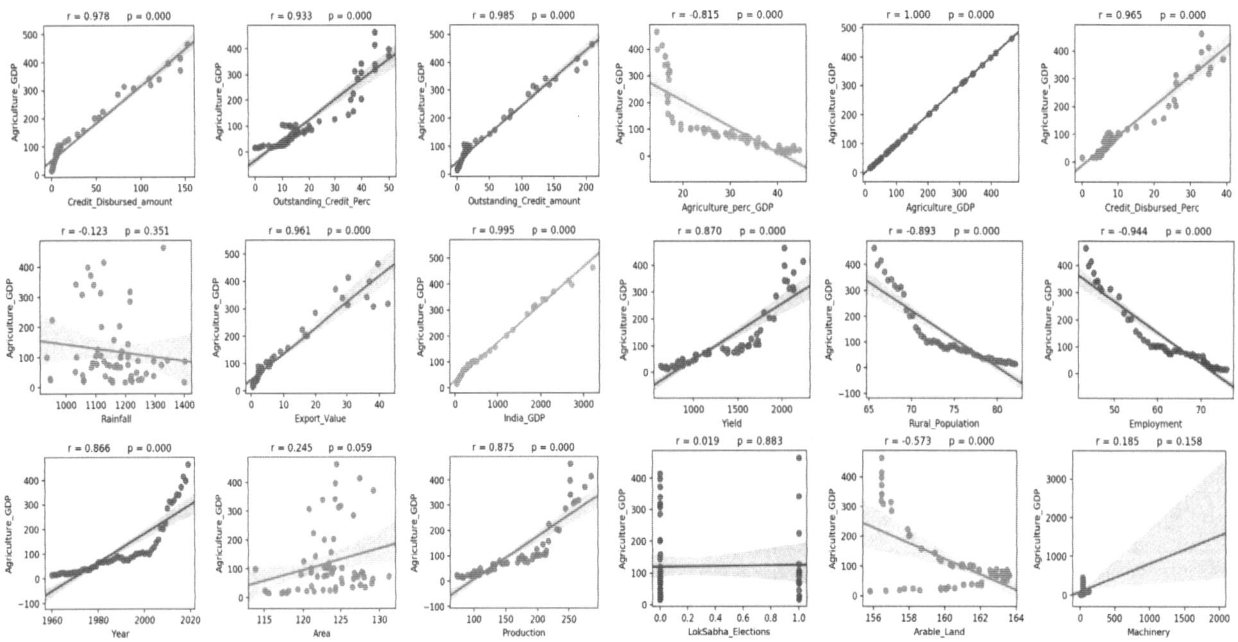

Figure 1: Correlation of features with Agricultural GDP

4.1 Credit disbursed vs. outstanding credit as a percentage of Agricultural GDP

The ratio of Agricultural Credit outstanding to Agricultural GDP peaked from 0.6 % in 1950-51 to 9.81 % in 1971-72. [7]The rate shows an escalated trend increasing to 21.76 %, from 1972 to 1986-87. The remarkable success of agricultural credit vs. agricultural GDP from the 1950s-1980s is based on banks' nationalization and the initiation of RRBs. However, the backward trend in the ratio commenced from 1991 onwards, and it fell to 13.34 % in 1998. Following 1999, the rate heightened steeply and reached up to 39.55 % in 2006-07, which indicates that the entrance of KCC was a big promoter for agricultural credit and induced a significant transformation in widening the reach of credit to the farming population. The loan waivers' announcement negatively influenced the borrowers' reparation behavior and made the banks reluctant to fresh lending.

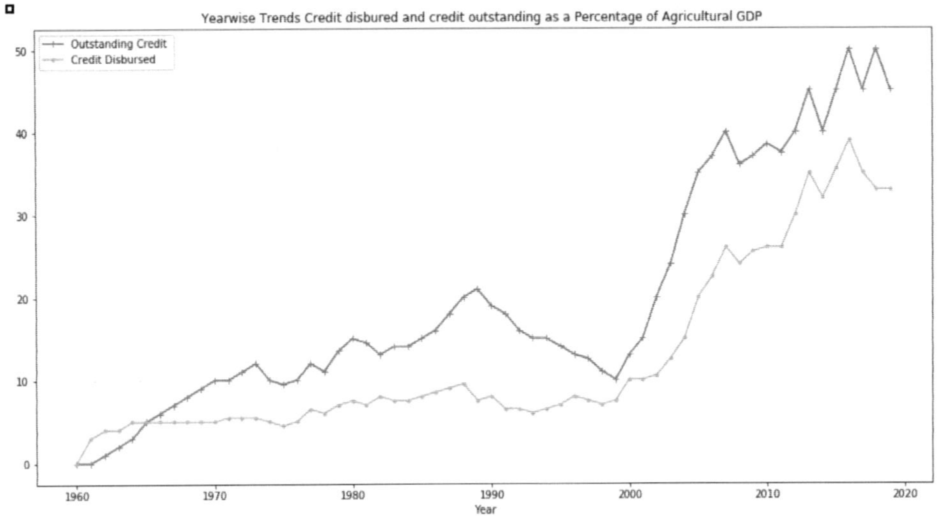

Figure 2: Credit disbursed vs. outstanding credit year wise trends

4.2 Export Value of crops

India is second in the global farming generation, but its share of world agricultural exportation is as low as 2%. Agricultural exports' benefaction to India's GDP is also as little as 2%, weaker than other developing agrarian country. Since the economy reopened in 1991, [27] India's agricultural business surplus listed more than ten-fold rise between 1991-92 and 2013-14. The brisk pace at which exports increased to balance the corresponding increase in imports. However, in the last few years, agricultural exports fell by 22% while imports increased by 62%, resulting in a significant drop in a trade surplus of 70%, consequently affecting the agricultural GDP.

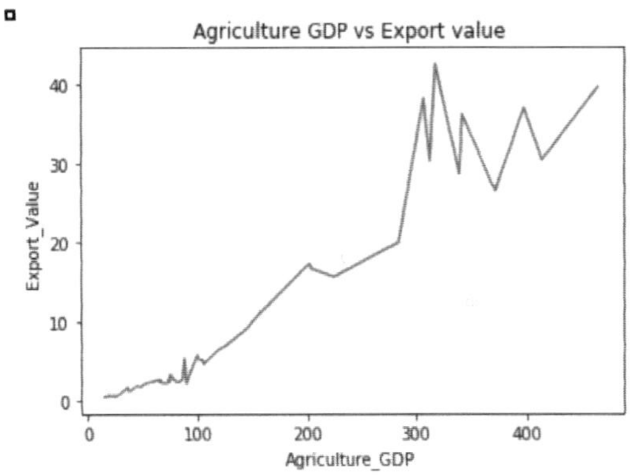

Figure 3: Export Value Trends vs Agri-GDP

4.3 Rural Population of India

[12.] In 2018, the rural population in India was 892,321,700. Over the past 58 years, it reached a maximum of 892,321,700 in 2018 and a minimum amount of 369,791,500 in 1960. In rural areas, agriculture plays an essential role in the economy. [21] Over 70 percent of rural families depend on agriculture for their daily sustenance. It employs over 60% of the rural population. Indian agriculture has registered remarkable growth over the last few decades, along with its inexorable rise in population. Hence, it is clear that the growth of the population is linearly related to the Agricultural GDP.

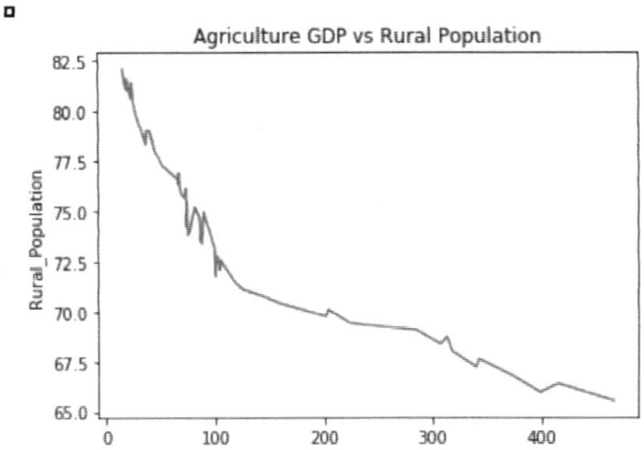

Figure 4: Population Trends vs Agri-GDP

5. Trends of Agricultural NPA

Loan enforcement of PSBs shows co-movement with rainfall deficiency, except in 2008-09 and 2009-10, where loan write-offs could have influenced the lessening of the Non- Performing Assets (NPA) level despite high rainfall scarcity. [7]In the recent years of 2016-17 and 2017-18, the NPA level has dramatically risen, possibly symbolizing strategic default stemming from the state-level loan waiver releases. The linkage between agrarian distress and unfavorable rainfall in India is well established, as the foundation for irrigation is still lacking. At the state level, rainfall performance in the years was typical, with no notable deviation from Long Period Average (LPA) between 2004 to 2008, after three straight years of drought between 2001 and 2003, causing a drop in GDP.

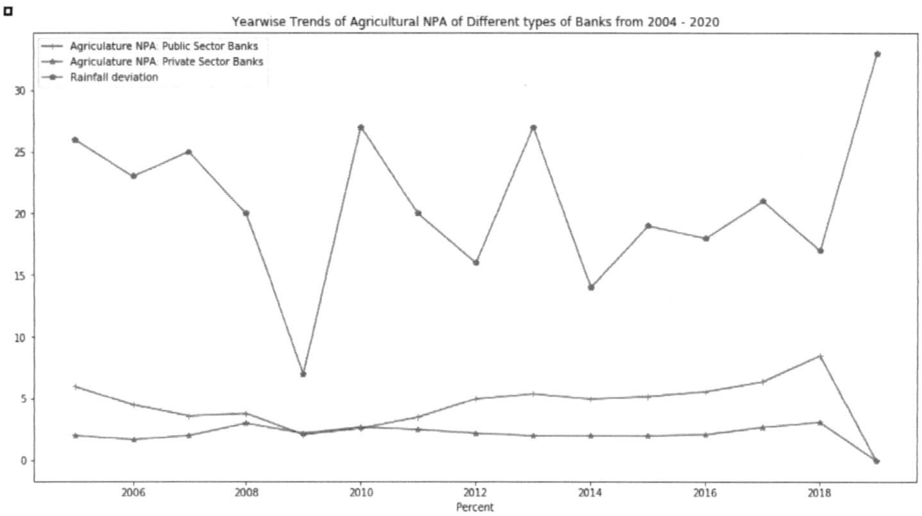

Figure 5: . Public sector NPA vs. Private sector NPA

5.1 State-wise data analysis

The data shows that the NPA level advanced for all states that have announced a farm loan waiver program in 2017-19. On the other hand, [7]almost all other states (except for Bihar, Odisha, and Haryana) have shown no significant change in their NPA level or have registered for a decline between 2016-18. Taken collectively, this could be suggestive of moral risk, with borrowers defaulting strategically in expectation of loan waiver. The states giving to the maximum amount of NPAs are Maharashtra (15.5%), Karnataka (10.3%), and Uttar Pradesh (10.2%)).

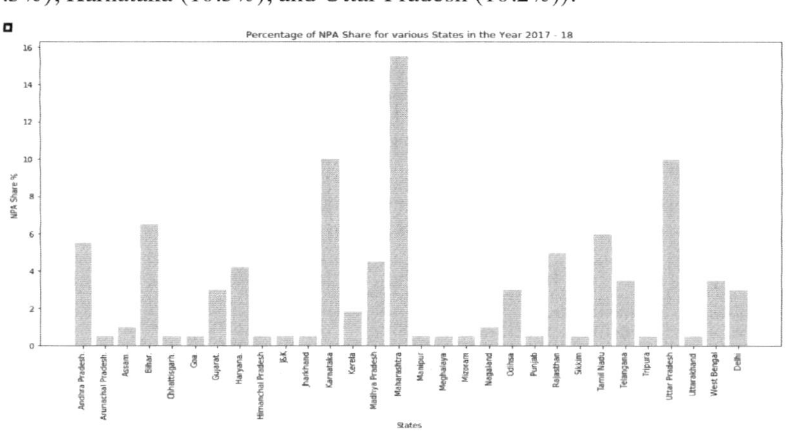

Figure 6: State-wise NPA

6. Proposed Model

Post data analysis, and feature selection, the dataset was fit into a Multiple Linear Regression algorithm to obtain the two prediction models. Since multiple independent variables are affecting the dependent variable, this model is used. The equation for multiple linear regression algorithm used in the model is:

$$y = b_0 + b_1 * x_{i1} + b_2 * x_{i2} + \ldots + b_n * x_{in} + e$$

In the above equation y is the dependent variable (here Agricultural GDP and NPA). Independent variables are x_{i1}, x_{i2} … x_{in} having coefficients/weights $b_1 \ldots b_n$, b_0 is the y-intercept and e is the model's error term. OLS method is used to obtain a line that is closest to the actual data points. The intent of regression is to determine the value of coefficients, such that the error is insignificant. OLS strategy helps us to achieve that. The prediction model is designed after reducing its multi-collinearity and selecting features having a maximum correlation to the dependent variable.

7. Results and Discussion

After analysis, we can state that the key features affecting Agricultural GDP from the model (based on p-value) are the Rural Population of India, Credit amount Disbursed as Agricultural Loans, Export Value of Agricultural Goods.

OLS Regression Results

Dep. Variable:	Agriculture_GDP	R-squared:	0.999
Model:	OLS	Adj. R-squared:	0.998
Method:	Least Squares	F-statistic:	1516.
Date:	Sun, 17 May 2020	Prob (F-statistic):	2.59e-34
Time:	19:38:28	Log-Likelihood:	65.611
No. Observations:	42	AIC:	-99.22
Df Residuals:	26	BIC:	-71.42
Df Model:	15		
Covariance Type:	nonrobust		

	coef	std err	t	P>|t|	[0.025	0.975]
const	0.2430	0.063	3.845	0.001	0.113	0.373
Year	-0.8429	0.209	-4.028	0.000	-1.273	-0.413
Area	-0.0111	0.022	-0.498	0.622	-0.057	0.035
Production	0.2178	0.070	3.116	0.004	0.074	0.362
Rural_Population	-0.9147	0.186	-4.922	0.000	-1.297	-0.533
Employment	0.0537	0.151	0.355	0.725	-0.257	0.364
Rainfall	-0.0222	0.012	-1.833	0.078	-0.047	0.003
Export_Value	0.0694	0.027	2.576	0.016	0.014	0.125
India_GDP	0.9596	0.114	8.383	0.000	0.724	1.195
Agriculture_perc_GDP	0.0943	0.051	1.862	0.074	-0.010	0.198
Credit_Disbursed_Perc	-0.0387	0.073	-0.530	0.601	-0.189	0.111
Credit_Disbursed_amount	0.3620	0.142	2.556	0.017	0.071	0.653
Outstanding_Credit_Perc	0.1627	0.072	2.256	0.033	0.014	0.311
Outstanding_Credit_amount	-0.6192	0.194	-3.192	0.004	-1.018	-0.220
LokSabha_Elections	-0.0024	0.004	-0.556	0.583	-0.011	0.006
Machinery	0.0161	0.015	1.056	0.301	-0.015	0.048

Omnibus:	11.975	Durbin-Watson:	1.455
Prob(Omnibus):	0.003	Jarque-Bera (JB):	13.893
Skew:	0.927	Prob(JB):	0.000962
Kurtosis:	5.122	Cond. No.	199.

Figure 7: Results for Agricultural GDP Model

Similarly, key features affecting Agricultural NPA from the model (based on p-value) are Area of agricultural land in India, Production of Agricultural crops, Employment rates and Credit Amount Disbursed as Agricultural Loans.

IOP Conf. Series: Materials Science and Engineering 1042 (2021) 012007 doi:10.1088/1757-899X/1042/1/012007

OLS Regression Results

Dep. Variable:	Outstanding_Credit_Perc	R-squared:	0.980
Model:	OLS	Adj. R-squared:	0.972
Method:	Least Squares	F-statistic:	131.8
Date:	Sun, 17 May 2020	Prob (F-statistic):	2.92e-22
Time:	19:38:30	Log-Likelihood:	2.8892
No. Observations:	42	AIC:	18.22
Df Residuals:	30	BIC:	39.07
Df Model:	11		
Covariance Type:	nonrobust		

	coef	std err	t	P>|t|	[0.025	0.975]
const	-0.3241	0.229	-1.416	0.167	-0.792	0.143
Year	0.1704	0.391	0.436	0.666	-0.629	0.969
Area	0.2702	0.076	3.555	0.001	0.115	0.426
Production	-0.7758	0.238	-3.263	0.003	-1.261	-0.290
Employment	-1.3162	0.541	-2.431	0.021	-2.422	-0.210
Rainfall	0.0501	0.046	1.085	0.287	-0.044	0.144
India_GDP	-0.0008	0.290	-0.003	0.998	-0.592	0.591
Agriculture_perc_GDP	0.0371	0.179	0.208	0.837	-0.328	0.402
Credit_Disbursed_Perc	0.8179	0.187	4.374	0.000	0.436	1.200
Credit_Disbursed_amount	-0.5982	0.288	-2.075	0.047	-1.187	-0.009
LokSabha_Elections	-0.0039	0.017	-0.228	0.821	-0.039	0.031
Machinery	-0.1274	0.057	-2.238	0.033	-0.244	-0.011

Omnibus:	4.849	Durbin-Watson:	1.981
Prob(Omnibus):	0.089	Jarque-Bera (JB):	3.628
Skew:	0.673	Prob(JB):	0.163
Kurtosis:	3.512	Cond. No.	84.2

Figure 8: Results for Agricultural NPA Model

7.1 Residual Analysis

Residual is computed by calculating the difference between the predicted output from the regression model and the measured output from the validation dataset. They represent the part of the validation data not defined by the model. Hence, we perform a residual analysis to check the accuracy of the model.

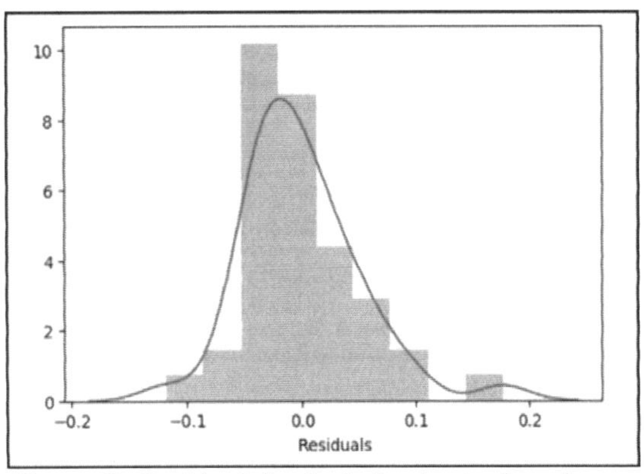

Figure 9: Residual Analysis for Agricultural NPA

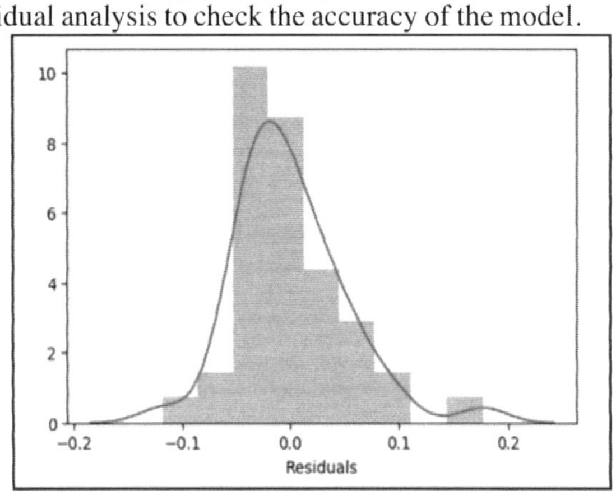

Figure 10: Residual Analysis for Agricultural GDP Model

As we can see, the residuals are normally distributed and have a mean of zero, which means that both the models are suitable.

7.2 Model Evaluation

Evaluation metrics used for the model are R Score and Mean square error to analyze its performance.

7.2.1 R score. The data points' closeness to the fitted regression line gives the R-squared statistical measure that ranges from (0 - 100 %). This value is also known as the coefficient of multiple determination for multiple regression. R-squared is f the percentage of the response variable variation that is explained by a linear model. A higher value of R-squared implies a better-fitted model for the data.

R-squared = Explained variation / Total variation

7.2.2 MSE (Mean squared error). MSE analyses how nearby a regression line is to the collection of data points. It calculates this value by taking the distances from the points to the regression line (these distances are known as "errors") and squaring them. Squaring the values helps to remove any negative signs. It also gives more weight to more significant differences.

Model	Accuracy (r score)	Mean Squared Error
Predict Agricultural GDP	0.995	0.011
Predict presence of bad loans	0.958	0.137

Table 1: Metrics for model evaluation

8. Application

The The intelligence from these models can be used to identify the growth and security of any sector. It can also be used to determine the number of threats any asset poses in terms of NPA. For example, should there be some agricultural sector assets, prone to turning into bad loans, then the necessary actions to manage them can be taken. The same holds for statistically predicting the presence and effect of these bad loans in the sector's GDP. Furthermore, should the data be analyzed in real-time, then the extracted intelligence will also be timely.

- The model notebook created can be further extended into an API for data-fetching applications or research extension purposes.
- The work can be deployed into a product with an interactive frontend to input and extract data resulting in an ever-growing dataset, to monitor the agricultural industry's performance.
- The concepts mentioned above can be released as an open-source repository for presenting contribution to data scientists worldwide to improve the accuracy and enhace the vision achieved in the current model.

8.1 Limitations of work

- The requirement to assemble a training dataset is achieved through manual labeling of the posts, which is resource-demanding, especially in the case of the multi-class dataset.
- Finally, the intricate features that have specific effects on the NPAs and the growth of a sector, as a whole, were not studied data-intensively.
- The model created cannot be reused as a starting point for creating models for other industries but only as a mark of citation due to similarity in use cases.
- The program does not provide a functionality to change the model algorithm as the nature and availability of data will change over the years to come.
- The Internet's available data is relatively inadequate since there is no proper record of data before 1960.

9. Conclusion

Agriculture is definitively one of the most influential contributors to our economy. [17] Despite such an overwhelming majority of the population engaged in agricultural activities, it is deplorable that the agrarian credit system is not yet well developed in India, and the existing systems are plagued with increasing numbers of NPAs. It is essential to realize that this is a systemic quandary that would not be fully solved by superfluous means such as farm loan waivers. It has to be tackled by resorting to some modern techniques and sophisticated mechanisms. These routines will help lessen the number of NPAs and further aid in boosting production levels, diminishing risks, and, consequently, leads to the overall growth of farmers' sustainable way. This work demonstrates the use of Multiple linear regression algorithms to predict economic growth in terms of GDP and bad loans as a percentage of GDP in the agricultural sector. [15] Even though most data analysts and financial practitioners have information about the NPAs and jeopardies involved in the industry's assets, distinguishing such assets and managing them has always posed a difficulty. Some features have been highlighted in this paper that affects the repayment of farm loans adversely like the rural population, the export value of crops, and crop production for that year. This consequently affects the GDP of the country. It also shows how identifying the features associated with a sector can help us determine its future growth and the risk involved in investing assets of those sectors. The acquired knowledge value also shows that this research also has a commercial development potential in the financial technology market that is yet unavailable.

10. Future Scope

This research work can provide a foundation for a more general framework for real-time data analysis from various sectors of the nation besides the agricultural industry. An interactive method that allows better visualization and exploration of the data would enable the analyst to understand better the risks and threats associated with assets. The relevance of this topic for future research lies in examining the extent of the adaptation of suggested technologies and mechanisms and further examining the impact of such adaptations on the NPA levels.

References

[1] Department of Agriculture, Cooperation and Farmers Welfare Annual Report 2018-29. http://agricoop.nic.in/sites/default/files/AR_2018-19_Final_for_Print.pdf
[2] Annual Report of the RBI 2018-19, Reserve Bank of India. https://www.rbi.org.in/Scripts/AnnualReportMainDisplay.aspx
[3] Economic Outlook. https://economicoutlook.cmie.com
[4] India, Knoema. https://knoema.com/atlas/India/topics/Agriculture
[5] Indiastat. https://www.indiastat.com/economy-data/8/union-budget/24469/2018-2019/1118397/stats.aspx
[6] Jayshree S, Focus on banking sector to tackle slowdown, Observer Research Foundation, 22 January, 2020. https://www.orfonline.org/research/focus-on-banking-sector-to-tackle-slowdown-60550/
[7] Reserve Bank of India's "Report of the Internal Working Group to Review Agricultural Credit"
[8] Jha, Vijay Shekhar, "Problems of NPA in Banking Sector in India & Debt Recovery Remedies", (May 1, 2018)
[9] Gopal, Sumathi, "NPA's-A Comparative Analysis on Banks & Financial Institutions and its Implications- (Doctoral Thesis)", (June 2010)
[10] S.K. Mishra, "Trends in growth of agriculture sector of the Indian economy", (2007)
[11] Sagarika Rastogi, "Developing Block Chain Smart Contract Platform for the Sale of Stressed Assets in Banks along with NPA Analysis for various sectors through Predictive Models and Visualization Techniques", (June 2019)

[12] India - Rural population, Index Mundi. https://www.indexmundi.com/facts/india/rural-population
[13] Pacha Malyadri, S. Sirisha, "Comparative Study of Non-Performing Assets in Indian Banking Industry"
[14] Kaveri.V.S., "Prevention of NPA suggested Strategies," Vinimaya, 23(8), (2001)
[15] Prashanth. K. Reddy. "A comparative study of non-performing Assets India in the global context–similarities and dissimilarities, remedial measures," (2002)
[16] JinWang, Zhiliang Chen, Kaiwei Sun, Hang Li, Xin Deng, "Multiple Target Regression via target specific features", (2019)
[17] Ahita P, Examining the rise of Non-Performing Assets in India, PRS Legislative Research 13 September 2018. https://www.prsindia.org/content/examining-rise-non-performing-assets-india
[18] Pacha Malyadri, S. Sirisha, "Comparative Study of Non-Performing Assets in Indian Banking Industry"
[19] Gülden Kaya Uyanık, Neşe Güler, "A study on Multiple Linear Regression Analysis", (2013)
[20] "An Analysis of Agriculture Sector in Indian Economy" IOSR Journal of Humanities And Social Science (IOSR-JHSS) Volume 19, Issue 1, Ver. X (Feb. 2014), PP 47-54
[21] "Impact of Agricultural Inputs on Agricultural GDP in Indian Economy" by Tamma Koti Reddy·, Madhubanti Dutta, Department of Economics, IBS Hyderabad (A Constituent of ICFAI Foundation for Higher Education), Hyderabad, India.
[22] "Estimation, Analysis and Projection of India's GDP, A Time Series Model" by U. R Daga, Rituparna Das and Bhishma Mahesh
[23] "High Ratio of Agriculture NPAS In Priority Sector Lending by Public and Private Banks in India –Reasons, Suggestions and Discussions" by Arpita Baijal - International Journal of Science and Research (IJSR) ISSN (Online): 2319-7064.
[24] "Non-Performing Loans and Terms of Credit of Public Sector Banks in India: An Empirical Assessment'' by Rajiv Ranjan and Sarat Chandra Dhal
[25] Abhini S, Feature Selection with sklearn and Pandas, Towards Data Science, 11 February 2019. https://towardsdatascience.com/feature-selection-with-pandas-e3690ad8504b
[26] RBI Working Paper Series No. 03 "Re-emerging Stress in the Asset Quality of Indian Banks: Macro-Financial Linkages" Shashidhar M. Lokare
[27] India Agricultural Trade: Expanding Export Opportunities Amid Persistent Limitations, USDA Foreign Agricultural Service 21 December 2016. https://www.fas.usda.gov/data/india-agricultural-trade-expanding-export-opportunities-amid-persistent-limitations

IOP Conf. Series: Materials Science and Engineering 1042 (2021) 012008 doi:10.1088/1757-899X/1042/1/012008

A survey on Pose Estimation using Deep Convolutional Neural Networks.

Manisha Patel*, Dr Nilesh Kalani

Research scholar, Electronics and Communication Department, School of Engineering, RK University, Rajkot, India.

Director, IQAC, RK University, Rajkot, India.

*m71074@ldce.ac.in

Abstract. Human pose estimation is a technique, which identifies the human body's landmarks in images and videos. Human pose estimation can be divided into single person pose estimation and multi-person pose estimation, also an estimated human poses in crowded places as well as in videos. Depends upon the application such as activity recognition, Animation, Sports, Augmented reality, etc., Pose estimation output can be in 2D or 3D coordinate format. 3D pose estimation is estimated considering joint angles in 2D. Some challenges like small and barely visible joints, strong articulations, occlusions, clothing, and lighting changes increase difficulty in estimating pose. Remarkable progress has gained in the field of human pose estimation using Deep learning-based CNN models. In this paper, we compare and summarized various deep learning models for pose estimation of a single person and multi-person.

1. Introduction

In computer vision, the main goal is to model and replicate the things which see by humans. To do so, there are interactions between humans and objects for such a high level of reasoning, the machine first has to interpret the posture from the given visual data. The task of identifying and determining the posture of a person in videos or images is reflected as human pose estimation [1]. Human pose estimation is an important and complex problem in a real-time application, such as sports, assisted living, assisted driver systems, games, etc. It is used to track a person's small body movement and do a biomechanical analysis in real-time.

Single Person Pose Estimation is easier to locate than Multi-Person Pose Estimation as an additional problem of inter-person occlusion occurs in it. As per the requirement of output dimensions, pose estimation can be classified in 2D or 3D.

Despite many years of research, articulated pose estimation of human is a challenging problem. Among the most significant challenges are strong articulations, small and barely visible joints, cluttered background, the variability of clothing in images, lighting conditions(variations), occlusion in the scene, motion blur, etc. To identify human pose estimation using an artificial intelligence technique, a convolutional neural network draws maximum attention nowadays. This review paper is organized as follows. Section 1 gives the introduction of human pose estimation and provides the purpose of the study. Section 2 describes the basic structure of the Convolutional neural network(CNN) with all elementary concepts and LeNet architecture. Section 3 describes various deep learning-based approaches for human pose estimation and Section 4 illustrates the Conclusion.

2. Convolutional Neural Network

A convolutional neural network is a deep learning approach that will take an input image or video, will assign weights and biases to extract and classify the given image. Compare to ANN (Artificial Neural Network), CNN can learn highly abstract features to identify objects very efficiently. CNN has certain beneficial advantages like sparse connection, parameter sharing and equivariance representation due to which training parameters reduced significantly, resulting in improved generalization. The architecture of the Convolutional neural network was inspired by the human nervous system and the organization of the visual cortex. To extract the features from images the network trains on thousands of images. This feature of CNN models makes extremely accurate for the computer vision tasks. In the deep convolutional neural network, there are tens or hundreds of hidden layers, so the network becomes more complex which make it able to learn to extract the feature from an image.

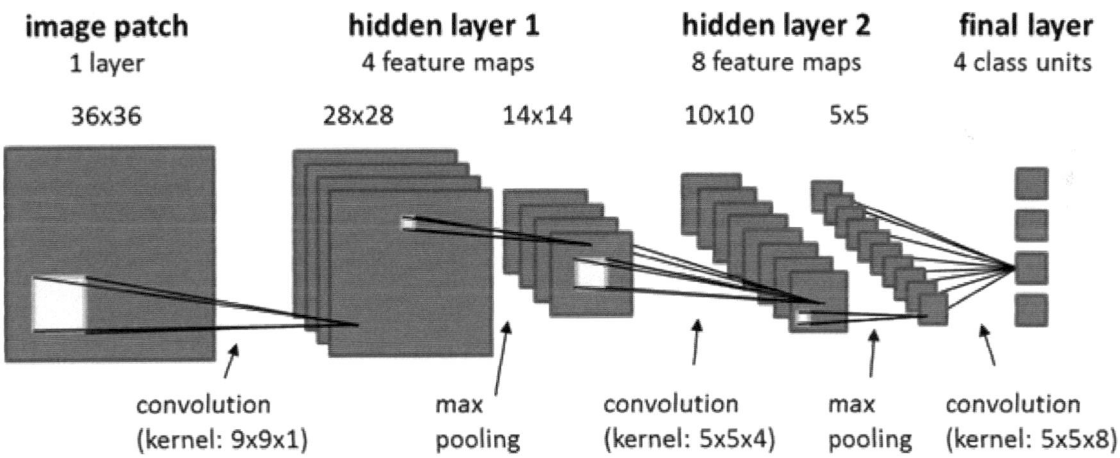

Figure 1. Convolutional Neural Network

In CNN fewer parameters that need to train, so the training is smooth and not suffer from over fitting. Further, the classification stage is incorporated with the stage of feature extraction as both use learning process. Implementation of CNN is compared to easy that large ANN. Due to all above remarkable performance CNN are widely used in image classification[8], object detection [9], face detection [10], vehicle recognition [11], speech recognition [12], diabetic retinopathy [13], facial expression recognition [14] and many more. Pose estimation incorporate with object detection, body key points location etc. so CNN is a highly advisable network and thus a majority of pose estimation model is comprised of CNN as a basic building block. A basic model of CNN consists of four components- Convolutional layer, Pooling layer, Activation function and fully connected layer.

2.1 Convolutional Layer [7]

Raw input data (images) is load and store at the input layer whose final classification will be done at the output layer. The input data - images can be black & white or colour. In black & white images two parameters- height and weight and in colour image three parameter- height, width and number of channels are specified. In RGB image number of channels will be 3. Convolution is a mathematical operation takes place between input image matrix and a filter or kernel. In convolutional layer, height and width of filter is smaller compare to input image matrix. The filter is slide across the width and height of input image matrix and compute dot products at every spatial position.

After applying convolution operation, the output a_{ij} for next layer can be

$$a_{ij} = \sigma ((w * x)_{ij}) + b$$

Where x is the input layer, w is a weight vector or filter, b is the bias and " * " represents convolution operation and σ is non- linearity introduced in the network.

2.2 Pooling Layer

When the features are detected its extracted location becomes less significant so the convolutional layer is followed by pooling or subsampling layer to reduces the number of learning parameters and introduce translation invariance [15]. As it reduces data representation progressively over the network, which directly helps to control over fitting. To resize the data, pooling layer uses certain operation like max, min and average. Pooling layer uses filters to perform a down sampling process on the input. The most common set up for a pooling layer is to apply 2×2 filter with a stride of 2, so will down sample each depth slide in the input volume by a factor of 2 on the spatial dimensions (width and height). Pooling layer has hyper parameters, which are not learnable.

2.3 Activation Function

To add nonlinearity, nowadays Rectified Linear Unit (ReLU) proved better than the former sigmoid function due to several advantages like – the calculation of partial derivatives of ReLU is easy [15], training time is lesser [8], and ReLU do not allow gradients to disappear. Dying ReLU problem can be resolved by using Leaky ReLU.

2.4 Fully Connected Layers

Fully connected layers have normal parameters and hyper parameters. In this layer feature map matrix will be flattened in to vector and feed to a fully connected layers.They perform a transformation on input data as a function of activations (weights and biases of neurons).

2.5 LeNet Architecture

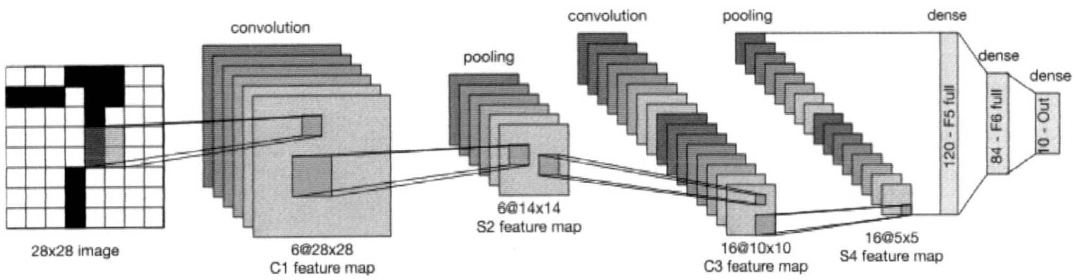

Figure 2. LeNet 5 Dataflow [18]

To understand CNN in a better way, LeNet architecture is mentioned above, which have a convolutional layer, 5×5 kernel and processes every output with a sigmoid activation function. The output channel of first convolutional layer is 6, which increase up to 16 in second layer. Height and weight are shrunk while the number of channels is increased considerably. So it makes the parameter sizes of two convolutional layer similar. To overcome the over fitting problem pooling layer down sample the size of matrix up to one quarter Three fully connected layers of 12, 84 and 10 outputs, added with bias is passed through softmax activation function.

3. Deep learning-based approaches

Deep Convolutional Neural Network techniques have shown outstanding performance in many fields of computer vision like visual classification tasks, object detection etc. In this section, some of the novel architecture is mentioned for pose estimation.

3.1 Deep pose[2]

Deep pose was the first paper that applied deep learning-based approach to the pose estimation of human. In this proposed network, pose estimation is formulated as a DNN based regression problem towards body joints. They define pose vector, which consists of all k body parts joints. To get better results they used a cascade of such regression refinements. They propose a simple but powerful approach in a holistic fashion, means even if certain joints are hidden, they can be estimated if the pose is reasoned about holistically.

Figure 3. DNN based Pose regression(Deep pose) [2]

This paper argues that DNN naturally provides this sort of reasoning. The input is a full image and location of each body joint can find out without the use of graphical models. They used AlexNet as their CNN architecture with an extra final layer. In this network seven layers are made up of **convolutional** layer**,** pooling layers and fully connected layer, in which only convolutional and fully connected layers have learnable parameters. They both contain linear transformation followed by a non-linear rectified linear unit (ReLU). The size of convolutional layers is defined as Width × Height × Depth, where the depth defines the number of filters. The model is trained using L2 loss for regression. The model takes an input image of size 220×220 via stride of 4 and the learning rate is set to 0.0005. The drop out regularization for the fully connected layers are set to 0.6.

The data sets used in this model are Frames Labeled in Cinema(FLIC) which has 4000 training and 1000 test images from popular Hollywood movies and second is LSD(Leeds Sports Dataset) having 11000 training and 1000 testing images from sports activities, which are quite challenging in terms of articulations.

3.2 Stacked Hour Glass [6]

In this landmark paper, a novel CNN architecture is presented which beat all previous method. The network called stacked hourglass- as it consists of successive steps of pooling and up sampling layers looks like as hourglass and these are stacked together to produce a final set of predictions. The motivation of this network is to capture information like person's posture and limb articulation at every scale.

When designing a network for Local evidence, it is essential to identify features like eyes, face, hands, etc., but to identify full- body pose estimation- need to require Global context. Hourglass capture all of these features and present pixel-wise predictions as an output. The Hourglass model performs repeated bottom-up (From high resolution to low resolution) and top-down (from low resolution to high resolution)processing with intermediate supervision.

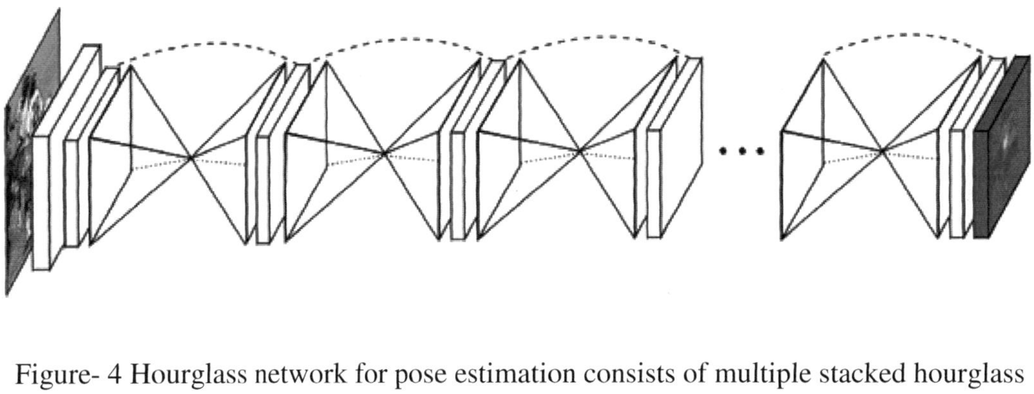

Figure- 4 Hourglass network for pose estimation consists of multiple stacked hourglass modules which allow for repeated bottom-up, top-down inference.

The hourglass model consists of convolutional and max-pooling layers to process features down to a very low resolution. At each max-pooling layers, the network branches off and applies more convolutions at original pre pooled resolution. When it reached the lowest resolution, the model begins the top-down sequence of up-sampling and combination of features across scales. They do nearest neighbour up-sampling of the lower resolution followed by an element-wise addition of the two sets of features. Hourglass has a symmetric topology, means for every layer present on the way down there is a corresponding layer going up.

The output of the model is a set of a heat map, for every given heat map the network predicts the probability of a joint presence at each and every pixel. The network operates at full input resolution of 256× 256, so it requires a significant amount of GPU memory, so the final output resolution (highest) is 64×64 but it is not affected the network's ability to produce precise joint predictions. The full network consists of 7×7 convolutional layer with stride 2 which followed by residual module and max-pooling which down resolution from 256 to 64. The hourglass network was tested on FLIC and MPII benchmark datasets. It achieves more than 2% improved accuracy on MPII across all joints and almost 4 to 5% on difficult joints such as knees and ankles.

3.3 Convolutional Pose Machine [4]

Convolutional pose machine [5] learns implicit spatial models via a sequential composition of convolutional architecture, such that it learns both image features and spatial models for prediction tasks. The architecture of this model is similar to the recurrent network as it generates the end result in multiple steps. It receives the images of resolution 368×368 as input, then applied few convolutions to the image in order to predict for each pixel map with a key point (head, neck, right elbow etc.)

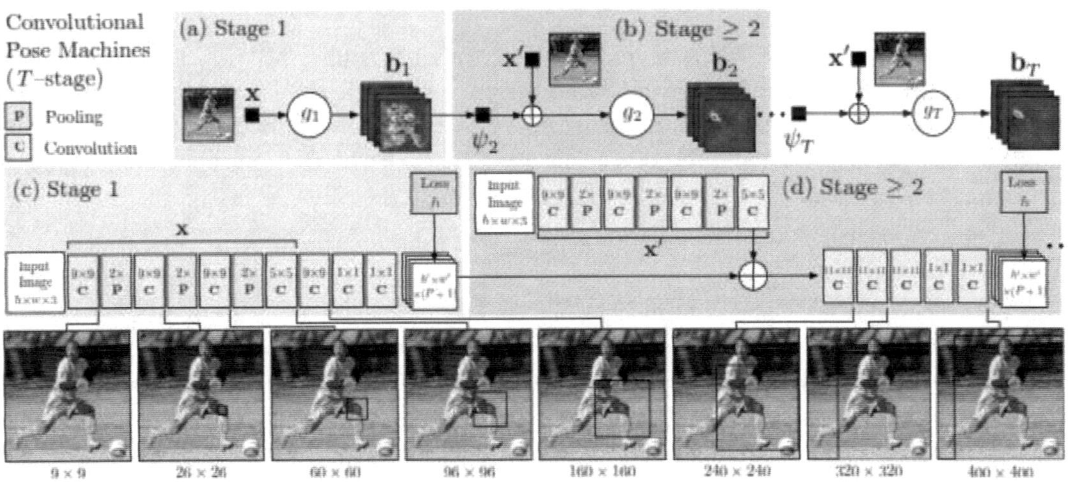

Figure 5. Convolutional Pose Machine

The loss function is a mean squared error. It gives the error between the expected output per keypoint and predicted one. A small Gaussian function applied to expected output maps to find the correct position of the keypoints. To overcome the problem of vanishing gradient, apply losses of after each step in the architecture. Benchmark datasets MPII, LSP and FLIC used in this model.

IOP Conf. Series: Materials Science and Engineering 1042 (2021) 012008 doi:10.1088/1757-899X/1042/1/012008

3.4 Open Pose [3]

Open pose, an efficient method for multi-person 2D pose detection is open-source. This paper proposes a real-time approach for detecting 2D human poses in images and videos. Bottom-up representation using Part Affinity Fields(PAFs) is presented in this paper. Most of the model uses a top-down approach for multi-person pose estimation. In top-down strategy, first, a person is detected and then have estimated its pose independently on each detailed region. This technique is directly applicable for single person pose estimation but in multi-person, it fails to capture the spatial dependencies across a different person that require global inference.

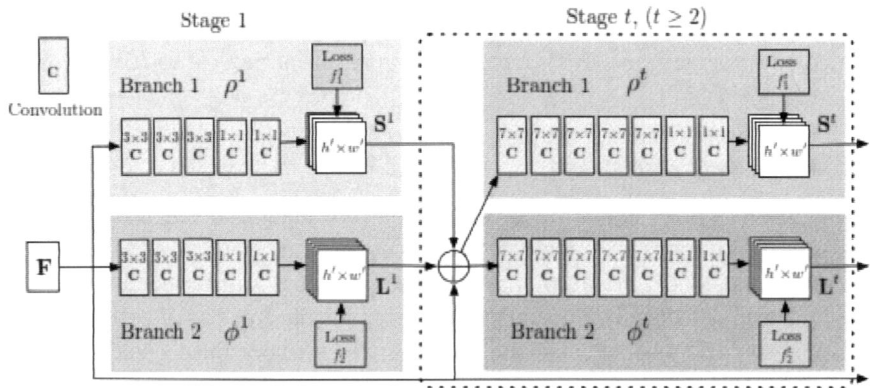

Figure 6. Two branch multilevel CNN [3]

The network first extracts features from an image using a few layers. The extracted features are then fed into two parallel branches of convolutional layers. Out of two, first predicts a set of 18 confidence map with each map represents a specific part of a human skeleton. The second branch predicts a set of 38 PAF which represents the degree of association between parts.

To understand and interpret multiperson 2D pose estimation is critical aspects in computer vision. In this paper several landmark contributions they present like, representation of keypoint includes both position and orientation of human limbs, second, they propose an architecture that jointly learns part detection and association. Regardless of the number of people in the image, high-quality parses of body pose can be demonstrated with Greedy parsing algorithm. They compare both PAF refinement and combined PAF and body part location refinement and conclude that only PAF refinement is more important as runtime and accuracy point of view. Finally, they estimate a single model which have combine body and foot estimation which reduces the inference time of training.

4. Conclusion

Deep convolutional neural networks are successfully applied to learn various computer vision tasks. Here we, thrown light upon the basic understanding and elementary concept of CNN to solve many complex problems and review deep learning approaches for human pose estimation. Deep neural network with a graphical model is introduced by Thomson. In this model the parameters are learned jointly with the given network. With the introduction of "Deep pose" [2] research on human pose estimation shift towards deep learning. A. Toshev

[2] proposed CNN based regression problem towards body joints in a holistic fashion. To resolve the problem of vanishing gradient Intermediate supervision was applied at every stage in stacked hourglass [6] architecture. Convolutional pose machine [4] is a sequential architecture for tackling structured predictions in the computer vision field without the need of a graphical model. Intermediate supervision enforced at every stage to resolve the problem of vanishing gradient. Open pose[3] is a real-time multi-person system to detect the human body key points on a single image. Its library is widely used for human re-identification, retargeting and human-computer interaction.

References

[1] M. Andriluka, L. Pishchulin, P. Gehler, and B. Schiele. 2D human pose estimation: New benchmark and state of the art analysis. In CVPR, 2014.

[2] A. Toshev and C. Szegedy, "Deeppose: Human pose estimation via deep neural networks," in CVPR, 2014.

[3] Z. Cao, T. Simon, S.-E. Wei, and Y. Sheikh, "Realtime multi-person 2d pose estimation using part affinity fields," in CVPR, 2017.

[4] S.-E. Wei, V. Ramakrishna, T. Kanade, and Y. Sheikh, "Convolutional pose machines," in CVPR, 2016

[5] V. Ramakrishna, D. Munoz, M. Hebert, J. Bagnell, and Y. Sheikh. Pose Machines: Articulated Pose Estimation via Inference Machines. In ECCV, 2014.

[6] A. Newell, K. Yang, and J. Deng. Stacked hourglass networks for human pose estimation. In ECCV, 2016.

[7] Sakshi Indolia, Anilkumar Goswami, S.p. Mishra, Pooja Asopa: Conceptual understanding of convolutional neural network- A deep learning approach. In ICCIDS 2018 science direct procedia computer science 132 (2018) 679-688.

[8] Krizhevsky, A., Sutskever, I., and Hinton, G. E. (2012). "Imagenet classification with deep convolutional neural networks. "In Advances in neural information processing systems (pp. 1097-1105).

[9] Szegedy, C., Toshev, A., and Erhan, D. (2013) "Deep neural networks for object detection." In Advances in neural information processing systems (pp. 2553-2561).

[10] Timoshenko, D., and Grishkin, V. (2013) "Composite face detection method for automatic moderation of user avatars." *Computer Science and Information Technologies* (CSIT'13).

[11] Luo, X., Shen, R., Hu, J., Deng, J., Hu, L., and Guan, Q. (2017) "A Deep Convolution Neural Network Model for Vehicle Recognition and Face Recognition." *Procedia Computer Science* **107**: 715-720.

[12] Sainath, T. N., Kingsbury, B., Mohamed, A. R., Dahl, G. E., Saon, G., Soltau, H., Ramabhadran, B. (2013) "Improvements to deep convolutional neural networks for LVCSR." *In Automatic Speech Recognition and Understanding (ASRU), 2013 IEEE Workshop on* (pp. 315-320).

[13] Pratt, H., Coenen, F., Broadbent, D. M., Harding, S. P., and Zheng, Y. (2016) "Convolutional neural networks for diabetic retinopathy." *Procedia Computer Science* 90: 200-205.

[14] Uçar, A. (2017, July) "Deep Convolutional Neural Networks for facial expression recognition." *In Innovations in Intelligent Systems and Applications (INISTA), 2017 IEEE International Conference on* (pp. 371-375).

[15] Zhou, Y., Wang, H., Xu, F., and Jin, Y. Q. (2016) "Polarimetric SAR image classification using deep convolutional neural networks." *IEEE Geoscience and Remote Sensing Letters* **13** (**12**): 1935-19.
[16] Lee, K. B., Cheon, S., & Kim, C. O. (2017) "A convolutional neural network for fault classification and diagnosis in semiconductor manufacturing processes."*IEEE Transactions on Semiconductor Manufacturing* **30** (**2**): 135-142.
[17] J. J. Tompson, A. Jain, Y. LeCun, and C. Bregler, "Joint training of a convolutional network and a graphical model for human pose estimation," in NIPS, 2014.
[18] LeCun, Y., Bottou, L., Bengio, Y., and Haffner, P. (1998) "Gradient -based learning applied to document recognition." *Proceedings of the IEEE* **86** (**11**): 2278-2324.

Reconstruction of Gene Regulatory Network using Bayesian Network

Abhishek Bhola[1], Sandeep Mittal[2]

[1]Chitkara University School of Engineering and Technology
Chitkara University, Himachal Pradesh, India
abhishek@chitkara.edu.in
[2]Department of Computer Science and Engineering,
R D Engineering College, Ghaziabad
sandeep.mittal2363@gmail.com

Abstract. Gene Regulatory Networks (GRN) are complicated govern systems, that refers to the intercommunication of genes and it also helps to understand and visualize all the biological interconnections. Reconstruction of GRN helps in finding out various hidden facts that can be helpful in solving various complexities related to living life, curing many diseases and inventing new drugs for the humans. There are number of computational methods available in literature for reconstructing GRNs. In this paper, a method is proposed for reconstructing GRN i.e. based upon Bayesian network. The mentioned method in this paper is considered as one of the most capable method for reconstructing GRN. The proposed method functions in two stages. In its first stage, there is two level clustering is employed to diminish the size of given dataset. In its second stage, Bayesian Network will be utilized to reconstruct the network for the sub-clusters that are gained from the first stage. The proposed method is also validated and evaluated using different parameters.

Keywords: Gene regulatory Network, K-mean Clustering, Bayesian network, Gene Expression Data

1. Introduction

Gene is the basic building block of living life and gene controls it's all biological processes. All the biological activities are directly or indirectly connected to genes. Due to its complicated and mysterious nature, study of gene regulatory networks (GRN) has utmost importance therefore numerous methods have been proposed in the past to reconstruct GRN[1-6].Ramesh Ram et al. have proposed a Fuzzy logic model that searches Microarray data for activator/repressor relationship. This method eradicates likely fallacious predictions possible from the traditional fuzzy model, so that permitting a larger search space for inferring regulatory relationships among number of Genes [7]. Young Hwan Chang et al. have developed a method i.e. based on the concept of repairing Sparse Low-Rank Matrices. This method is important from the view that it considers perturbations can affect the result inferred by GRNs. It estimates the error, occurred due to these perturbations and identifies the common graph structures among biased inferred structures [8]. Sudip Mandal et al. have specified Neural Model for reconstructing GRN, which is effective at that particular time to specify relationship among number of Genes [9]. This model is implemented in the form of Weighted Matrix. Further, another method has been specified i.e. proposed by Nilanjan Santra et al. They have known the importance of GRN, because it helps in revealing the disease motifs and also in solving the complexities related to living life. They have used Firefly Algorithm for reconstructing and learning GRN effectively. Although the method supposed by this work has accuracy rate i.ee above 50 % [10].Raed I. Hamed et al. have discussed a fuzzy Petri net (FPN) method i.e. used for reconstructing Gene Regulatory Networks. The approach specified by their work is more propitious than the Fuzzy Logic. The concept used by this work is better as compared to simple Fuzzy Logic approach because it removes the plausible untrue predictions from the classical Fuzzy model so that enabling a larger search space for inferring regulatory relationship [11].

This paper has been organized in four main sections i.e. Bayesian Networks, datasets & methodology, results and conclusion. Section 2 discusses about Bayesian Networks and how they are inferring GRN effectively. Section 3discusses data sets and methodology used in this work. Next Section 4 analyses result and discuss it. Finally, section 5highlights conclusion based on the proposed work and future directions.

2. Bayesian Network Method

Bayesian Network methods are one of the mostly used inference architectures for reconstructing GRNs. In this paper, Bayesian Network Methods have been used for reconstructing GRNs. These methods are also called probabilistic models, because they make use theorem of probability i.e. each node is connected to another on the basis of some probabilistic value. Bayesian Methods are directed acyclic graphs, which consist of nodes describing characteristics and acyclic edges present between them highlighting independancy. Each of the nodes is joined to the other on the basis of Probability Distribution Table, highlighting the chances of connecting each node to the other. Bayesian methods are considered to be highly capable because they make use of both graph concept and probability concept. These methods are quite understandable and have been used successfully in many of the applications and research works. These methods are mathematically capable because of having concept of probability and reconstruct the GRNs effectively by giving accurate results.

3. Proposed Methodology& Datasets

The reconstruction of GRN can be said as one of the key tasks in the area of Bioinformatics because for reconstructing GRN we need calculative and capable computational method that can infer GRN effectively. Bayesian Computational method makes use of Genomic Datasets as input those are obtained from the Microarray analysis. But handling of complete Datasets obtained from Microarray Gene Expression Data is impossible that's why Clustering technique has been used for monitoring a large number of Genes.

There are some of the authenticated source like gene expression omnibus (GEO)repository of NCBI that are considered for obtaining the needed gene expression datasets[12-13].In the proposed framework the Gene Expression Dataset of E. coli is used and the methodology that has been used to used and implemented to reconstruct the GRN is shown in Fig 1.

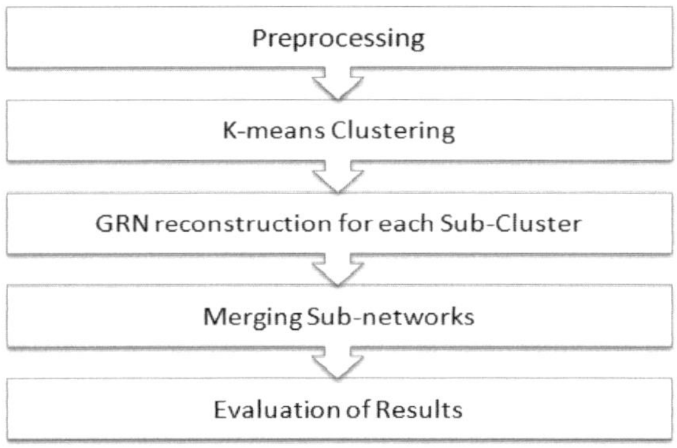

Fig. 1 Proposed methodology

Step 1: **Pre-processing**: In this step, pre-processing is done on the dataset obtained from the GEO data repository. It eliminates the Genes those are having missing names from the Gene Expression Data. It also removes the gene profiles that have low variance.

Step 2: **Clustering**
In this step Clustering is implemented on the gene expression data [14]. Clustering is a technique of grouping the similar things into one group and different objects into different Groups. This technique can group the Genes having similar expressions under various conditions. There are number of Clustering techniques present in literature, but in the proposed method K-mean clustering has been used [15]. It is a clustering technique that classifies the objects (Genes) based on some characteristics into K number of groups (clusters). In this clustering, computational approach begins, by positioning K points into the space represented by the objects that are being clustered.

Step 3: **GRN Reconstruction for sub-Clusters**
Now, clusters have been created in the previous step. In this step, Bayesian Network is used for reconstructing GRN for each of the sub-clusters [16]. Each sub-cluster will be treated as an individual, independent and complete Dataset and proposed method will be implemented and GRN will be reconstructed for each of the sub-Clusters.

Step 4: **Merging**
After conducting the previous step, there has been number of sub-networks or network is created for each of the sub-clusters [17]. Now, in this step merging of sub-networks is done. This step enables the making of main network. For creating main network, all sub-networks (related to each of the cluster) are concatenated and main network is produced.

Step 5: **Evaluation and validation of Results**
In this step the results are evaluated based on the number of parameters e.g. True Positive Rate, True Negative Rate, F-score and Total Time taken.

For validation Gene-Mania is used to validate accuracy of results.

4. Result and Discussion

The datasets used in this work has been downloaded from publicly available Gene Expression Omnibus (GEO) data repository of NCBI [18].In this section, the results obtained from the proposed methodology, has been discussed. At the end of first step i.e. pre-processing step, the obtained gene expression data will be free from most of the anomalies [19-20].

After pre-processing, the clustering technique has been applied so that various clusters of genes can be formulated. After clustering technique, GRN is constructed for each of the sub-clusters individually as shown in Fig 2. These all sub-networks may vary in shape and size all though this work has applied same method for reconstructing GRN for each of the sub-clusters.

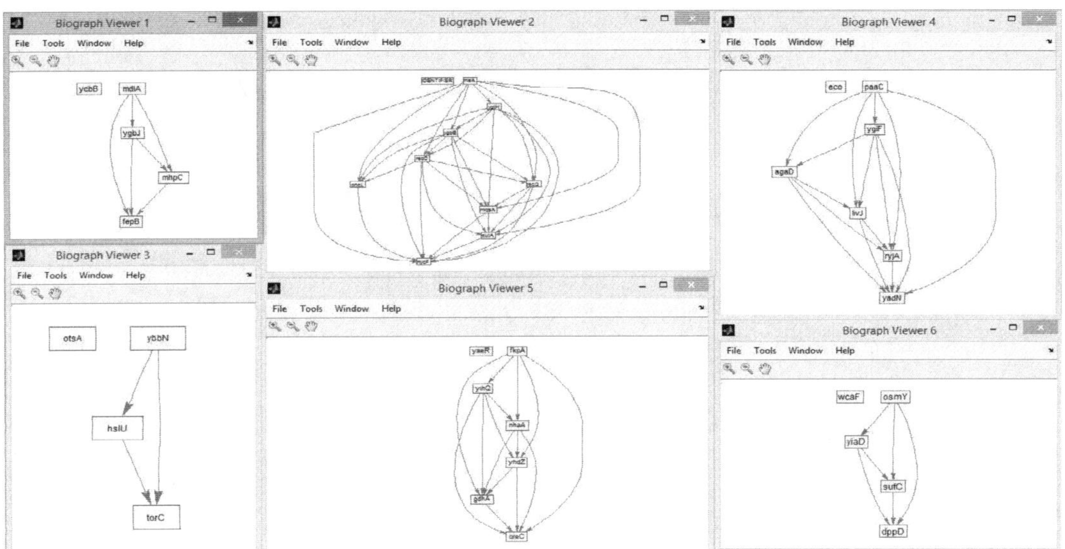

Fig. 2 Sub-network for each of the sub-clusters

All the GRN of sub-cluster reconstructed in previous step are merged into a single Network as shown in Fig 3.

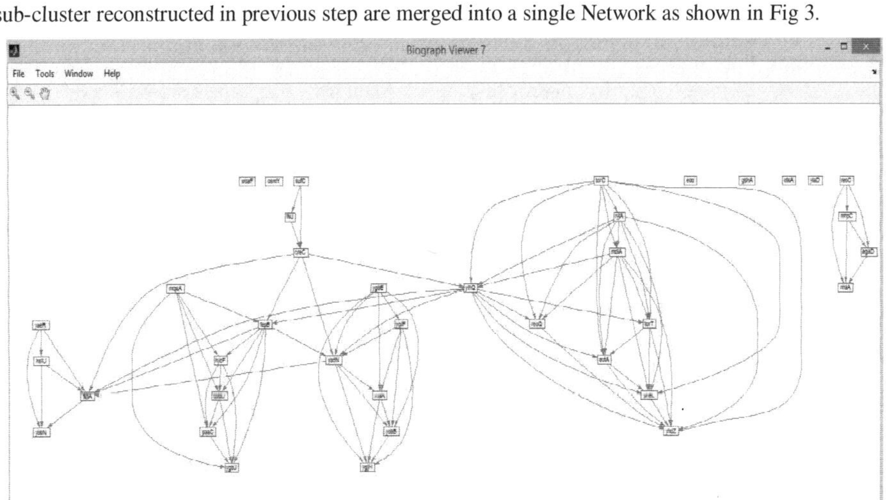

Fig. 3 Mcrgcd GRN

In the previous step all the sub-networks belonging to each of the sub-clusters are merged to form single network. Now results are evaluated and validated using Gene-MANIA platform [21].

Various parameters used for the evaluation of results includes: true positive rate, True Negative Rate, F-score and Total time taken. The results for each parameter that are taken into consideration for evaluating performance, is shown in the Table 1.

Table.1 Evaluation of Results

Parameter	Calculated Value
True Positive Rate	0.3
True Negative Rate	0.285
F-score	0.29
Total Time taken (in Seconds)	102

Based on the above parameter's performance measure of the proposed method for different number of genes has been also done and shown in Table 2.

Table.2 Performance assessment using various numbers of genes

Experiment No.	Number of Genes	True Positive Rate	True Negative Rate	F-score
1	7	0.80	0.57	0.66
2	8	0.69	0.60	0.64
3	9	0.60	0.60	0.60
4	10	0.52	0.52	0.52
5	11	0.50	0.52	0.50

5. Conclusion and Future Scope

This work has discussed all the facts related to GRN and stated how previous work lack in reconstructing GRN effectively. The work done in this paper, we tried to eradicate some of the gaps which are present while reconstructing GRN efficiently. For future perspectives more work needs to be done and investigated to improve the result. It is one of the challenging fields to which needs more intense works as it helps in uplifting the standard of human beings by inventing personalized drugs and diagnosis tools. The study of GRN also helps in inspecting the Gene's behavior towards changing environment and solving the complexities related to living life. Therefore, the work to reconstruct GRN must be continued so that results can be further improved.

References

[1] Fernando M. Delgado, Francisco Gomez-Vela: Artificial Intelligence in Medicine, Volume 95 (April 2019)
[2] Sudip M, Goutam Saha,Rajar KrPal :Reconstruction of Dominant Gene Regulatory Network from Microarray Data using Rough Set and Bayesian Approach (2013)
[3] Paras B, Shailendra S, Abhishek, A Review, On Bayesian Network Techniques for Inferring Gene Regulatory Networks (2015)
[4] Jure Bordon, MihaMoskon, Nikolaj Zand MihaMraz: Fuzzy logic as a computational tool for quantitative modelling of Biological Systems with uncertain Kinetic data (2015)
[5] A Brazma, P Hingamap, J Quackenbush, G Sherlok: Minimum Information about a microarray experiment (MIAME) – towards standards for Microarray Data (2001)
[6] R Seidel- Discrete and Computational Geometry: The nature and meaning of perturbations in Geometric Computing (1998)
[7] Ramesh Ram, Madhu Chetty, Trevor I. Dix: Fuzzy Model for Gene Regulatory Network (2006)

[8] Young Hwan C, Roel Dobbe, Palak Bhushan, Joe W. Gray, and Claire J. Tomlin: Reconstruction of Gene Regulatory Networks based on Repairing Sparse Low-Rank Matrices (2016)
[9] Sudip Mandal, Goutam S, Rajat K Pal: Neural Network Based Gene Regulatory Network Reconstruction (2015)
[10] Nilanjan S, Surama Biswas, SriyankarAcharyya: Neural Modeling of Gene Regulatory Network using Firefly Algorithm (2015).
[11] Raed I. Hamed, Syed I. Ahson, Rafat Parveen: Fuzzy Petri Nets Approach (2010)
[12] A Kolinski, J Skolnick, JS Fetrow: Structural Genomics and its importance for gene function analysis (2000)
[13] JJ Goeman, P Buhlmann- Bioinformatics: Analyzing Gene Expression Data in terms of Gene sets: methodological issues (2007)
[14] AR Deshpande, L Lobo-International Journal of Engineering: Text Summarization using Clustering Technique (2013)
[15] L Morisette, S Chartier- Tutorials in Quantitative Methods for Psychology: The K-means clustering technique: General considerations and implementation in Mathematics (2013)
[16] ABhalerao, A Katkar, J Gahlot: Systems and Methods for managing sub-clusters within dependent clustered computing systems subsequent to partition events (2015)
[17] C Fontaine, PR Guimaraes Jr, S Kefi, N Loeuille: The Ecological and Evolutionary implications of merging different types ofnetworks (2011)
[18] H O, MLazaar, Y Tabii: Effect of Feature Selection on Gene Expression Dataset Classification Accuracy (2018)
[19] ST Sherry, MH Ward, M Kholodov, J Baker: dbSNP: the NCBI database of Genetic variation (2001)
[20] E Clough, T Barret- Statistical genomics: The Gene Expression Omnibus Database (2016)
[21] J V, K Zuberi, H Rodriguez, R Arnold: Novel Function discovery with Gene-MANIA: a new integrated resource for gene function prediction in E. Coli (2015)

INTELLIGENT TRAVEL ROUTE SUGGESTION SYSTEM BASED ON PATTERN OF TRAVEL AND DIFFICULTIES

Nikhitha Pulmamidi, Rajanikanth Aluvalu, Uma Maheswari V

Vardhaman College of Engineering, Hyderabad, Telangana, India

Nikhithareddy595@gmail.com, rajanikanth.aluvalu@gmail.com, umamaheshwariv999@gmail.com

Abstract. People travel a lot to new places for career, business, education, travel, health and for various other reasons. Travellers always search to find shortest and safest route to reach the destination. They majorly consider distance, modes of transport and safety as major parameters. Various algorithms, mechanisms are proposed earlier to serve the purpose. However existing algorithms, mechanisms suffer from various shortcomings like unable to analyze the suggestions and filtering the database. The existing models use collaborative or content filtering algorithms. These methods do not collect the past user experience. The main objective of this paper is to help the travellers from their source to target from a database with best possible way. Proposed method provides the information from the travellers and problems faced by them during their travel. Based on the reported issues and experiences, new path will be enrooted to the destination. In the proposed model C4.5 decision tree algorithm is used to collect data from previous user experience, minimum redundancy maximum relevance (MRMR) and features selection algorithm is used to reduce execution time and increase system accuracy. The experimental outcome of this paper is to improve the system accuracy and performance.

Keywords: Recommendation System; Tourist Destination, Feature Selection; Filtering methods; Mutual information; Classification; Decision Tree.

1 INTRODUCTION

Travels and Tourism sector the role of planning effective routes is complex, dynamic and includes various decision-making factors such as quality of attraction, travel time, cost, safety, accommodation, tourist numbers, leisure activities, weather and so on. Recently, ICT and Internet technologies started playing an important role in tourism. The tourists and tour operators can review, pick, evaluate and take decisions easily by implementing decision support systems known as "Recommendation Systems" (RS). The majority of earlier Travel Route Suggestion (TRS) relied on estimations based on the user's expectations and requirements to decide the location, events, attractions, tourism facilities(for example, restaurants, hotels, transportation). Lacking in technical aspects such as (e.g. scalability, reliability, system accuracy, personalization theories etc.) and functional aspects like (e.g. customer retention, accessibility, etc.) fail. These aspects are not accessible. One of the greatest problems is increasing a TRS that provide personalized recommendations of visitor targets and develop the visitor decision-making method. In order to complete this, it needs a deep understanding of the travellers' decision-making and develops new models form their information seeking process. In addition, the efficiency of the recommendation and the customer satisfaction level of the program can be improved. By reducing extra attributes in the system the complexity could be decreased. This paper proposes a new TRS, which suggests tourist destination to solve the difficulties mentioned. Three main developments are included in the proposed TRS. First of all, unwanted (like redundant) inputs have been eliminated and the complexity of the interface is reduced by using two feature selection methods. Secondly, a decision tree C4.5 is used as a classifier to see the traveller destination selection process. Lastly, the future method uses real world information that has been collected from Chiang Mai, Thailand [1].

The paper is majorly organized into 6 sections. Section 2 provides background on recommendation systems in the tourism domain. Section 3 describes the data collection process used in this paper. Section 4 presents the proposed TRS framework using the DM approach. The experiment setup for this study is demonstrated in Section 5. Section 6 shows the results and the evaluation analysis of the proposed TRS. Finally, presented some tentative conclusion and future scope of this work.

2. LITERATURE SURVEY

Travellers choose a destination based on tourist or traveller guide. Thailand Annual Report 2013 stated that , the individual demographics that influence the actions are identified. 4,000 Information-seeking Questionnaires are distributed and collected in five tourist destinations of Chiang Mai, Thailand. The favourite destinations lists were taken from the website of the Trip Advisor1. All worldwide (60%) and domestic tourists (40%) participated in the study. The data pre-processing phase was assisted by 3,696 valid questionnaires and 146 variables, although 35 samples not fully supplied. The task includes artificial intelligence to study from travellers' knowledge and to use this information to direct future travellers to their own destination. You may also sell similar videos by using this form for a video watchmaker based on the tickets that you previously booked. Therefore, one of the previous visitor meetings and one of the patterns will be taught two times by the actual consumer. An-Jung Cheng et al [2] proposed travel route recommendation based on photographic elements such as gender, time, and event. People attributes are useful for making decisions on travel landmarks and paths and also helpful for route planning and personalized travel recommendation. The development of recommendation systems is based on user interests. Social data, including check-in behaviour, social contacts, feedback and recent studies, lead to making highly accurate travel recommendations. The author provides a site advisory programmer: the UI module, the idea module, the rating module and the Location Based Social Network (LBSN). The position scores are 0 to 1 value and are used by a recommendation application to match the LBSN data with the user attributes. Pantano et al proposed a way to plan a trip using the history of Flickr[3][4]. The author suggests a photographer behaviour model path, which estimates the probability of the photographers visit. Ge Cui et al. [5] propose two approaches. [6, 7, 8, 9] is the two form of collaborative travel route recommendation are collaborative travel route recommendation CTRR, and an expanded version of CTRR. Both methods employ personalized routes followed by GPS-based users. Users are calculated using collaborative filtering techniques to optimize travelling speeds and to achieve the greatest likelihood of travel using the Naive Bayes model [10, 11, 12, 13].

3 PROPOSED SYSTEM

Choosing a tourist destination using the information available on internet and from other sources is one of the difficult tasks for visitors in tour planning before and during trips. In order to solve this problem TRSs is proposed earlier. However, some of the system accuracy, usability and satisfaction have been ignored. In order to solve this issue, it needs a full understanding of the visitor's decision-making and new models of the process of seeking information. This paper proposes C4.5 decision tree algorithms which take experiences of previous users and then build a model. Users provide their requirements to the system, and then decision tree will predict best location based on his given input. Sometimes datasets will have empty or garbage values and gives bad effect on decision tree model. It can be eliminated by applying pre-process techniques. To predict or build model no need to use all attributes values from dataset and this unnecessary attributes can be eliminated by applying features selection algorithms.

4 METHODOLOGIES

The process of identifying tourist destination is done through C4.5 decision tree approach. The suggested approach uses data from the real world from Chiang Mai, Thailand.

4.1 Recommendation system

A suggestion framework RS, a subset of Decision Support Systems (DSS), is a technique that can suggest an item based data collected for the user preferences [14,22]. It helps customers to make decisions based on their preferences and concerns by providing them with helpful information. RS plays a significant job and is common in popular online business sites, for example, Amazon, Netflix, Pandora, and so forth [15, 16]. The web based business RSs recommend item to the client which includes news, papers, individuals, URLs, etc[17,21].

4.2 Systems of Travel Advisory

Tourism is a leisure activity and difficult to make decision processes, for example, the choice of destinations, attractions, events and services. So many researchers in both academic and industry receive a TRS attention. Different TRSs (e.g. web, browser, mobile) have been developed in and on several platforms. For the purposes of assessing user interest, TRS suggest results to customer for use the of user preferences, selecting point of interest (POIs), defining or classifying services or routes or as a holistic journey. Most of the existing TRSs aim is to help single tourist, there are some systems that support travel agencies as well. They share similar structures but vary in the choice of technologies, personalization philosophies, data inputs, types of interaction and suggestion techniques. Data is embedded in and keyed in the server from many resources (e.g. sensors, GPS, surveys, comments etc.). The recommendation engine may consist of a several subsystems, including an optimization subsystem, a numerical subsystem and an intelligent subsystem, etc. In general, before and during the journey, most TRSs are used to display the end result. Fig 1 shows framework of travel route recommendation systems.

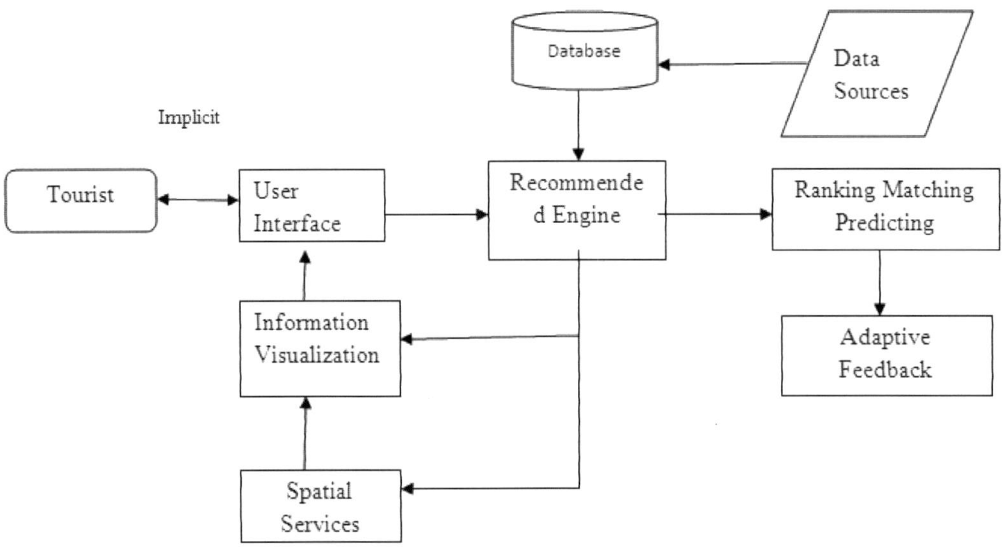

Fig- 1. Framework of travel route recommendation system.

4.3 Recommendation techniques

The effectiveness and consistency of the proposed technique can be measured by the degree of personalization. This can be designed from low to high, with non personalization, short term customization and long term customization. The non-customized RS is a very simple program that does not come into account user needs when making suggestions. For example, the RS produces only a list of the common products based on the number of transactions or reviews. Customized RS has been progressed more than a non-customized RS due to its limited decision-making ability. This means that each user will see another set of guidelines, according to his / her preferences. For example, a target based on user socio demographic information is being suggested by tour-advisors. In addition, several forms of Personalized RSs have been classified and analyzed in previous studies using the method of information filters. In the next section, we will briefly investigate the recommendation engine Fig.2 which is composed of several recommendation techniques based on findings

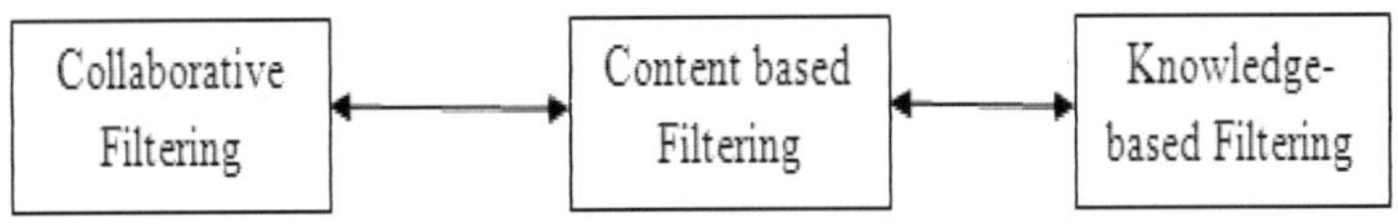

Fig.2. Recommendation Engine

4.3.1. Collaborative filtering: This approach is widely adopted by the most implemented recommendation systems. It suggests the items to the customer based on comment given by new visitors, who distribute the same features. There is an issue with a cold-start approach [18], which will have to be classified before a new item or consumer would make a suggestion.

4.3.2. Content-based filtering: This technique suggests items to the user based on his/her previous searches or queries for items. The main drawback is the cold-start problem for the user, where the user wants to present a lot of information earlier than a decision can be made. Another popular trouble is over-expectation, as the method would most probably recommend the item was mainly liked by the user with less variance, the quality results are produced through the archive of extensive historical data set [19].

4.3.3. Knowledge-based filtering:
Based on domain awareness this recommends the items to the user. In other words, the system has some data of how the particular item relates to a particular customer. Predominantly, this method can be achieved by using case-based analysis or ontological methods [20]. This suggestion technique is designed to benefit from past experiences between travel agencies and groups.

4.3.4. Hybrid filtering: The following technique has some strengths and limitations. The purpose of the hybrid recommendation technique is to achieve the best performance and to remove the weaknesses/disadvantages of one technique by complementing (which technique rewrite the sentence) it with the advantages of another. Many hybridization methods are also possible, for example by combining weight, flip page, mixture, function combinations, cascades, feature increases and metric rates.

4.3.5. Methods

I) the frequencies of user visits to different destinations the program will begin learning from scratch. The most popular destinations can be categorized and shown to the visitor, indicating that he / she travels to the most commonly visited spot. This program will slowly follow the needs of a consumer who visits somewhere else. Therefore, for training the program in two classes, it requires a machine learning algorithm similar and different destinations. The number of visitors to each other can be counted and averaged from one source to both locations by complete tourists. This counter will then have a value of 0 to 1. 1 if the same people visit both locations. 0 if the same people do not visit two locations. The user's favourite routes from his travel experience are based on this similarity score system.

(II) The program can also calculate the difficulties in any direction based on a comment of users and forecast a route that is complex or simple to achieve from feedback score (1-5) from visitors to these sites. Use of these scores can alert users of difficult routes. Once the next visitor enters the program, the previous score and number of visitors will be updated to get a new ranking. The program can also be educated on the basis of texts on the traveller's path. When the route is considered to be complicated, the findings can be preserved in a chart concerning the trip. This can be saved as a nice ride on a smooth and fun route. This method can then be regressed logistically to teach the input nodes identical to the possible amount of terms in a sentence, say 100. For each one, the input and output of which document class are complex and which class are nice can be identified and supplied to the Neural Network. It immediately takes a challenging or simple trip, when we offer a new feedback on place travellers. Unlike this person, any chosen route may display different rates.
Any route's complexity can be educated using input parameters such as:

1. Road conditions
2. Weather conditions
3. Transport
4. Number of accidents
5. Human personality (e.g. age, individual gender)

A controlled approach is training on routes that are already established as difficult and easy. 3-4-layer network can be used for this project with each parameter between 0 and 1 in value.

5 EXPERIMENTAL PROCESSES

I. Road suggestion based on travel guide

Phase 1: To store the number of visitors to destinations, a database is developed.

Phase 2: The user can choose between the number of visits and the device.

Phase 3: The similarity score is determined from a data base on a pair of routes that are separated by the total number of visitors to each route.

Phase 4: Based on traveller's history related routes are displayed for choosing next targets
The fig 3 is road suggestion based on travel guide

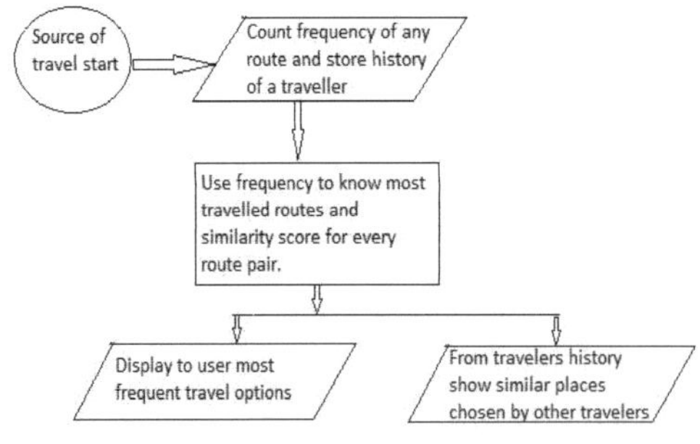

Fig. 3. Route suggestion based on travel guide of other user history

IOP Conf. Series: Materials Science and Engineering 1042 (2021) 012010 doi:10.1088/1757-899X/1042/1/012010

II . The complexity path to the user

Phase 1: Store visitor difficulty input in a score (1-5) and/or input text for all routes in the database.
Phase 2: Compute the problem score by dividing score by total guests. Apply 1 to the number of problems and change the problem score each time.

Phase 3: Train a neural network for traveller textual feedback to predict routes that are difficult and successful.
Phase 4: Show new users both the challenge and their simple understanding of the difficulties of travelling. The fig 4 knows the difficulty of a route

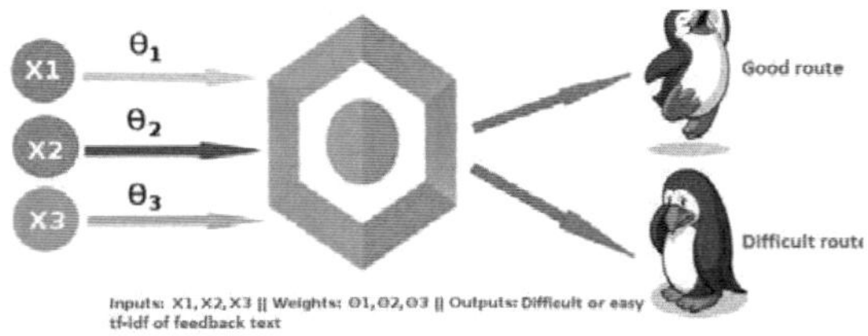

Figure 4. Knows the Difficulty of a route

Measures for assessment:

- Latest ways is based on travel habits of others and visitor's history are shown properly.
- For known difficult paths, a high difficulty score is correctly displayed.
- Below are the columns or attributes of the dataset used by previous users.

This collection of data is complemented by TripAdvisor.com crawling. Reviews are considered on destinations in 10 groups in East Europe. -- Rating is classified as Excellent (4), Really Good (3), Great (2), bad (1) and Terrible (0). - Group is labeled with average ratings per user.

Columns and meanings of the data set

Theatres, religious establishments, venue, user ids, gallery-art, salsa clubs, liquid bars, restaurants, museums, spa, place and picnic spots.

The column names above and the column values below are

User1,0.93,1.8,2.29,0.62,0.8,2.42,3.19,2.79,1.82,2.42,Amsterdam_Heining_2

User2,1.02,2.2,2.66,0.64,1.42,3.18,3.21,2.63,1.86,2.32,Amsterdam_Jachthaven_ijbur

User3,1.22,0.8,0.54,0.53,0.24,1.54,3.18,2.8,1.31,2.5,Amsterdam_Bert_Haanstra_Kad

User4,0.45,1.8,0.29,0.57,0.46,1.52,3.18,2.96,1.57,2.86,Amsterdam_Ruigoord_Ker

For each user, the meaning of the first column is USER ID and the second column is ART GALLERIES. The third column is DANCE CLUB etc.
Now using above values we can build C4.5 decision tree and prediction will be done using below test values

'User 122', 0.93, 1.8, 2.29, 0.62, 0.8, 2.42, 3.19, 2.79,1.82, 2.42,?
'User 222', 1.02, 2.2, 2.66, 0.64, 1.42, 3.18, 3.21, 2.63,1.86, 2.32,?
'User 3222', 1.22, 0.8, 0.54, 0.53, 0.24, 1.54, 3.18, 2.8,1.31, 2.5,?
'User 4222', 0.45, 1.8, 0.29, 0.57, 0.46, 1.52, 3.18,2.96, 1.57, 2.86,?
'User 522', 0.51, 1.2, 1.18, 0.57, 1.54, 2.02, 3.18, 2.78,1.18, 2.54,?

New users have provided new test values for location with above-average ratings but new users don't know where such services are available, so when we upload test values into the decision tree they determine so predict the user's best position and notify him.

6 EXPERIMENTAL RESULTS

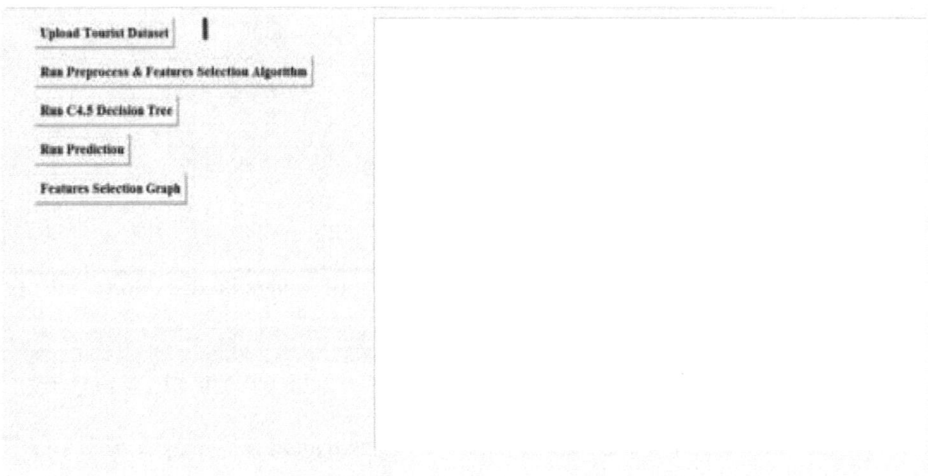

Fig. 5. Upload tourist data package

In the above tab press the button "Upload tourist data package" after uploading, all the data set information are given below the screen (write in detail)

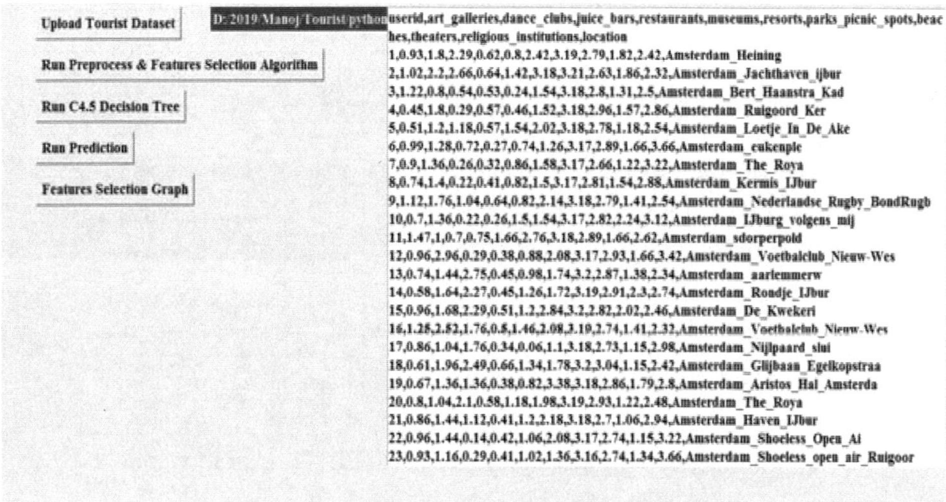

Fig. 6. Displaying Datasets of Previous Users Experiences

The above screen contains all users that have been loaded with the previous experience and a total of 12 attributes.
To delete empty values and reduce the size of attributes, press 'Run Pre-process Selection Algorithm'.

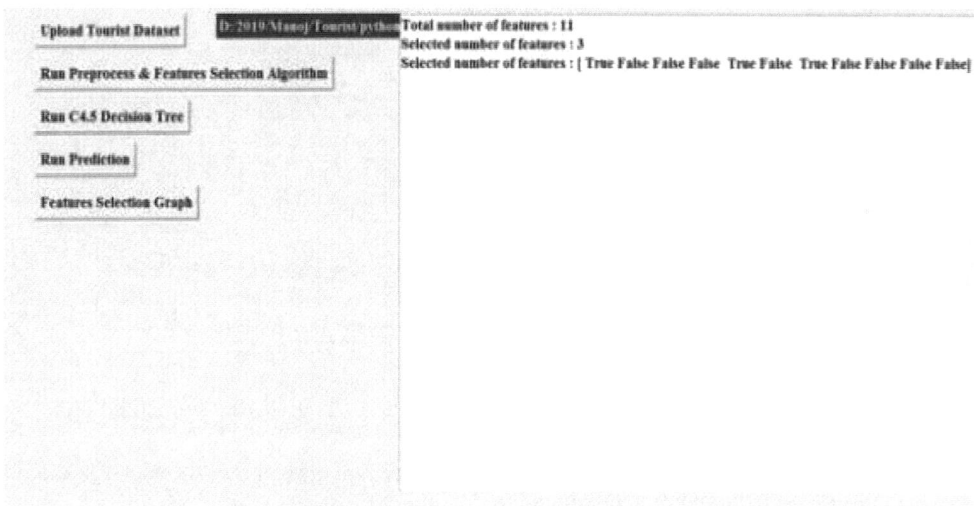

Figure- 7. Displaying the Selection Feature Attributes

All those attributes which have column TRUE and column FALSE are ignored in the above screen after applying the MRMR scope size to 3.
Click on the "Template Tree for Generating C4.5 Decision"

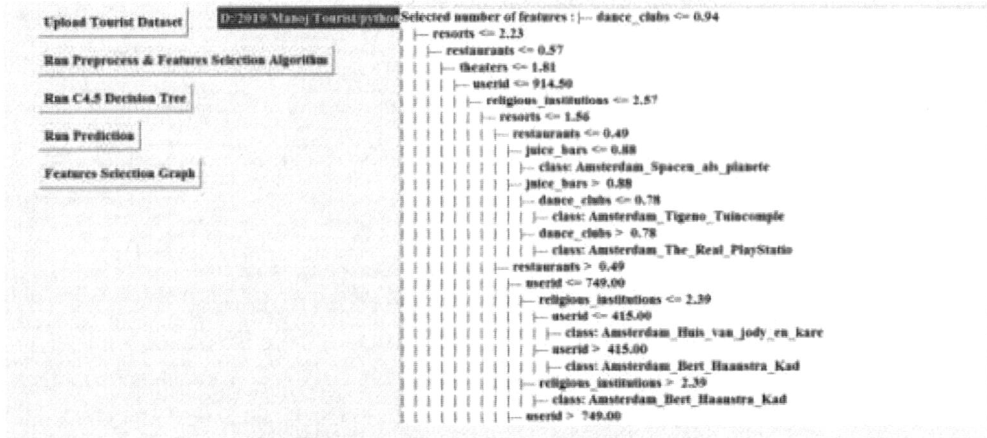

Fig. 8. Display the IF and ELSE declaration tree

We can see in the above screen the models created by the IF and ELSE declaration tree. If > any decision will be made, if < another decision will be selected. Click on the 'Tourist Recommendation' button to upload the test file without the place name.

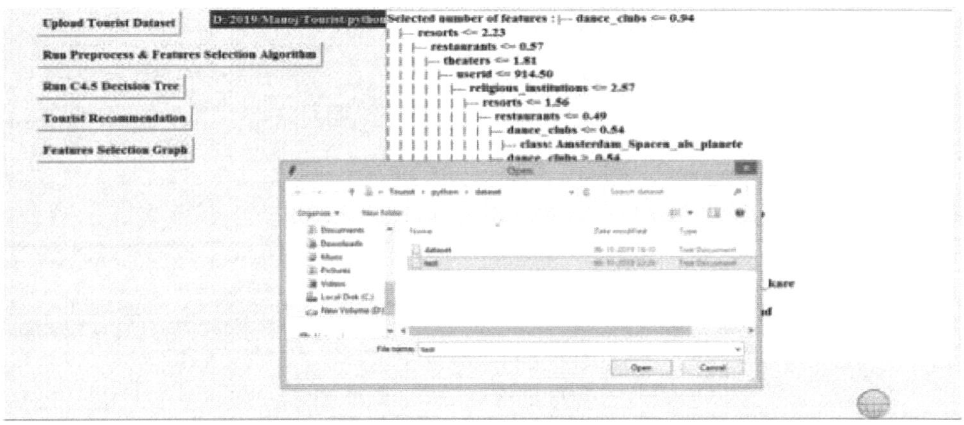

Figure- 9. Uploading the Test File in Tourist Recommendation

I have uploaded the test file on above screen and now click on open to get the location that I suggest. In the name of the test file, there is no appeal

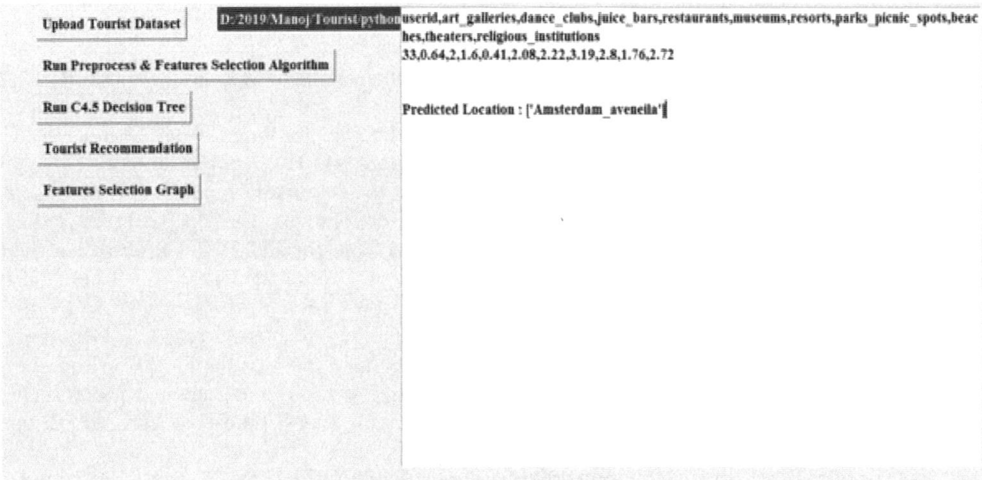

Figure- 10. Test Files

Once the test data are uploaded, all the values are in the test data but the position name and base on the application of the test values are predicted or the place name is not suggested in the above screen.

Click the Set of Features button below the diagram

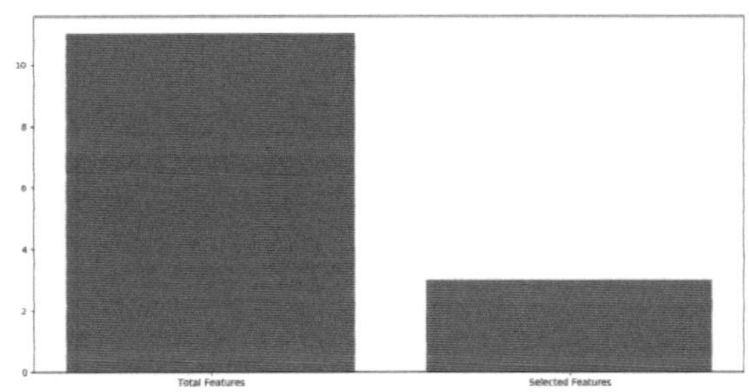

Figure- 11. Feature Selection Graph

In the Figure-11. the x-axis is total characteristics are shown and the MRMR selected and y-axis is countless and in above graph the size of characteristics decreases to 3 after the application of the MRMR technique

7 CONCLUSION

This Paper proposes a decision tree based visitor recommendation method to address the existing TRS problem. The data set was broken down into two subsets using specific knowledge of the tourism. This is used in order to increase the system accuracy and to decrease the decision tree complexity. The optimal decision trees of NMIFS were optimized for the option of the destination, with the maximum precision and ease (i.e. less leaf and tree sizes). Decision guidelines derived from decision-making bodies. The NMIFS is the optimum method because it uses less number of features than MRMR for both of the data sets. Finally, the test results indicate that the proposed decision tree is valid. The proposed decision tree satisfies the travellers' requirements who plan to visit or during their visit the city. In addition, different types of classifiers can be considered to increase the classification accuracy rate for the data sets. Moreover, front-end web application and an interactive and adaptive user interface will be designed and implemented.

REFERENCES

[1] "Economic Impact of Travel & Tourism 2014 Annual Update: Summary." World travel & tourism council. "Thailand Annual Report 2013."

[2] Personalized Travel Recommendation by Mining People Attributes from Community-Contributed Photos, An-Jung Cheng et al., Copyright 2011 ACM 978-1-4503-0616-4/11/11

[3] E. Pantano and L. D. Pietro, "From e-tourism to f-tourism: emerging issues from negative tourists' online reviews," J. Hosp. Tour. Technol., vol. 4, no. 3, pp. 211–227, 2013.

[4] B. Pan and D. R. Fesenmaier, "Semantics of Online Tourism and Travel Information Search on the Internet: A Preliminary Study," Inf. Commun. Technol. Tour. 2002 Proc. Int. Conf. Innsbr. Austria 2002, pp. 320–328, Jan. 2002.

[5] Personalized travel route recommendation using collaborative filtering based on GPS trajectories, Ge Cui, Jun Luo & Xin Wang, International Journal of Digital Earth Volume 11, 2018 -Issue 3

[6] E. Pitoska, "E-Tourism: The Use of Internet and Information and Communication Technologies in Tourism: The Case of Hotel Units in Peripheral Areas," Tour. South East Eur., vol. 2, pp. 335–344, Dec. 2013.

[7] Maheswari, V. Uma, G. Varaprasad, and S. Viswanadha Raju. "Local Directional Maximum Edge Patterns for facial expression recognition." *Journal of Ambient Intelligence and Humanized Computing* (2020): 1-9.

[8] G. Häubl and V. Trifts, "Consumer Decision Making in Online Shopping Environments: The Effects of Interactive Decision Aids," Mark. Sci., vol. 19, no. 1, p. 4, Winter 2000.

[9] F. Ricci, L. Rokach, and B. Shapira, "Introduction to Recommender Systems Handbook," in Recommender Systems Handbook, F. Ricci, L. Rokach, B. Shapira, and P. B. Kantor, Eds. Springer US, 2011, pp. 1–35.

[10] P. Resnick and H. R. Varian, "Recommender Systems," Commun ACM, vol. 40, no. 3, pp. 56–58, Mar. 1997.

[11] G. I. Alptekin and G. Buyukozkan, "An integrated case-based reasoning and MCDM system for Web based tourism destination planning," EXPERT Syst. Appl., vol. 38, pp. 2125–2132, 2011.

[12] R. Anacleto, L. Figueiredo, A. Almeida, and P. Novais, "Mobile application to provide personalized sightseeing tours," J. Netw. Comput. Appl., vol. 41, pp. 56–64, May 2014.

[13] L. Sebastia, I. Garcia, E. Onaindia, and C. Guzman, "e-TOURISM:: A TOURIST RECOMMENDATION AND PLANNING APPLICATION," Int. J. Artif. Intell. Tools, vol. 18, no. 5, pp. 717–738, Oct. 2009.

[14] J. B. Schafer, J. A. Konstan, and J. Riedl, "E-Commerce Recommendation Applications," in Applications of Data Mining to Electronic Commerce, R. Kohavi and F. Provost, Eds. Springer US, 2001, pp. 115–153.

[15] R. Burke, "Hybrid Recommender Systems: Survey and Experiments," User Model. User-Adapt. Interact., vol. 12, no. 4, pp. 331–370, Nov. 2002.

[16] D. Jannach, M. Zanker, A. Felfernig, and G. Friedrich, Recommender Systems: An Introduction. New York: Cambridge University Press, 2010.

[17] Chennam, Krishna Keerthi, and M. Akka Lakshmi. "Cloud security in crypt database server using fine grained access control." *International Journal of Electrical and Computer Engineering* 6.3 (2016): 915.

[18] M. Montaner, B. Lopez, and J. L. de la Rosa, "A taxonomy of recommender agents on the Internet," Artif. Intell. Rev., vol. 19, pp. 285–330, 2003.

[19] F. M. Santiago, F. A. López, A. Montejo-Ráez, and A. U. López, "GeOasis: A knowledge-based geo-referenced tourist assistant," Expert Syst. Appl., vol. 39, no. 14, pp. 11737–11745, Oct. 2012.

[20] D. R. Fesenmaier, K. W. Wöber, and H. Werthner, Destination recommendation systems [electronic resource] :behavioural foundations and applications / edited by Daniel R. Fesenmaier, Karl W. Wöber, Hannes Werthner. Wallingford, UK ; Cambridge, MA : CABI Pub., c2006., 2006.

[21] N. Leiper, "Tourist attraction systems," Ann. Tour. Res., vol. 17, no. 3, pp. 367–384, 1990.

[22] K. L. Andereck and L. L. Caldwell, "The Influence of Tourists' Characteristics on Ratings of Information Sources for an Attraction," J. Travel Tour. Mark, vol. 2, no. 2–3, pp. 171–190, Feb. 1994.

VMATE – Intelligent E-Learning Management System for Educational Institutions

Pudiyadatha Anjali[1], Nunnabhatla Harshita[2], M.D.N. Akash[3], Patlolla Vinith[4] and Mr. G.S Prasada Reddy[5]

[1] Computer Science and Engineering, Vardhaman College of Engineering

[2] Computer Science and Engineering, Vardhaman College of Engineering

[3] Computer Science and Engineering, Vardhaman College of Engineering

[4] Computer Science and Engineering, Vardhaman College of Engineering

[5] Assistant Professor, Computer Science and Engineering, Vardhaman College of Engineering

E-mail[1]: nairpanjali2000@gmail.com

Abstract. The world is facing a huge health crisis which is demanding to reorganize the way it functions from square one. The spread of the pandemic, the shutdown of education institutions, and the shift to online learning were so fast that it barely gave any time to consider the difficulties faced by students. This is a great opportunity where educational institutions will shift to blended learning; hence it will become a new way of learning where there will be a need to find new ways to deliver the best content by using learning management systems. Therefore, the virtual learning world will create a great impact on the current education system due to the diverse effect of the epidemic. Hence different organizations are tremendously advancing their technology, and it will become more crucial over time. VMATE –Intelligent E-Learning Management System provides an interface for learners, which is of great significance and practical value for the improvement of learning and teaching quality. The use of Computers and the Internet has introduced technological conditions for teachers and students who can make good use of online information provided and communicating with others. This can be used by institutions and for different environments, where they can have computer-based instructions, blended-learning, and increased quality and quantity learning. Besides some basic management functions, the system focuses on assignment grading, assessment for various learning courses which makes it suitable to provide personalized grading and detailed feedback.

1. Introduction

The world today is facing the largest public health risk which is resulting in one of the biggest and fastest reorganizations of the global order. By the end of March 2020, the epidemic has unfolded to over 185 countries and resulted in the closure of over ninety percent of all schools, colleges, and universities impacting near to 1.38 billion students [1].

The closure of schools and universities has widespread individual, organizational, and learning and teaching implications for the students, faculty, directors, and also the institutions. The initial stage of speedy adaption during 2020 contained 3 primary responses to COVID-19: minimal legal response, delayed commencement of study periods, and speedy digitalization of curriculum [2] [3]. The closure of schools and colleges not solely interrupts the teaching for the students around the world; the closure additionally coincides with a key assessment period and a lot of exams have been rescheduled or canceled. In India, more than thirty-two crores of scholars and students have been affected by the several restrictions and the worldwide lockdown.[4]. As per the United Nations Educational Scientific and cultural organization report, about 14 crores of primary and 13 crores of secondary students are principally affected which are the two most affected levels in India [5]. On the bright side because of the epidemic, the world has stayed connected globally through the internet. Earlier it was exceedingly difficult to spread knowledge globally through the net. Earlier it was extremely tough to unfold information, but today it is very easy to share and learn by using the Internet. The Internet has helped the education system a lot by introducing a concept referred to as e-learning based Learning Management System (LMS). Several institutes started using these systems for serving different department needs. Generally, an online learning management system contains different components or modules [22],[23],[24]. A course management module provides facilities for adding new courses, managing, or updating existing courses, assigning teachers to courses, and other course-related details. [6] The student management module contains student enrolment, student registration for regular and elective courses. This is an especially important module for other modules to work properly with accuracy. In the online assessment module, the student can upload the assignment electronically so there is no need for submitting it by using paper [3]. Therefore, it saves a lot of papers. [6] A teacher can upload tutorials, videos, and other useful materials using an online course material management module. With the help of using this module, the student can view and download these files for study. In the feedback management module, students can give feedback for each subject, and teachers and only authorized persons can view the feedback. This is an immensely helpful module for teacher assessment.[6]. It provides prompt and efficient output as compared to a manual feedback system [8]. E-learning has rapidly evolved from just a thing of the future to a very practical approach towards education. It will continue to be an extremely useful classroom teaching tool as well as a self-study platform for learners. [7] E-learning approaches offer a vastly more engaging classroom atmosphere as well as the option for a fully tailored and self-paced online curriculum for home learning. The new advancements in learning technology have helped to support learners. Hence in this paper, we propose one such platform for improving the experience of learning effectively, especially during the pandemic like COVID-19 and other crisis. Hence it provides us an opportunity to improve our present education system.

2. Related work

This year 2020 has been the widely effected year where the world is at various points in pandemic rate, overall, there are about 1.2 billion children in 186 countries affected by shut down of institutions due to the epidemic. Hence, the education system has changed a lot, with the rise of E-learning, which is being delivered on online platforms. Even before the pandemic, there was high growth in online earning, with overall investments for online education. There has been a drastic growth in E-Learning when services like online learning software, conferencing tools, and online tutoring platforms were introduced. Now, many online learning interfaces are providing free resources, including some platforms like BYJU'S, which is an online learning application. Hence a single platform is being provided by other firms for both the teachers and students. For example, Lark, a Singapore company took the initiation for providing intelligent planning, auto-translation tools, and unlimited video conferencing time, with many other functionalities. Keeping in mind the functionality and usability, Lark has upgraded its global server and engineering techniques to make sure the connection is reliable [9]. To experiment using a virtual laboratory Mihaela M. Albu et al. introduced a system based on the online web. The system is used for multiple purposes and developed in the LabView environment. [6] The main benefit of this system is that the students can access their laboratory equipment's from anywhere across the globe. The main goal of the system is to conduct practical power engineering using this virtual laboratory.[6] [25] Since there are various e-learning software's available in the market. There is confusion about choosing a system for an institute. A web-based learning system introduced by Nadire is an easy way to evaluate Learning Management Systems [10]. It is a decision support system, which gives the users a choice to select between the best learning system based on their requirements. Almost all learning systems are based on open source technologies. Raja Maznah Raja Hussain and Hamsiah Mohd Dahalan proposed an online learning management system to help with the teaching and learning of institutions [11]. Some Learning Management Systems are developed for providing an online platform to users for multiple purposes [12]. An LMS is based on multi-instance learning which was introduced by Amelia Zafra et al [13]. To categorize students in LMSs. It mainly focuses on three functions. They are Assignment, Forum, and Quizzes for the students. Ardeshir Bahreininejad and Mahkameh Yaghmaie proposed an effective learning system is based on sequencing and adaptation, Sharable Content Object Reference Model (SCORM), semantic

Web ontology, and multi-agent system [14]. Amidst the pandemic, there have been many successful changes observed in many institutions. For example, The Imperial College in London introduced a class for the science of coronavirus, which is now the most taken class on Coursera this year. Zhejiang University managed to get more than 5,000 courses online in two weeks using "DingTalk ZJU" [9]. A detailed comparison of different LMS is shown above. There are many Learning platforms that provide various purposes. The features of these systems are generally course management, virtual classroom, feedback management, and other common features. However, none of these systems is the unification of learning and gaining information on the area of interest of the users.

2.1. *A detailed comparison of different LMS*

Table 1. Previous LMS Works

Technologies	Features	Platforms and tools
MS- Visual Studio.Net, MS-SQL Server	Tracking of learning Testing, Registration, Online course material, Delivery, Offline course material, communication.	Not Reported
Moodle	Assessment	Moodle
Moodle	Moodle features	Moodle
Not Reported	Quizzes, Assignment, Forum.	Not Reported
C, C++, Java, XML, SOAP	Account Management, Security Protection, Collaborative Learning, Student Activity, Feedback Collection, Online Electronic Experiment.	Pentium IV 2.80GHz, 512MB RAM, Window 2003 Server.
Stand-Alone, Web Based, Adaptive hypermedia, Semantic Web	Person and group information, person's profile and learning history, assessment, groupings of learning content, dynamic content sequencing, learner competencies, learning activities, searching across federated databases, connecting diverse eLearning tools.	Not Reported
Not Reported	Not Reported	CSCL tools; Collage and Gridcole.
PHP5, MySQL, HTML, Javascript, XHTML, CSS	Carry out Practical, assignments in astronomy, automatic control, chemistry, image processing, robotics and real-time operating systems.	Ciclope software, Apache
Not Reported	Face to Face(F2F) Learning over internet, Discussing on forum, studying lessons, solving self-test, creating seminars, course creation, examination, evaluation.	AHyCo Tool
Moodle	Student motivation and Interaction patterns. Apache Moodle share education institutes learning object repositories.	Apache LOP2P Plug-in, Java language with JXTA framework.
Not Reported	Course access, video and animation, quiz, virtual classroom.	Adobe Connect Professional, Adobe

		Presenter, Adobe Captivate software, Adobe Flash, 3ds Max Studio.
PHP, XML	Laboratory experiments in automatic control, biomechanics and fluid mechanics.	SysQuake Remote.
Not Reported	Learning management system features, online virtual experiment.	Active Document (AD) system.

There are many Learning platforms used for different purposes. The features of these systems are generally course management, virtual classroom, feedback management, and other common features. However, none of these systems is the unification of learning and gaining information on the area of interest of the users.

3. Proposed approach

During a crisis like COVID -19, sharing knowledge becomes very much necessary. Today it is extremely easy to share knowledge due to evolution in technology. In this paper, we have explained our approach to an e-learning based learning management system. Keeping in mind the basic parameters such as technology, features, and platform-tools. Based on these parameters a new unified system is proposed with all these parameters and other new features. The main idea is to provide an online user interface, to increase the efficiency of Virtual Learning Management, and to organize e-learning content in one location. We have proposed a learning platform that would work both for e-learning and learning during normal situations. Since there are different modes of learning, we have selected asynchronous learning as our main way of teaching and learning. By doing this we provide both the faculty and students equal time and opportunity to organize a class. The instructor can manage the lectures according to his or her schedule or the timetable set by the organization and the student also has the flexibility to check out the classes anytime. This overrules the problem of network connectivity and time management. Another feature that we have proposed is students can see the number of views of an uploaded course, and by analyzing these views, appropriate suggestions are provided to students taking the courses. Our proposed model also follows the general norms of a basic e-learning platform that is learning end teaching is completely computerized and thus decreases manual power. It does not consume a large amount of time and paperwork. This system has no scope for human contact and hence exceptionally reliable during pandemics and other crisis. Apart from these basic features, we provide a single platform for the users to get to know their associated major more proficiently, by providing them with a lobby area where the alumni and students can post their views, suggestions, and experiences in the industry. We also have included a feature of attendance check-in where the student attendance is taken care of. unlike in the existing system there no single platform for acquiring information on various aspects of different domains and evaluating student performance. Apart from this our approach also proposes flexible searching using Machine Learning. Hence the Outcomes are mainly focused to increase Instructor-Student time and to provide a flexible interface for online learning, which reduces learning and development time. Hence this approach makes learning flexible for the community.

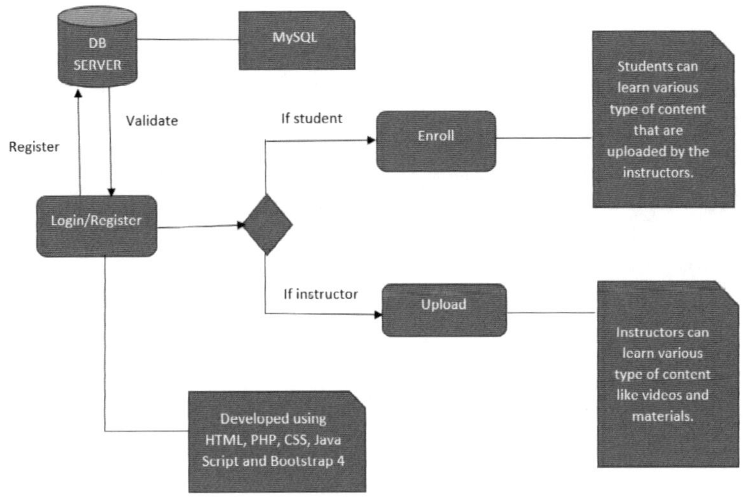

Figure 1. General process flow of the proposed approach.

4. Experimental setup

Web development is undergoing continuous evolution over some time and it is expanding its reach. This advancement has made way for web technologies. Web development is the task of developing dynamic web applications. Two major segments of web development front-end development are the client end and back-end development that is server-end. [15] Front-end development is what a user can see when they load a web application which includes the content, design, and how one can interact with it. This task is performed with the help of three codes mainly HTML, CSS, and JavaScript. Back-end development is the charge to control all the actions that take place behind the scenes of the web application. The back-end development uses a database that will generate front-end [16] [15]. To make a web application, we need

to make the server, application, and database communicate with each other, which is done by using server side languages like PHP, Ruby, Python, Java, and .Net, and tools like Oracle, SQL, and MySQL to save, change, or find data and send it back to the user in front-end code these tools are used to create or contribute to web applications with clean, portable, well-documented code. [17] But before writing that code, we need to know the system requirements, then translate those into technical requirements and come up with the most effective and efficient solution for the technology. When a new user accesses our website, he or she can register with our services, and all the credentials of the registered users are stored in our database server. Once the user enters valid details, they either upload or enroll in courses. We have embedded different areas on the website for the flexibility of users using HTML, CSS. If the instructor wants to upload their course and materials, initially they have to select a category from the provided catalog, if the required category is already provided in our service, they can proceed to upload their content, where all the uploads are stored in the database for future accessing. On the other hand, if the student wants to enroll and start learning a course, he can select the category and start learning by enrolling in the course. For making our website reliable, we are using various Machine learning algorithms that can receive the input data and use statistical analysis to predict the output, based upon the type of data available. By using supervised and Unsupervised Machine learning algorithms such as linear regression, clustering, the association we are providing applications such as search engine, user behavior analysis to give personalized recommendations. For example, the content suggestion to the students based on the type of video he or she watches the most and gives personalized recommendations. Using the Machine learning algorithms, we provide a search engine that enables the search option for the students which provides the best results. Also, whatever the page is being opened by the students for a particular topic frequently will remain at the top of the page for a long time. The architecture of our approach shows the interaction between the user, middleware system, and database to ensure multiple processes work together. With our application, we have a server and client-side, two programs are running simultaneously. The code developed with HTML, CSS, JS which lives in the browser and responds to user input. The code developed using PHP, Python, JS lives on the server which response to the HTTP request. When the user tries to access our web application, through the internet browser, the website lives on and loads the requested page. The server then responds by sending files over the browser, after this action, the browser executes the HTML, CSS, and PHP files to show the requested page to the user. in case of any stored information is to be loaded, the server accesses the database and sends back the required information which is again sent over to the user through the browser.

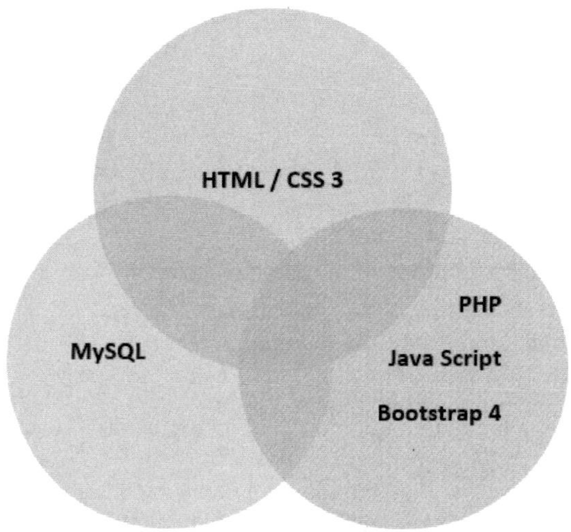

Figure 2. Technology ecosystem for the proposed approach.

5. Results and discussion

The digital revolution has resulted in exceptional changes in how and what way the content is accessed, consumed, discussed, and shared.[18] In this proposed paper educational courses are uploaded by the experienced instructors that can be taken up by the students which suit them depending on their availability and comfort. In contrast to classroom teaching, with online-based learning students can access the courses numerous amounts of times. This is especially required at the time of revision when preparing for an exam and unexpected situations, for example, COVID-19 and a similar crisis. In the traditional form of the learning system, if the students cannot attend any lecture, then they must prepare for that topic on their own which might be difficult. In e-learning, students can attend the lectures whenever they want with ease and even during normal times, e-learning platforms can be used along with face-to-face learning for better student-teacher interaction. As compared to the present system, the online learning model has relatively quick delivery cycles. This indicates that the time which is required to learn is reduced to 30%-65% of what is required in the existing system. Some of the reasons why the learning time is reduced with our approach are that students can define their speed of learning instead of following the speed of the whole group. It saves time as the student does not need to travel to the training venue. They can learn at their comfort and in their place. They can also choose to study specific and relevant areas of the learning material without focusing on each area. For example, they can skip a particular area that they do not want to learn. [19] There are also many benefits to online learning. It enables educators to get a higher degree of coverage to communicate their content in a consistent way for their target audience. This makes sure that all learners receive the same type of training. It is also cost-effective as compared to traditional forms of learning. The reason for this reduction is because learning through online mode happens quickly and very easily. [20] A lot of training time is reduced concerning trainers, travel, course materials, and accommodation.[19] This cost-effectiveness also helps in enhancing the profitability of the organization. [20] Also studying at one's place relives paying for travel expenses when training happens in another city, state, or any external learning materials. E-learning is a paperless way of learning; it protects the environment to a lot of extents and hence the highly eco-friendly way of learning. Due to the pandemic the learning was at a halt for few days, this became highly inconvenient for students as well as the administration but thanks to technology, many professional institutions have started developing their own LMS to reflect on the current educational system.

6. Conclusions and Future Work

One of the many questions raised in this research was whether e-Learning and traditional face-to-face learning should be measured based on the same definitions of and approaches to effectiveness during pandemics and another similar crisis. To find appropriate answers to this question, the authors had to consider the identified definitions, measures, and factors when designing for an effective E-learning platform. Whatever e-learning demands, an LMS is there to deliver. Without it, the online world would be a huge textbook filled with unorganized knowledge, there would be very little order to our studying efforts, and even little methodology to help us gain, understand, and retain all the information. Learning would still be possible, of course, but not in an effective, systematic manner that LMS provides [21]. No E-learning platforms can be alike for choosing the best for education. The complex process of teaching and learning requires complex, models of accomplishment. A universal idea behind the selection of LMS is that E-learning is organized and managed within an integrated system. The learning management system normally offers discussion forums, file sharing, management of assignments, lesson plans, syllabus, and other features. A sustained process requires aligning context changes with changes and approaches to learning methods and available technology. The Institutions must adapt, using technologies and models of understanding, with the change in environment and global situations.

References

[1] https://www.indiatoday.in/education-today/featurephilia/story/covid-19-4-negative-impacts-and-4-opportunities-created-for-education-1677206-2020-05-12, last accessed 12/09/2020.
[2] https://en.wikipedia.org/wiki/Impact_of_the_COVID-19_pandemic_on_education, last accessed 29/08/2020.
[3] Crawford, Joseph; Butler-Henderson, Kerryn (2020). "COVID-19: 20 countries' higher education intra-period digital pedagogy responses". Journal of Applied Learning and Teaching. 3 (1): 9–28. doi:10.37074/jalt.2020.3.1.7. Retrieved 24 August 2020.
[4] https://scroll.in/latest/957955/covid-19-india-now-has-1834-cases-after-biggest-single-day-jump-global-count-nears-a-million, last accessed 30/08/2020.
[5] Jena, Pravat. (2020). Impact of Pandemic - COVID-19 on the Education in India. International Journal of Current Research. 12. 12582-12586.10.24941/ijcr.39209.07.2020.
[6] Patel, Chirag & Gadhavi, Mahesh & Patel, Atul. (2013). A survey paper on e-learning based learning management Systems (LMS). International Journal of Scientific and Engineering Research. 4. 171-176.
[7] https://www.quora.com/Can-I-tutor-online-and-earn-online, last accessed 18/09/2020.
[8] Ellis, Ryann K. (2009), Field Guide- Learning Management, ASTD Learning Circuits are archived from the original on 24- August- 2014, retrieved in 5- July -2012.
[9] https://www.weforum.org/platforms/covid-action-platform/articles/coronavirus-education-global-covid19-online-digital-learning, last accessed 13/09/2020.
[10] Nadire. Cavus, the evaluation of Learning Management Systems using an AI fuzzy logic algorithm, Advances in the Engineering Software, Volume- 41, Issue- 2, February- 2010, Pages - 248- 254.
[11] Hamsiah Mohd Dahalan, Raja Maznah Raja Hussain, Development of Web-Based Assessment in Teaching and Learning Management System (e-ATLMS), Procedia - Social and Behavioral Sciences, Volume 9, 2010, Pages 244-248, ISSN 1877-0428, 10.1016/j.sbspro.2010.12.144.

[12] Tufan Adiguzel, Ozge Yasar, A working successor of learning management systems: SLOODLE, Procedia - Social and Behavioral Sciences, Volume 2, Issue 2, 2010, Pages 56825685, ISSN 1877-0428, 10.1016/j.sbspro.2010.03.928.

[13] Amelia Zafra, Sebastián Ventura, Cristóbal Romero, Multiple instances learning for classifying students in learning management systems, Expert Systems with Applications, Volume 38, Issue 12, November–December 2011, Pages 15020-15031.

[14] Ardeshir Bahreininejad, Mahkameh Yaghmaie, A context-aware adaptive learning system using agents, Expert Systems with Applications, Volume 38, Issue 4, April 2011, Pages 32803286, ISSN 0957-4174, 10.1016/j.eswa.2010.08.113.

[15] https://thenextscoop.com/technology-build-website/, last accessed 15/09/2020.

[16] https://ellencrichton11gmail.wordpress.com/category/software-development/, last accessed 17/09/2020.

[17] https://www.fourbrick.com/technologies, last accessed 17/09/2020.

[18] https://www.elearningscript.com/e-learning-script-best-online-training-software/, last accessed 19/09/2020.

[19] https://elearningindustry.com/9-benefits-of-elearning-for-students, last accessed 20/09/2020.

[20] https://divinelinksoft.wordpress.com/2018/10/08/benefits-of-elearning-for-students/, last accessed 20/09/2020.

[21] https://www.betterbuys.com/lms/root-of-all-elearning-lms/, last accessed 10/09/2020.

[22] A.M. Payne, J.E. Stephenson, W.B. Morris, H.G. Tempest, A. Mileham, D.K. Griffin, The use of an e-learning constructivist solution in workplace learning, International Journal of Industrial Ergonomics, Volume 39, Issue 3, May 2009, Pages 548-553, ISSN 0169-8141, 10.1016/j.ergon.2008.10.019.

[23] Salem Alkhalaf, Steve Drew, Thamer Alhussain, Assessing the Impact of e-Learning Systems on Learners: A Survey Study in the KSA, Procedia - Social and Behavioral Sciences, Volume 47, 2012, Pages 98-104, ISSN 1877-0428, 10.1016/j.sbspro.2012.06.620.

[24] Daniel Palacios-Marqués, Rocío Cortés-Grao, Clemente Lobato Carral, Outstanding knowledge competences and web 2.0 practices for development of successful e-learning project management, International journal, volume 47 – 2012, pages-98.

[25] K.E.; Heydt, Albu, M.M.; Holbert, G.T.; Trusca, Grigorescu, S.D.; V.; "Embedding remote experimentating in power engineering education,", IEEE Transactions on Power Systems, vol.19, no.1, pp. 139- 143, Feb. 2004.

IOP Conf. Series: Materials Science and Engineering 1042 (2021) 012012 doi:10.1088/1757-899X/1042/1/012012

A Multifarious Diagnosis of Breast Cancer Using Mammogram Images – Systematic Review

M.Swapna[1], Dr Nagaratna Hegde[2]

[1] Asst Professor, Dept of CSE, Vardhaman College of Engineering.
[2] Professors, Dept of CSE, Vasavi College of Engineering
1swapna0801@gmail.com
2nagaratnaph@gmail.com

Abstract. Breast Cancer is most common disease in worldwide leads to high rate in mortality. Detection of symptoms at early stage is difficult to identify the breast cancer. Diagnosis through General Medical Examination cannot detect the disease. This paper represents various common methods to detect the Breast Cancer and detection of tumors using mammogram images with Artificial Intelligence (AI). Algorithms helps Images in preprocessed using various feature extraction methods and classification algorithms to predict the class of tumors is benign or malignant. This article insight the recent approaches used in detection of tumors and stage of breast cancer.

KeywordsDetection Breast Cancer, Volatile Organic Compounds, Urinary Exosomal, CT scan, Mammography, Ultra Scan.

1. Introduction

One of the leading diseases, which lead to more death in women, is cancer. When new patients are diagnosing with cancer, one of the four detected as breast cancer [1]. As per the ICMR (Indian Council of Medical Research), every year 70,000 people die from this disease and 1.5 lakhs are new cases [4]. Breast cancer is most common at age of above 40 years [5]. The p53 gene act as a tumor suppressor, which will regulates the normal cell proliferation and in p53 mutations obtains an output of nonfunctional protein, which accumulates in tumor cell nuclei leads to breast cancer [6]. Patients delay in regular screening and ignoring symptoms of the disease can cut the survival rate of current stage [7] and it is also observed and proved that tumor size also depending on the cause of the prognosis of the patient, as tumor size accumulate, less in survival rate [8]. Breast cancer occurs in Male and female, in few countries men has more incident rate but most countries female have more incident rate [9].

The radiologists identify the calcification (break down chips) and lesions (cotton balls) using mammograms [11]. Benign disorder breasts Asymmetric breast development and discharge of nipples, breast cysts, breast pain can because of breast cancer [12], The presence of receptors like Estrogen, the progestin in cells lead to cancer, it is called as HER2 Positive [13]. When Estrogen and progestin receptor are tested negative and excess proteins are said to be Triple-negative Breast Cancer [14]. Lymph was built on the skin of the breast, symptom of breast cancer, swollen lymph are present under the armpits [15].

In this article, section 2 describes the extensive literature survey on various approaches used to predict the breast cancer, comprehensive study and performance is summarized in Table 1, section 3-detailed review on various preprocessing, segmentation, Feature extraction and section-4 reviews on classification methods.

2.Literature survey

2.1 Multifarious Methods for Breast Cancer Examination

Test for breast-screening method is very simple and attainable [18]. **Breast Cancer SelfExamination** has to perform at regular menstrual cycle from day three to five. Observations are Dimpling, Shape, and outline are deformed, nipple pin inside lumps near armpits to know the abnormality of breast [18]. **A breath Analysis Test** can be the primary screen for breast cancer disease, this test performed by analyzing the breath samples using the analytical technique is gas chromatography and measures the mass-to-charge ratio of ions in mass spectroscopy. The breath methylated alkenes contour values are determining breast cancer. Which determines a woman has breast cancer or not. The negative-positive values are more accurate than mammograms [37]. VOCs will help to identify the existence of breast cancer or not. Componentslike heptanes, 1-phenyl-ethanone, and isopropyl myristate, 2, 3-dihydro-1-phenyl-4(1H)-quinazolinone, 2-propanol. [34]. the novel approach for Breast Cancer diagnostics applies to premenopausal, younger, pregnant women who are Liquid biopsy-based on **Biomarkers**. Urinary miRs are used for breast cancer detection [22].

Ultrasound is an important approach for the detection of breast cancer, to avoid the needle biopsy and needle aspiration. Apart from lesion characteristics and reporting of lesions it also helps us to identify the lesion is benign or malignant. **MRI Examination** is used to detect and study the breast more accurately than any another method, even it is more costly and it is preferred for high-risk women's and it does not expose the human body to heavy radiations and the duration of the test is for 15 to 30 minutes, the human body is passed into a narrow tube structure and create a magnetic field and radio frequency wave environment [16]. MRI Screening will result in effective Positive prediction values when we compare with mammography, ultrasound, and mammography with MRI. A patient lay on the machine, An X-Ray beam is projected only on the Breast portion of the body and it is a circular scan on different projection views and builds the 3D Images of the breast (300 to 500 projections in a 360-degree rotation of X-ray Tube). The disadvantage of CT scan is radiations observed are 10 times more than mammography [51].**CT scan** helps to the detection of benign breast lesions and primary and secondary malignancies, it observed that the detection rate various from 24% to 70%, and it is an alternative examination. The radiologist should concentrate while performing regular chest CT scans carefully [2].

3.Related Work

3.1 Breast Cancer Detection using Mammogram

Mammogram images are used to detection of abnormal growth of cells. It uses a low dose of X-Ray and High quality with low radiations on Human breast [23]. In figure 1, the model diagram represents process flow to detect breast cancer. These Digital Images are meant to observe calcification or Lumps look like cotton balls. Mammogram images are preprocessed after reading an image, smoothing an image by removing noise data, identify the boundaries, Region of Interest, detection methods applied to the image [24].

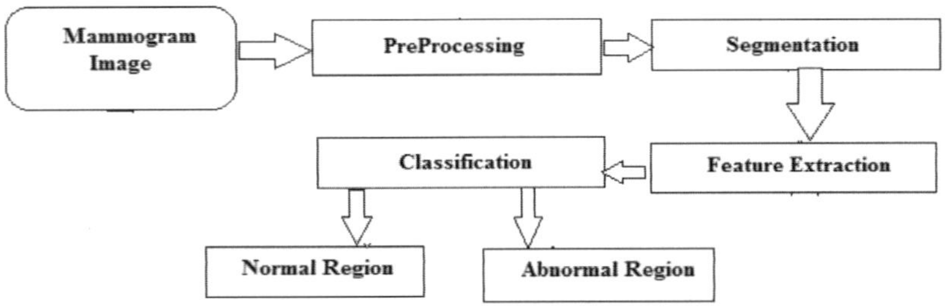

Fig 1.Flow Chart of Mammogram Image Detection Model

Breast images were screened in two views one is the CC view another is the MLO view. The images are X-Ray Film records in either Digital or Film Screen Mammography. Digital Mammogram is viewed [26]. Images are stored on the storage devices with DICOM file format [27]. DICOM images are viewed with Open- Software like OSIRIX [28], ImageJ[29], Micro View, MIPAV[30]. **Smoothing Image** is a Images are preprocessing by Removal of noise like Salt & Pepper noise, Gaussian noise using Median Filter, Min and Max Filters to change the intensity of the image for clarity of the organ and denoising image is done by Winer filter [25].

Table 1: literature Survey on comparative Analysis of Detection of Breast Cancer

S. No	Method of Detection	Examination Procedure	Factors for Observation	Size of the Dataset	Results
1	Breast Self Examinatio n	Self Examination of breast and observe of Armpits.	Lumps, Deformation of the shape, Dimpling, Nipple pins	None	AUC 81.40%
				None	AUC 89.70%
2	Biomarkers Analysis	Breath Analysis (VoC). analyze gas chromatography and mass spectroscopy[37]	VoC's such as alkanes, ketenes, halogenated, hydrocarbon, aldehydes, esters [36].	101	93.8% sensitivity and 84.6% specificity
		Urinary Exosomal MicroR	miR-Factors	109	98.6% is sensitivity 100% specify
3	Magnetic Resonance Imaging	human body is passed into a narrow tube structure and create an magnetic field and radiofrequency wave enviourment	BRCA Gene Mutation Analysis	529	sensitivity is 91% Specificity of MRI is 97.2%
4	Ultrasound Images	sound waves are transmitted into body and bounce back ,they are collected by computer to create images[46]	Shape of lesion, margins, boundaries and orientation.	167	71.3% Accuracy
5	Computer Tomograph y	Humandy body is passed an X-Ray Beam on 360-degree projection.	Abnormal Edges, shapes and RIM enhancement on breast	-	24% to 70% Variance with radiologist

TheEdge detection method is aImages are divided into (Region of Interest) RoI using edge Detection techniques will able to identify the boundaries of the breast and analyze the shape of the breast, it also helps to observe both breasts are symmetry in shape or not. There are many proved efficient methods for edge detection like Robert Edge Detection use for horizontal and vertical filters, Sobel Edge Detection used for edge detection, Prewitt Edge Detection for horizontal and vertical edges [31], Zero-Cross, and Canny uses mathematical functions [32].

Feature Extraction Methods is a Detection of masses and calcification, identification of the region of interest. This phase is most important in breast cancer detection can be obtained by the classification of regions and clustering algorithms [17] [33].

4. ClassificationTechniques used for Breast Cancer Detection

Classification Techniques helps to classify the input data into a labeled class. Class features are defined based on the training dataset. This will help new facts to predict the class label. The broadly categorized algorithms are quadratic classifier, Random Forest Method, Neural Network, Support Vector Machine, Linear Classification and Logistic Regression.

4.1 Sequential Minimal Optimization (SMO) is methods used to train the Support Vector Machine (SVM), it is invented by John Carlton Platt. This approach is simple and fast, the concept used to solve the problem is dual quadratic optimization, which will include two elements are at each iteration. The large quadratic program divides into a small optimization problem. The smallest problem solved individually and parallel to decrease the time. The attributes like clump thickness, Cell Size, Cell Shape, Marginal Adhesion, Single Epithelial Cell Size, Bare Nuclei, Bland Chromatin, Normal Nucleoli, Mitoses are input variable X1,X2,.... Xn and Y1, Y2...Yn are binary integer value has positive or negative values for classification. Support Vector Machine has trained by a quadratic optimization algorithm to create the margin. SMO method able to classify the instances correctly is 657 and incorrectly classify are 26, the overall accuracy achieved 96.19% [23].

4.2Radial Basis Function Neural Network (RBFNN). This method is an artificial feed-forward neural network, which consists of three layers, Input Layer has four neurons Use to represent features like ASM, Correlation, Sum Entropy, and Sum Variance of the GLCM features extraction approach. The second layer is a hidden layer, which has a total number of neurons less than the no of epochs in the dataset; they depend upon the total no of datasets available. The third layer is the output layer, which is the linear sum of the radial basis function. The x value considers based on the threshold. Here 'k' is the upper threshold value and 'j' is the lower threshold value. If 'x' value is greater than 'k', then x=k or if 'x' is less than 'j' then x=j. This Work, There are two different classifications are performed, first time RBFN used to classify normal or abnormal and the second time RBFN used to classify the data is benign or malignant. 'Y' is the outputs of the network refer in equation (1).

$$Y = \sum_{j=1}^{p} w_j \phi_j \qquad (1)$$

w_j are the weights from j[th] neuron to output layer.

ϕ_j is a kernel function, i.e. Gaussian Function refer in equation (2)

$$\phi_f = \exp\left(-\frac{(x - c_j)^2}{2\sigma_j^2} \right) \qquad (2)$$

$x = (x_1, x_2, x_3, x_d)^T$ is input vectors.

$c_j = \left(c_{1j}, c_{2j}, \ldots . c_{dj}\right)^2$ is center vector, $'\sigma'$ is a spread.

4.3 Region Extraction by Clustering and Classification Algorithms (RECC)

In this work, mammogram images are giving as inputs and they are dividing into three steps. In step-1, removal of noise and enhancement of image using medium filter because it can efficiently remove noise for non-linear digital image and Gaussian Filter used to enhance the images. In step-2, images have segmented using a **k-Means** algorithm, every pixel compared with the centroid value of the cluster and the nearest cluster has mapped to the pixel. **CART** is a Classification and Regression Trees, which is a machine-learning algorithm, which helps to build a binary tree using the GINI Index to split the attributes. The recursive splitting of attributes stops at min-count at leaf node condition. The outcome is to predict the disease of class. This method has been achieved an accuracy of 93.75 %. **J48** algorithm has used for classification in the data-mining tool Weka. This algorithm will help to generate a binary tree based on the training data. It is an extension of the ID3 algorithm. It will generate a decision tree, which may be pruned or unpruned. the training set is generated based on information entropy. The attributes are divided into subset and entropy used to find the optimum attribute until the leaf nodes are pure. This algorithm is able to obtain an accuracy of 92.85%. **JRIP** is a repeated incremental pruning. The basic concept is to reduce the errors and it is one of the fast learning and efficient methods to generate a decision tree. It has different steps a) **Growing Rule** b) **Pruning** c) **Optimization** d) **Selection.In Growing Rule** will add the antecedents until it is 100% **pure**, the incremental inspecting the rule with previous rule of class, **Optimization** of rule, after generating the initial rule, and in the **selection phase** it will regulate the classes. This method is able to achieve an accuracy of 95.48%. **Navi Bayesian Algorithm** is a classification algorithm based on statistical attribute values, members of the class dependence on the attribute dependency of class. The training of the data has to perform and noisy data is to be replaced with the most frequent occurred value, a transformation of data into relational tables. Finally the predication of the class is obtained based on the attribute and class table values. This algorithm has achieved an accuracy of 93.61%. **SVM** (Support Vector Machine) in a machine-learning algorithm, it the most commonly used method used for binary classification, which separates the group of data into classes. The decision line drawn between the data objects falls. It will separate them into two classes. This approach has obtained an accuracy of 92.40%.

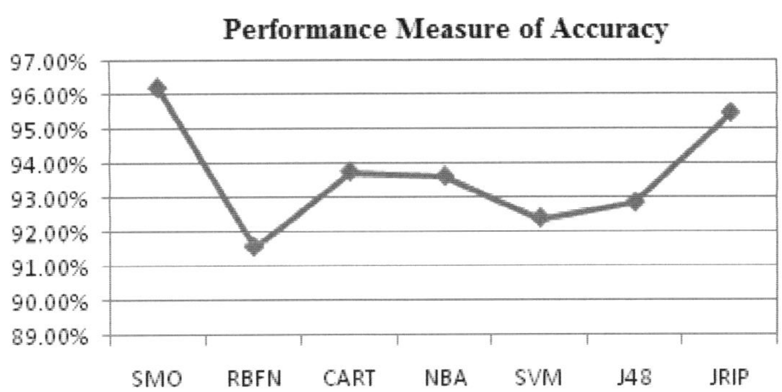

Fig 2 Comparative Analysis of Performance Measure of Accuracy

Table 2: Multifarious Algorithms used to classify the Classes

S. No	Author	Feature Extraction	Attributes	Classification Algorithm	No. of Image	Accuracy (%)
1	Vikas et. al [23]	Numeric values as input	clump thickness , Cell Size, Cell Shape, Marginal Adhesion, Single Epithelial Cell Size, Bare Nuclei , Bland Chromatin, Normal Nucleoli, Mitoses are attributes numerical values	Sequential Minimal Optimization	683	96.19%
2	Mellisa et.al[33]	Gray-level co-occurrence matrix	ASM, Correlation, Sum Entropy and Sum Variance	Radial Basis Function Neural Network	330	91.57%
3	Venkatesan et.al[37]	Gray-level co-occurrence matrix	Clustering images into K cluster using K-means algorithm. Each pixel is comparing with centric of cluster to consider in nearest cluster.	Classification and Regression Trees(CART)	50	93.75 %
				Naive Bayes Algorithm		93.61 %
				Support Vector Machines		92.40 %
				J48 Algorithm		92.85 %
				JRecursive Incremental Pruning		95.45 %

5. Conclusion

In this paper, Different approaches to detection of breast cancer and brief knowledge of characteristics and symptoms of BC. Preliminary image profile screening is best option for detection of breast cancer. Mammogram image analysis proven to be more accurate compare to any another approach, even patient will undergoes low radiations ,cheaper in cost and most important is highly availability of technology. Images are preprocessed and identify the ROI with feature extraction, In this work images are preprocessed and segmented using Gray-level co-occurrence matrix and one of the example numeric values are considered. This paper discuss on various method used to classify normal/abnormality of patients using different classification methods SMO,RBFN,JRIP,Navi Bayesian,SVM,J48 and CART. The Sequential Minimal Optimization is achieving maximum accuracy in comparison of other methods.

References

[1] American Institute for Cancer Research, Published an article on"world cancer statistics for the most common cancers" in 2018.

[2] O'Connell, A.M., Karellas, A., Vedantham, S. and Kawakyu-O'Connor, D.T., 2018, February. Newer technologies in breast cancer imaging: dedicated cone-beam breast computed tomography. In Seminars in Ultrasound, CT and MRI (Vol. 39, No. 1, pp. 106-113). WB Saunders.

[3] Kanimozhi, G., Shanmugavadivu, P. and Rani, M.M.S., 2020. Machine Learning-Based Recommender System for Breast Cancer Prognosis. Recommender System with Machine Learning and Artificial Intelligence: Practical Tools and Applications in Medical, Agricultural and Other Industries, p.121.

[4] Jemal, A., Ward, E. and Thun, M.J., 2007. Recent trends in breast cancer incidence rates by age and tumor characteristics among US women. Breast Cancer Research, 9(3), pp.1-6.

[5] Allred, D.C., Clark, G.M., Elledge, R., Fuqua, S.A., Brown, R.W., Chamness, G.C., Osborne, C.K. and McGuire, W.L., 1993. Association of p53 protein expression with tumor cell proliferation rate and clinical outcome in node-negative breast cancer. JNCI: Journal of the National Cancer Institute, 85(3), pp.200-206.

[6] Rauscher, G.H., Ferrans, C.E., Kaiser, K., Campbell, R.T., Calhoun, E.E. and Warnecke, R.B., 2010. Misconceptions about breast lumps and delayed medical presentation in urban breast cancer patients. Cancer Epidemiology and Prevention Biomarkers, 19(3), pp.640-647.

[7] Garg, M., Nagpal, N., Sidhu, D.S. and Singh, A., 2016. Effect of lump size and nodal status on prognosis in invasive breast cancer: Experience from Rural India. Journal of clinical and diagnostic research: JCDR, 10(6), p.PC08.

[8] Ly, D., Forman, D., Ferlay, J., Brinton, L.A. and Cook, M.B., 2013. An international comparison of male and female breast cancer incidence rates. International journal of cancer, 132(8), pp.1918-1926.

[9] Karmilasari, S.W., Hermita, M., Agustiyani, N.P., Hanum, Y. and Lussiana, E.T.P., 2014. Sample K-means clustering method for determining the stage of breast cancer malignancy based on cancer size on mammogram image basis. IJACSA) Int J Adv Comput Sci Appl, 5(3), pp.86-90.

[10] Nguyen, Q.D., Tenreiro, A., Roberts, J.T., Tavana, A. and Robinson, A.S., 2020. Hematoma Mimicking Breast Cancer on CT Scan and Breast Ultrasound. Cureus, 12(7

[11] Santen, R.J. and Mansel, R., 2005. Benign breast disorders. New England Journal of Medicine, 353(3), pp.275-285.

[12] Harrell, J.C., Dye, W.W., Allred, D.C., Jedlicka, P., Spoelstra, N.S., Sartorius, C.A. and Horwitz, K.B., 2006. Estrogen receptor positive breast cancer metastasis: altered hormonal sensitivity and tumor aggressiveness in lymphatic vessels and lymph nodes. Cancer research, 66(18), pp.9308-9315.

[13] Anders, C. and Carey, L.A., 2008. Understanding and treating triple-negativebreastcancer. Oncology (Williston Park, NY), 22(11), p.1233.

[14] Marino, M.A., Avendano, D., Zapata, P., Riedl, C.C. and Pinker, K., 2020. Lymph Node Imaging in Patients with Primary Breast Cancer: Concurrent Diagnostic Tools. The oncologist, 25(2), p.e231.

[15] Bakic, P.R., Barufaldi, B., Pokrajac, D.D., Weinstein, S.P. and Maidment, A.D., 2018, July. Optimized simulation of breast anatomy for virtual clinical trials. In 14th International Workshop on Breast Imaging (IWBI 2018) (Vol. 10718, p. 107181X). International Society for Optics and Photonics.

[16] Kuhl, C.K., Schrading, S., Leutner, C.C., Morakkabati-Spitz, N., Wardelmann, E., Fimmers, R., Kuhn, W. and Schild, H.H., 2005. Mammography, breast ultrasound, and magnetic resonance imaging for surveillance of women at high familial risk for breast cancer. Journal of clinical oncology, 23(33), pp.8469-8476.

[17] Prasad, G.V. and Raju, S.V., 2018. A survey on local textural patterns for facial feature extraction. International Journal of Computer Vision and Image Processing (IJCVIP), 8(2), pp.1-26.

[18] Semiglazov, V.F., Moiseyenko, V.M., Bavli, J.L., Migmanova, N.S., Seleznyov, N.K., Popova, R.T., Ivanova, O.A., Orlov, A.A., Chagunava, O.A., Barash, N.J. and Matitzin, A.N., 1992. The role of breast self-examination in early breast cancer detection (results of the 5-years USSR/WHO randomized study in Leningrad). European journal of epidemiology, 8(4), pp.498-502
[19] Costantini, M., Belli, P., Lombardi, R., Franceschini, G., Mulè, A. and Bonomo, L., 2006. Characterization of solid breast masses: use of the sonographic breast imaging reporting and data system lexicon. Journal of ultrasound in medicine, 25(5), pp.649-659.
[20] Sehgal, C.M., Weinstein, S.P., Arger, P.H. and Conant, E.F., 2006. A review of breast ultrasound. Journal of mammary gland biology and neoplasia, 11(2), pp.113-123.,
[21] Thaineua, V., Ansusinha, T., Auamkul, N., Taneepanichskul, S., Urairoekkun, C., Jongvanich, J., Kannawat, C., Traisathit, P. and Chitapanarux, I., 2020. Impact of regular Breast Self-Examination on breast cancer size, stage, and mortality in Thailand. The Breast Journal, 26(4), p.822.
[22] Sood, R., Rositch, A.F., Shakoor, D., Ambinder, E., Pool, K.L., Pollack, E., Mollura, D.J., Mullen, L.A. and Harvey, S.C., 2019. Ultrasound for breast cancer detection globally: A systematic review and meta-analysis. Journal of global oncology, 5, pp.1-17.
[23] Singh, A.K. and Gupta, B., 2015. A novel approach for breast cancer detection and segmentation in a mammogram. Procedia Computer Science, 54, pp.676-682.
[24] Makandar, A. and Halalli, B., 2015. Breast cancer image enhancement using median filter and CLAHE. International Journal of Scientific & Engineering Research, 6(4), pp.462-465.
[25] Moreira, I.C., Amaral, I., Domingues, I., Cardoso, A., Cardoso, M.J. and Cardoso, J.S., 2012. Inbreast: toward a full-field digital mammographic database. Academic radiology, 19(2), pp.236-248.
[26] Vachon, C.M., Fowler, E.E., Tiffenberg, G., Scott, C.G., Pankratz, V.S., Sellers, T.A. and Heine, J.J., 2013. Comparison of percent density from raw and processed full-field digital mammography data. Breast Cancer Research, 15(1), p.R1.
[27] Yakami, M., Ishizu, K., Kubo, T., Okada, T. and Togashi, K., 2011. Development and evaluation of a low-cost and high-capacity DICOM image data storage system for research. Journal of Digital Imaging, 24(2), pp.190-195.
[28] Hermena, J., Molla, C. and Jahnena, A., 2008. Managing and viewing DICOM images with ImageJ. Proc. ImagesJConf.
[29] Haak, D., Page, C.E. and Deserno, T.M., 2016. A survey of DICOM viewer software to integrate clinical research and medical imaging. Journal of digital imaging, 29(2), pp.206-215
[30] Singh, N., Mohapatra, A.G. and Kanungo, G., 2011. Breast cancer mass detection in mammograms using K-means and fuzzy C-means clustering. International Journal of Computer Applica1.
[31] Ponraj, D.N., Jenifer, M.E., Poongodi, P. and Manoharan, J.S., 2011. A survey on the preprocessing techniques of mammogram for the detection of breast cancer. Journal of Emerging Trends in Computing and Information Sciences, 2(12), pp.656-664.
[32] Chaurasia, V. and Pal, S., 2017. A novel approach for breast cancer detection using data mining techniques. International Journal of Innovative Research in Computer and Communication Engineering (An ISO 3297: 2007 Certified Organization) Vol, 2.
[33] Pratiwi, M., Harefa, J. and Nanda, S., 2015. Mammograms classification using gray-level co-occurrence matrix and radial basis function neural network. Proceeding Computer Science, 59, pp.83-
[34] Hirschfeld, M., Rücker, G., Weiß, D., Berner, K., Ritter, A., Jäger, M. and Erbes, T., 2020. Urinary Exosomal MicroRNAs as Potential Non-invasive Biomarkers in Breast Cancer Detection. Molecular Diagnosis & Therapy, pp.1-18.
[35] Phillips, M., Cataneo, R.N., Ditkoff, B.A., Fisher, P., Greenberg, J., Gunawardena, R., Kwon, C.S., Tietje, O. and Wong, C., 2006. Prediction of breast cancer using volatile biomarkers in the breath. Breast cancer research and treatment, 99(1), pp.19-21.
[36] Speiser, D., Schneider, A., Staeck, O., Freitag, C., Lanowska, M. and Mangler, M., 2012.

Volatile organic compounds (VOCs) in exhaled breath of patients with breast cancer in a clinical setting. Ginekologia polska, 83(10).
[37] Venkatesan, E. and Velmurugan, T., 2016. Extraction of cancer affected regions in mammogram images by clustering and classification algorithms. Indian Journal of Science and Technology, 9(30), pp.1-9.

A contingent review on cloud computing trends predicting viable possibilities for future of computing

Ravula Arun Kumar[1(a,b)], **Vinuthna Kambalapally**[2]

[1(a)]Research scholar, Department of CSE, Koneru lakshmaiah Education foundation, Vijayawada,India.
[1(b)]Department of IT, Vardhaman College of Engineering, Hyderabad, India
[2]Associate professor, Department of CSE, Koneru lakshmaiah Education foundation, Vijayawada,India.

*arunravula12@gmail.com

Abstract. Cloud computing was maybe the wildest stage in innovation as well as business press all through 2019. This's absolutely nothing unexpected as the cloud sector has been creating quickly throughout the final several years. Synergy Research by Group a short while ago detailed a thirty-seven % by and huge development year-over-year within the public cloud. They as well bear in mind it's taken just 2 years for the open PaaS and IaaS marketplaces to twofold in size as well as their conjecture provides them multiplying once again within the following 3 years. As year that is New begins with targets in the lives of ours, likewise, for that IT business additionally constantly accompanies an objective in Cloud Computing. Cloud is cost-productive, robust, and scalable. Cloud innovation is beneficial for program development, using the cloud for specific application development has to be common. We're just beginning to view the improvement of this concept into a transformation. Cloud computing changes the fashion in that we consider information, the manner where businesses consider the operations of theirs as well as the manner where engineers consider building. Let us take a look at several of the cloud computing trends which will take place in deep 2020 in terms of serverless computing, omni cloud, Quantum computing. Kubernetes, digital natives and here try to predict the future scope of computing in terms of compute, network, store and mark analytics.

1. Introduction

Cloud computing as an adage have been around after the very first 2000s, although the thought of computing-as-a-service has been around for a lot, a lot more time -- as long ago even though the 1960s, when laptop bureaus would permit companies to lease phase on a mainframe, rather than have to buy one themselves. These' time-sharing' treatments had been commonly overtaken by the rise on the Pc that made running a laptop computer or maybe pc a great deal much more economical, and subsequently consequently by the rise of business information centers where companies will save big numbers of info. Though the thought of renting usage of computing power has resurfaced once more as well as once again -- from the application plan application suppliers, power computing, in addition to grid computing of late 1990 and soon 2000s. This was applied by cloud computing, which really took hold with the progress of a system as being a service along with hyperscale cloud computing vendors as Amazon Web Services [1]-[2].

Creating the infrastructure to assist cloud computing today accounts for more than a third of nearly all IT investing globally, based mostly on investigation from IDC. Meanwhile committing on standard, in house IT is going to continue to glide as computing workloads stay switching to the cloud, even if that is public cloud services offered by vendors or maybe perhaps private clouds created by enterprises themselves [three]. The major Research predicts this around a single third of enterprise IT having to spend is actually be on web host as well as cloud services this season "indicating a growing reliance on external sources of infrastructure, software, management in addition to security services". Analyst Gartner predicts that one half of globally enterprises using the cloud today would have gone all in on it by 2021[4]-[7].

According to sources that are reliable, globally having to pay on cloud offerings will attain 1dolar1 260bn this season up through 1dolar1 219.6bn. It's also growing at a faster speed instead of the analysts expected. Though it's not totally clear exactly how much of that need is really coming from businesses that truly want to shift to the cloud and precisely how much is truly being created by vendors which currently just provide cloud types of the products of theirs (often since they are sharp to move to out from advertising one off licences to marketing probably more successful and also predictable cloud subscriptions).Here the article compares the future of computing via serverless compute, Kubernetes, Quantum computing, omni cloud and digital natives and will discuss the pros and cons of computing in detail in next sections, to introduce what the sections, let us dive into first component serverless computing.[8]-[11]

Serverless Computing is actually a cloud computing execution design which offers freedom to create, deploy, and operate applications without stressing about provisioning servers. Developers are helped by it to concentrate on their core item rather than passing time to provision, scope, and control servers. This cloud computing execution airer does not eliminate the demand for a server. Though it provides a total infrastructure that usually requires to run the application of yours by including a level of abstraction over the cloud infrastructure. Several of the cloud service providers which offer a serverless framework are actually AWS, Microsoft Azure, and also the Google Cloud Platform (GCP).[12]-[14]

In the start days of the web, any individual that needed to produce a web plan had to experience the physical hardware needed to operate a server, that's a cumbersome and expensive undertaking. After that delivered cloud computing, in what fixed amounts of servers or amounts of server room could be rented remotely. Businesses as well as designers which lease these fixed products of server home commonly over purchase to ensure that a spike in traffic or activity will not go over their month boundaries and bust the apps of theirs. This means which a great deal of the server location which will get paid for may go to waste. Cloud vendors have released automobile scaling models to contend with the problem, but despite automobile scaling, models to contend with the issue, but despite car scaling an undesirable spike in training, such as a DDoS Attack, might wind up being incredibly costly.

When nearly all of the businesses had been withdrawing the services of theirs from multi cloud, Omni cloud computing proved to become the last resort. Already IaaS (Infrastructure being a Service) providers had been gaining popularity. But businesses discovered they deserve faster moving uses for the new generation. Making uses much more powerful and lightweight was everyone's main concern. Omni-cloud computing brought all of these characteristics into the apps along with first-rate connectivity. This world class connectivity allows numerous platforms to streamline the data of theirs and incorporate them in the most effective way possible. Hence, the very best way to describe Omni cloud happens when barriers between the various platforms shrink then any multi cloud converts into an Omni cloud system. The data deployment process is much more accessible and organized when using the Omni cloud system.

All computing depends on bits, probably the smallest unit of info that's encoded as an on condition or even an off state, much more commonly called an one or perhaps a zero, in a few physical medium or perhaps another. The majority of the time, a little takes the physical type of an electric signal going over the circuits within the computer 's motherboard. By stringing several bits together, we are able to represent useful and complex more things as textual content, music, and other things. Quantum bits (or perhaps qubits) are actually produced of subatomic particles, specifically specific photons or perhaps electrons. Because these subatomic debris conform far more to the guidelines of quantum mechanics compared to classical mechanics, they display the bizarre qualities of quantum particles. Probably the most salient of these attributes for computer scientists is actually superposition. This's the thought that a particle is able to occur in a number of states simultaneously, at minimum until that state is actually calculated as well as collapses into a single status. By utilizing this particular superposition property, personal computer scientists are able to create qubits encode an one and a zero at the very same time.
Kubernetes is actually a container management technology produced in Google lab to handle containerized uses in different sort of locations like physical, virtual, as well as cloud infrastructure. It's an open source process which helps in producing and controlling containerization of application. This tutorial offers an overview of different sort of functionalities and features of Kubernetes and also shows you how you can handle the containerized infrastructure as well as program deployment.

2. Literature survey:

Born from a would like to produce platform as a system (PaaS) a lot more affordable, fine grained, and accessible, serverless computing has garnered curiosity from each and every academia as well as industry. The effort [nine] is actually created to create an understanding of the very first days or maybe days of serverless computing: what it is, exactly where it comes from, what is the existing state of serverless know how, and precisely what are its main obstacles and opportunities. [ten] has an evaluation and multilevel feature evaluation of seven business serverless computing os's. It reviews extant exploration on these platforms also as identifies the development of AWS Lambda as a true base platform for searching on company serverless cloud computing.
A novel design of performance focused serverless computing wedge deployed around Microsoft Azure, and making use of Windows canisters as features delivery locations [eleven]. There are actually metrics suggested evaluating the delivery performance of serverless platforms and hold out assessments with the suggested prototype. The measurements showed considerable advancement in acquiring significantly greater throughput than a number of other platforms at nearly all concurrency levels.
The substitute platform alternatives [twelve] to AWS Lambda, absolutely no discrete academic investigations using Azure Functions, Google Cloud Functions, IBM Bluemix OpenWhisk, Webtask, Iron.io Ironworker, Galactic Fog Gestal Laser had been identified. [thirteen] presents three demonstrators for IBM Bluemix OpenWhisk. Event-based programming is actually evidenced by them brought on by weather forecast info, Apple WatchOS2 program info, also speech utterances. Furthermore, it demonstrated a chatbot by utilizing IBM Bluemix OpenWhisk that calls on IBM
Watson services such as for instance information that is current, jokes, dates, clean water, music tutor,and a security alarm program. [fourteen] conducted a survey on the present serverless os's from company, academia, and next open source projects, primary abilities and use cases, and also describe open issues and complex challenges. This work provided a handson knowledge of making use of the serverless systems accessible from various cloud providers as IBM, Amazon, Google and Microsoft.

ServerlessOS, comprised of three important components: (a) a novice driver desegregation model, and that leverages desegregation for abstraction, but enables assets to move fluidly in between servers for performance; (b) a cloud orchestration level which manages good grained source placement in addition to allocation through the application' s lifetime via local and global option making; in addition to (c) an isolation ability that enforces info or resource isolation[15].

A good resource management process for serverless cloud computing frameworks with all of the goal to boost helpful resource with an emphasis on brain allocation among containers. The design and style included a level top of an open sources serverless wedge, OpenLambda. It's grounded uponsoftware workloads and serverless function' s brain requires happenings are essentially triggered. The mind limits the same lead to variants in the quantity of pots spawned on OpenLambda[16].

Kubernetes has considerably decreased the complexity to create brand new developer experiences, and a range of completely new experiences have been designed or perhaps are actually in the works which offer simplified or perhaps targeted developer experiences as Functions-as-a-Service, in addition to core Kubernetes-as-a-Service. By embedding core performance like lumber and checking in the bunch itself and allowing designers to make the most of such services mainly by deploying the application of theirs into the bunch, Kubernetes has cut back the learning needed for developers to create scalable reliable applications [17].

Finally, Kubernetes has supplied a new, typical vocabulary for articulating the patterns as well as paradigms of sent out system development. This basic vocabulary implies that we are able to more quickly describe and talk about the typical methods in what our distributed systems are actually built, and moreover we are able to create standardized, re usable implementations of such methods. The net result of this's the improvement of higher quality, dependable distributed programs, more easily [18]-[21].

3. Comparative study on future ready computing with pros and cons:

S.No	Computing sources	Advantages	Cons to Work on
1	Serverless Computing	The provider scales and also manages the required resources	No access to virtual devices, operating system or perhaps runtime environments
		Rapid provision of materials in real time, even for unexpected peak lots and disproportionate growth	Implementing serverless structures is incredibly labor-intensive
		Customers are just charged for the materials used	Lock-in effect - for instance, when modifying provider, you commonly have to recode all event based functions
		High error tolerance because of flexible hardware infrastructure within the provider's computer system centers	Relatively complicated monitoring and debugging procedure, as in depth functionality, and error analyses are usually not possible
2	Omni cloud	Save cost of storage in cloud	Backups

		Data redundancy avoided	Restore
		Tiering cost for saving of data	Internet utilization
		Compliance	-
		Protection	-
3	Quantum Computing	Faster computing	-
		Exponential fastup in calculations	Inconsistent electron
		Classical Algorithmic calculations	No positive progress
4	Kubernetes	Open source	Uses different YAML files
		Scalable	Sometimes it have to start Manually
		In production new deployment containers	Degree of reorganization
		Grouping tasks	
		Fault tolerant	
5	Digital Natives	Internet etiquette	Wide usage of internet
		Crap detection	

4 Common comparison between cloud providers with respect to serverless computing:

S No	AWS	AZURE	Google
1	A FaaS offering which belongs to Amazon Web Services was launched in 2014. Because the release of its, Lambda started to be associated with what serverless implies, having the role of the top item on the market place with probably the widest range of services out there. Probably, the most known instance of public serverless adoption is actually by Netflix.	The system launched in 2016 to participate with AWS Lambda. Azure Functions provides a comparable set of expert services to Amazon, with an emphasis on the Microsoft family of tools and languages. Among the instances of using Azure Functions is actually Have I Been owned. If you're keen on the application structure and just how it does on Azure, you might look at the volume report that contains detailed info on expenses and analytics.	One of the 4 largest, Google released the solution of its only in 2017. GCF service utilized to lag behind Lambda and Azure, but during 2018, Google managed to resolve earlier errors as evidenced by GCF release paperwork.

5 Conclusion:

Computing today typically favors small, self contained devices of computation to permit it to be simpler to manage and range in the cloud. A computation, which may be interrupted or perhaps even restarted, cannot depend on the cloud platform to maintain its state. This inherently influences the serverless computing programming cellphone models. There is, nevertheless, just no equivalent thought of scaling to zero when you're looking at express, since a continuous storage amount is really necessary. Nevertheless, whether or perhaps not the put in place of a stateful system must have continual storage space, a provider can make a pay-as-you-go pricing model that would make express handling serverless. Amazon Lambda, Azure functions and Google Kubernetes engine provides a vast scope for PAAS and in coming days the future could be cloud and still we have a future scope for optimizing the services in cloud providers side for usage in terms of compute where in for easy support in terms of there technicality of launching the services in PAAS.

References:

[1] Barrie Sosinsky, Cloud Computing Bible, 1st ed., Wiley Publishing, 2011.
[2] R Arokia Paul Rajan, "Service Request Scheduling based on Quantification Principle using Conjoint Analysis and Z-score in Cloud," International Journal of Electrical and Computer Engineering, vol 8, issue 2, pp.1238-1246, 2018.
[3] L.A. Barroso, J. Clidaras and U. Holzle. "The Datacenter as a Computer: An Introduction to the Design of Warehouse-scale Machines," Synthesis Lectures on Computer Architecture, vol. 8(3), pp. 1-154, 2013.
[4] Cisco, "Cisco global cloud index: Forecast and methodology, 2015- 2020,". CA: Cisco Public, 2016.
[5] Telerik, Accessed on 2nd August 2018, https://www.telerik.com/blogs/introduction-to-serverless-computing
[6] Evolution of Serverless computing, Accessed on 10th August 2018, https://www.ca.com/us/modern-software-factory/content/the-evolutionof- serverless-computing.html
[7] Amazon, Accessed on 16th August 2018, https://aws.amazon.com/lambda/serverless-architectures-learn-more/
[8] Iron, Accessed on 06th September 2018, https://www.iron.io/docs/Whitepaper_Serverless_Final_V2.pdf
[9] E. van Eyk, L. Toader, S. Talluri, L. Versluis, A. Uță and A. Iosup, "Serverless is More: From PaaS to Present Cloud Computing," in IEEE Internet Computing, vol. 22, no. 5, pp. 8-17, 2018.
[10] T. Lynn, P. Rosati, A. Lejeune and V. Emeakaroha, "A Preliminary Review of Enterprise Serverless Cloud Computing (Function-as-a- Service) Platforms," 2017 IEEE International Conference on Cloud Computing Technology and Science (CloudCom), pp. 162-169, 2017.
[11] G. McGrath and P. R. Brenner, "Serverless Computing: Design, Implementation, and Performance," 2017 IEEE 37th International Conference on Distributed Computing Systems Workshops, pp. 405- 410, 2017.
[12] P. Castro, V. Ishakian, V. Muthusamy and A. Slominski, "Serverless Programming (Function as a Service)," 2017 IEEE 37th International Conference on Distributed Computing Systems, pp. 2658-2659, 2017.
[13] J. Short, S. Ennis, and B. Judson. "Cloud Event Programming Paradigms: Applications and Analysis," In Proceedings of the 9th IEEE International Conference on Cloud Computing (CLOUD), pp. 400-406, 2017. 92
[14] I. Baldini, P. Castro, K. Chang, P. Cheng, S. Fink, V. Ishakian, N.Mitchell, V. Muthusamy, R. Rabbah, A. Slominski and P. Suter, "Serverless Computing: Current Trends and Open Problems," Research Advances in Cloud Computing, Springer, pp 1-20 2017.

[15] Z. Al-Ali, "Making Serverless Computing More Serverless," 2018 IEEE 11th International Conference on Cloud Computing (CLOUD), pp. 456-459, 2018.
[16] A. Saha and S. Jindal, "EMARS: Efficient Management and Allocation of Resources in Serverless," 2018 IEEE 11th International Conference on Cloud Computing (CLOUD), pp. 827-830, 2018.
[17] Comparison of Serverless architectures, Accessed on 2nd August 2018, https://dzone.com/articles/4-use-cases-of-serverless-architecture
[18] Microsoft, "Azure Functions Documentation", Accessed on 15th September 2018, https://docs.microsoft.com/en-us/azure/azure-functions/
[19] Google, "Google Cloud Functions Documentation". Accessed on 17th September 2018 https://cloud.google.com/functions/
[20] Ravula Arun Kumar, Dr Vinuthna kambalapally, "A study on privacy preservation techniques and scope of applying deep learning concepts in security: a survey", International Journal of Engineering & Technology, 7 (3.27) (2018) 586-592.
[21] Ravula Arun Kumar "A Review on various secure data access schemes and techniques in Fog for Internet of Things" IJCSE, volume 6, issue 1, 2017.

A Neural Network trained to perform as an Imprecise 4:2 Compressor using Machine Learning Algorithm

Lavanya Maddisetti[1,2,*], Ranjan K. Senapati[1] , JVR Ravindra[2]

[1]Department of ECE, KL Deemed to be University, Vijayawada, India
[2]Center for Advanced Computing Laboratory, Vardhaman College of Engineering, Telangana, India

([*]Corresponding author: lavanya.maddisetty@gmail.com)

Abstract: In many fields, such as speech recognition, image processing, Internet of Things (IoT) systems, arithmetic circuits, etc., artificial neural networks have become a popular solution for a wide range of problems. In this research work, a 4:2 compressor-based arithmetic circuit is considered and an approximation technique called probabilistic pruning is applied to it. This proposed compressor is compared with various existing approximate 4:2 compressors. Training of a neural network as proposed inexact 4:2 compressor is performed using a supervised machine learning algorithm. The accuracy values of the proposed compressor are acceptable at architectural and neural network levels. The absolute difference between the train set and the predicted test set accuracy values obtained is very less (1.73%) when compared with the other (inexact 4:2 compressors) trained deep neural network models.

1. Introduction

Low power, high speed with smaller silicon area is the demanding parameters from decades to design arithmetic circuits like 4:2 compressor [1]. The inexact compressor is the one that consumes less power and is accurate than the other existing inexact 4:2 compressors.
Two analyses have been shown in this work; the first is the proposed architecture level approach where simulations of compressors are done to find the area, power, and energy consumption. The second approach is the training of Neural Network (NN) as the proposed inexact 4:2 compressor. The truth table of the proposed 4:2 compressor is given as the dataset to train the NN and it is split into train and test sets to predict the output of the compressor circuit.

Our contributions:

- The inexact 4:2 compressor is proposed with the least number of transistors and observed the power consumption, energy, and area of it by comparing it with the existing inexact compressors.
- Error analysis is performed for the existing and proposed inexact 4:2 compressors.
- Training of NN as an inexact 4:2 compressor is done using supervised Machine Learning (ML) algorithm by giving the truth table of the proposed compressor as the dataset.

2. Related work

The authors of [1] proposed a high speed, low power compressors of different orders which operate at ultra-low voltages with multiplexers in the proposed architecture. Two CMOS+ gate-based circuits were proposed in [2] to reduce power, area, and delay of 4:2 compressor. In [3], the authors used a compressor structure in a multiplier design. Also, a sparse adder was proposed to use it in the final addition of a multiplier. M.Ben has given a general introduction to approximate computing in [4] and proposed a taxonomy to show the merits of different approximation techniques. The author discussed the opportunities and challenges of non-deterministic approximate computing. The authors of [5] have presented a study on combining three approximation techniques on more than one layer of system stack to get more benefits and said that benefits can be applied to different domains.

The authors of [6] have presented the way they have developed the spanning circuits, architecture, and software by applying techniques of approximate computing. In [7-8], 4:2 compressors were proposed and they were placed in a Dadda multiplier to get low power multipliers. Finally, these multipliers were used in image processing applications.

Due to approximations on digital circuits, the difference in the outputs of exact and inexact circuits can be calculated from the equations proposed in [9]. The work of [10] also resembles the same as [7-8] but the authors have used different operand lengths of approximate multipliers and image filtering application was shown in their research work. The ML techniques applied to CAD algorithms were given in [11] to make automation in the operation in the tools. In [12], the authors have proposed an approximate compressor using inexact logic minimization and embedded it in the multiplier for power efficiency and finally, an algorithm was given to simulate them in Google Colaboratory.

3. Proposed method

This section presents a proposed inexact 4:2 compressor employing a probabilistic pruning type of approximation. A feed-forward neural network is trained to behave as the proposed 4:2 compressor.

3.1 Architectural approach of proposed 4:2 compressor

The probabilistic pruning type of approach is applied to the exact 4:2 compressor by removing a multiplexer present at the output of Sum [1]. Thus, the Sum output for the proposed 4:2 compressor can be taken from the multiplexer whose select line is connected to the output of the second XOR [1]. With this approximation, there is no difference in the outputs of C_{out} and Carry. The proposed approximate method gives the least number of errors with less transistor count. The Sum expression for the proposed 4:2 compressor is obtained as equation (1).

$$\mathrm{Sum} = (X_3 \oplus X_4)' (X_1 \oplus X_2) + (X_3 \oplus X_4) (X_1 \oplus X_2)' \quad (1)$$

The transistor-level circuit of the proposed 4:2 compressor is shown in Figure 1.

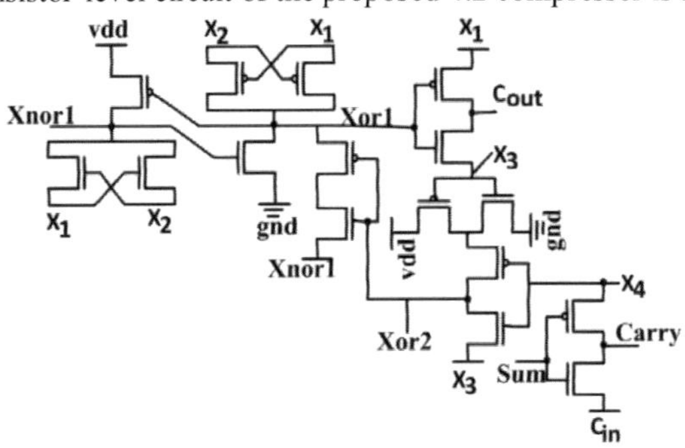

Figure 1: Proposed Inexact 4:2 Compressor with 16 Transistors

Minimum numbers of errors are present in the above circuit. i.e., C_{out} and Carry have `0' errors and the Sum output has `8' errors which give a 25% error rate considering all the three outputs. When the approximate design of [12] is examined, it also has a 25% error rate, but it is obtained considering only Sum output. The C_{out} and Carry outputs of [12] are giving 8 and 4 errors respectively when compared with the exact 4:2 compressor. Thus, the total errors of the proposed inexact circuit have `0,

0, 8' whereas [12] has `8, 4, 8' for C_{out}, Carry, Sum outputs respectively. With this least error rate, the inexact design was embedded in a high speed, parallel Dadda multiplier to reduce the power consumption of the multiplier, and intern the Arithmetic Logic Unit (ALU) and processor.

3.2 Neural Network model of an Inexact 4:2 Compressor

A supervised machine learning type of algorithm is used for training the neural network. The truth table of the proposed inexact compressor is given as dataset to the NN with C_{in}, X_4, X_3, X_2, X_1 inputs, and C_{out}, Carry and Sum target labels. The architecture of the NN with inputs and outputs is shown in Figure 2.

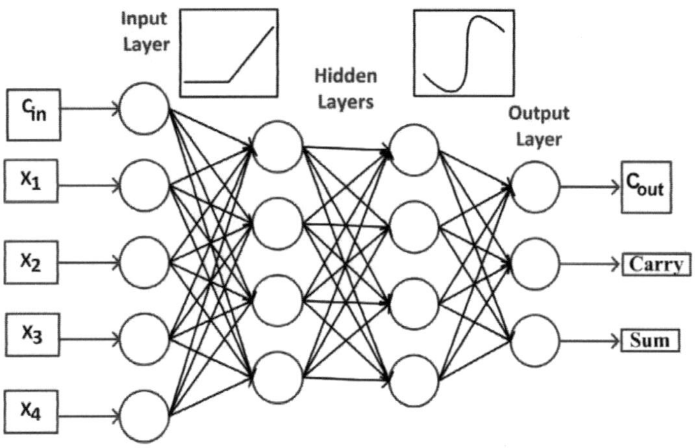

Figure 2: Neural Network Model as Inexact 4:2 Compressor with the activation functions, ReLU at input and Softmax at the output

The truth table is divided into train and test sets to find the accuracy of the NN model. The accuracy value obtained from the train and test sets describe how best the NN model is performing as the proposed inexact compressor. The difference between the trained accuracy and the predicted test accuracy should be equal in ideal cases for the NN model to be complete. When these accuracy values are equal then the model is said to be completely validated.

4. Results and Discussion

The experimental work of 4:2 compressors was performed using the Spectre simulator of Cadence Virtuoso for schematic simulations. To find the area of these architectures Cadence Layout XL was used in 45nm CMOS technology node. The error analysis of compressors was performed using MATLAB. The training part of the NN was done using the Jupyter notebook with Pandas, Numpy, and Keras libraries.

4.1 Simulation Results of 4:2 compressors

In this subsection, the proposed inexact 4:2 compressor is compared with the exact and existing approximate 4:2 compressors. The comparison has been done in terms of transistor count, power, energy, and area for exact, existing inexact, and proposed inexact 4:2 compressors. The simulation results of the compressors are tabulated in Table 1.

Table 1: Simulation Parameters of 4:2 Compressors

4:2 Comp.	Transistor Count	Power (nW)	Energy (pJ)	Area (nm^2)
Exact [1]	50	178.9	1861	29.893
Inexact [8]	40	157.4	1348	23.522
Inexact [12]	34	110	836.9	20.34
Inexact [7]	28	102.4	650.5	16.728
Proposed	16	83.23	240.3	9.641

When the proposed compressor is compared with the design1 of [7], the improvement in power, energy, and the area is 18.72%, 63.05%, and 42.36% respectively. From Table 1, it is observed that the power, area, and energy parameters are varying according to the transistor count of each 4:2 compressor. The graph is shown in Figure 3 also explains that these parameters are varying for the transistor count.

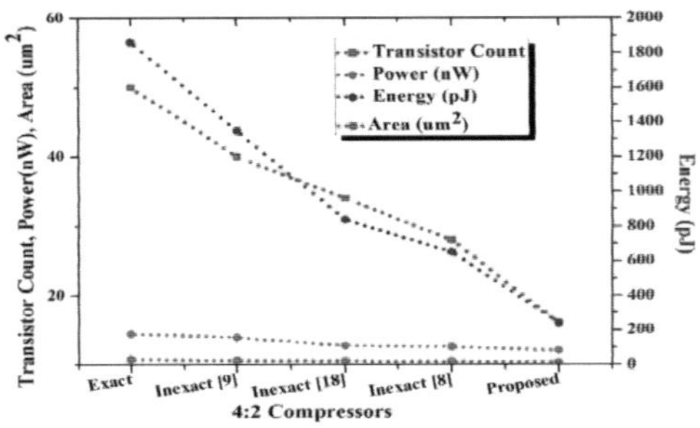

Figure 3: Transistor Count, Power, Energy and Area of Exact and Inexact 4:2 Compressor

Table 2: Error Metrics of Compressors

4:2 Comp.	Errors	ER (%)	MED	NMED	RED
Exact [1]	0	0	0	0	0
Inexact [8]	5	31.25	3.3432×10^6	1.044×10^5	0.0604
Inexact [12]	8	25	5.535×10^7	1.729×10^6	1
Inexact [7]	12	37.5	7.551×10^7	2.359×10^6	1.3641
Proposed	8	25	1.203×10^3	37.6025	2.1736×10^{-5}

ER is Error Rate, MED is Mean Error Distance, NMED is Normalized mean error distance

The observations from Table 2 are

- The proposed inexact 4:2 compressor and [12] have the same error rate. This is because in [12] only sum is considered for error rate whereas in the proposed circuit the error rate is 25% considering all the outputs (0+0+8 for C_{out}, Carry, and Sum).
- Although the transistor count is very less, the design1 of [7] was giving a 37.5% error rate which is higher than the rest of the circuits, .
- The design of [8] has only 4 inputs C_{in}, X_4, X_3, X_2, X_1 inputs, and two outputs (Carry and Sum). Hence error analysis has been performed for 16 inputs instead of 32 inputs. Therefore, the error rate is 31.25%.

IOP Conf. Series: Materials Science and Engineering 1042 (2021) 012014 doi:10.1088/1757-899X/1042/1/012014

4.2 Simulations of a deep neural network model

Training of NN has been done by splitting 66% of the dataset into a train set and 34% of the dataset into a test set. The parameters used for building the NN is given in Table 3.

Table 3: Neural Network Parameters used in Training

NN Parameters	Values
Test size	34% of the dataset
Classifier	Sequential
Kernal initializer	Uniform
Input activation function	ReLU
No. of Hidden Layers	2
Output activation function	Softmax
Optimizer	adam
Loss function	Categoricalcross-entropy
Batch Size	10
No. of Epochs	1000

For the proposed inexact compressor, the train set accuracy is obtained as 61.90%. To assess the performance of the trained model, the prediction of its accuracy has been done by the test set. After prediction, the confusion matrix is obtained as $\begin{pmatrix} 5 & 0 & 0 \\ 4 & 1 & 0 \\ 0 & 0 & 1 \end{pmatrix}$

$$\text{Test set accuracy} = \frac{\text{No. of correct outputs}}{\text{Total outputs}}$$

$$= \frac{5+1+1}{11} = \frac{7}{11} = 63.63\%$$

The absolute difference between the accuracies of train and test sets is 1.73%. This difference suggests that the NN model is perfectly built as a proposed inexact 4:2 compressor. The accuracy and loss graphs for the number of epochs are shown in Figure 4.

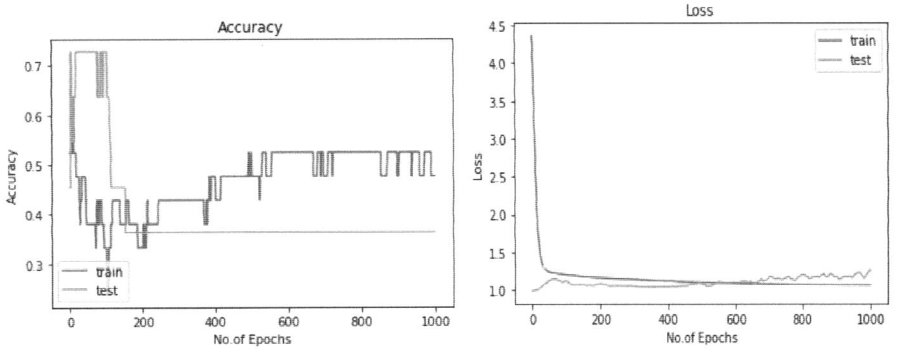

Figure 4: Accuracy and Loss graphs for the Train and Test sets of the proposed inexact NN model

To compare train and test accuracies of the other approximate designs, NN models are built using datasets, and accuracies are obtained as shown in Table 4.

Table 4: Train and Test set Accuracies

DNN Model	TrainSet Accuracy (%)	Confusion Matrix	Testset Accuracy (%)	Absolute Difference (%)
Exact [1]	85.71	$\begin{bmatrix} 1 & 0 \\ 3 & 2 \end{bmatrix}$	81.81	3.9
Inexact [8]	60	$\begin{bmatrix} 1 & 0 \\ 3 & 2 \end{bmatrix}$	50	10
Inexact [12]	57.14	$\begin{bmatrix} 2 & 4 & 0 \\ 0 & 4 & 0 \\ 0 & 1 & 0 \end{bmatrix}$	54.54	2.6
Inexact [7]	57.14	$\begin{bmatrix} 6 & 0 & 0 \\ 4 & 0 & 0 \\ 1 & 0 & 0 \end{bmatrix}$	54.54	2.6
Proposed	61.9	$\begin{bmatrix} 5 & 0 & 0 \\ 4 & 1 & 0 \\ 0 & 0 & 1 \end{bmatrix}$	63.63	1.73

From Table 4, it is clear that the proposed inexact 4:2 compressor is modeled perfectly since the absolute difference between train and test set accuracies obtained are higher for other models when compared with the proposed one.

5. Conclusion

Our findings suggest that the proposed inexact method can be applied to any arithmetic combinational circuit in terms of the architectural and training of neural network approaches. The proposed inexact 4:2 compressor model has the least absolute difference between a train and test accuracies which proves that the NN model is validated completely. The power and area consumption of the proposed compressor is also less with the least error rate.

Acknowledgments

This research project was carried out at C-ACRL, Vardhaman College of Engineering. The authors would like to thank the management and faculty for their constant support throughout.

References

[1]. Sreehari Veeramachaneni, Kirthi M Krishna, Lingamneni Avinash, Sreekanth Reddy Puppala, M.B. Srinivas, (2007), Novel architectures for high-speed and low-power 3-2, 4-2 and 5-2 compressors, Proceedings of International Conference on VLSI Design (VLSID), pp. 324-329.

[2]. Raphael Dornelles, Guilherme Paim, Bianca Silveira, Mateus Fonseca , Eduardo Costa , Sergio Bampi, (2017), A Power-Efficient 4-2 Adder Compressor Topology, 15th IEEE (NEWCAS), Strasbourg, France, pp. 281-284.

[3]. Sreehari Veeramachaneni, M. B. Srinivas, 2013, Design of optimized arithmetic circuits for multiplier realization, IEEE Asia Pacific Conf. on Postgraduate Research in Microelectronics and Electronics (PrimeAsia), India.

[4]. M. Ammar Ben Khadra, 2017, An introduction to approximate computing, December, [Online]: https://arxiv.org/abs/1711.06115.

[5]. Ankur Agrawal, Jungwook Choi, Kailash Gopalakrishnan, Suyog Gupta, Ravi Nair, 2016, Approximate Computing: Challenges And Opportunities, IEEE International Conference on Rebooting Computing (ICRC), San Diego, USA.

[6]. S.Venkataramani, S.T. Chakradhar, K. Roy, A. Raghunathan, 2015, Computing approximately, and efficiently, in Design, Automation & Test in Europe Conference & Exhibition, pp. 748-751.

[7]. Amir Momeni, Jie Han, Paolo Montuschi, Fabrizio Lombard, 2015, Design and analysis of approximate compressors for multiplication, IEEE Transactions Computers, 64(4), April, pp. 984–994.

[8]. Suganthi Venkatachalam, Seok-BumKo, 2017, Design of Power and Area Efficient Approximate Multipliers, IEEE Transactions on Very Large Scale Integration (VLSI) Systems, 25 (5), pp.1782-1786.

[9]. Jinghang Liang, Jie Han,and Fabrizio Lombardi, 2012, "New Metrics for the Reliability of Approximate and Probabilistic Adders, IEEE Transactions on Computers, 62(9), pp. 1760–1771

[10]. Darjn Esposito, Antonio Giuseppe Maria Strollo , Ettore Napoli , Davide De Caro , Nicola Petra, 2018, Approximate Multipliers Based on New Approximate Compressors, IEEE Transactions on Circuits and Systems I: Regular Papers, PP(99), pp. 1-14.

[11]. Peter A. Beerel, Massoud Pedram, 2018, Opportunities for Machine Learning in Electronic Design Automation," IEEE International Symposium on Circuits and Systems (ISCAS), Italy.

[12]. Lavanya Maddisetti, JVR Ravindra, 2019, Machine Learning based Power-Efficient Approximate 4:2 Compressors for Imprecise Multipliers, in 32nd International Conference on VLSI Design, New Delhi, India.

Performance Analysis of Activation Functions for Wheat Crop Yield Prediction

[1]**Shital H. Bhojani, Dr. [2]Nirav Bhatt**

[1]Research Scholar, RK University & Assistant Professor, AAU, Anand-Gujarat

[2]Associate Professor, MCA Department, RK University, Rajkot-Gujarat

shitalbhojani@aau.in, nirav.bhatt@rku.ac.in

Abstract. Activation function (AF) plays a vital role in the neural network. This research study focuses on the performance evaluation of the various AFs in multilayer perceptron (MLP) algorithm for the wheat crop yield prediction at the regional level. Nowadays researchers are publishing the new AFs and also proposing the improvement in existing AFs for more accurate results. Sigmoid AF is the default function used in the MLP algorithm. In this research we have applied the various AFs in MLP using the WEKA java libraries and performance analysis has been carried out for the selected districts of the Gujarat state. Based on the mean absolute percentage error (MAPE) performance measure output is discussed. Therefore scientists and researchers can select the appropriate AF to improve the prediction accuracy and hence improve the performance of the network as per the research data.

Abstract. Neural Network, Activation Function (AF), MLP, Yield Prediction

1. Introduction

Crop yield is affected by many factors such as weather parameters, rainfall, soil type and its fertility, atmospheric conditions, management practices, etc. Data mining is the process to extract significant knowledge, patterns and trends from datasets. Data mining and machine learning techniques are making their roots stronger in the field of agriculture. Neural networks can be considered as an approach to the problem of computation. We can apply the neural network approach to diverse fields such as speech and text recognition, image processing, various scientific, engineering and commercial disciplines, prediction, modeling etc. Therefore, it is essential to implement the new technologies in farming and have to follow the scientific farming methods for better output. This research paper was focused on various activation functions and its assessment used for the wheat crop yield prediction. The WEKA time series forecasting package was used with the help of WEKA JAVA libraries. Various activation functions were applied for forecasting using the MLP algorithm of classification technique.

2. Activation functions

The activation function, also referred to as the transfer function, is applied on the inner product in the neural network. To improve the performance of the neural network selection of activation function is very much important. Activation functions manipulate the data through gradient descent and generate an output for the neural network. We can categorise the activation function as linear or nonlinear. When we have the linear input we get the linear output. So activation functions are applied to the outputs of the linear models to produce the non-linear outputs. The input and output parameters of the neural network decide the type of activation function to be used for further processing. When activation function applies to input values are generally a sum of products formed and if bias exists then it is also included for the computing. Bias is an integer value associated with each node in NN as an extra input. In neural network derivatives of AFs are used for updating the curve and to know in which direction and how much to change. The derivative also called a gradient of an AF is really important for training the neural network. In neural network different activation functions exist like sigmoid, tanh, softmax, ReLU, Leaky ReLU (LReLU), ELU, loglog, logsigm (sigmoid square), cloglog etc.

2.1. Sigmoid
The Sigmoid is a non-linear activation function. Its value lies between 0 and 1. This function is majorly used in NN. Sigmoid activation function is providing good results with my research datasets. It is used in binary classification problems. It is mainly used in the feedforward neural network.

2.2. logsigm
This is the square of the sigmoid function and provides good results for the current research study.

2.3. The hyperbolic tangent – tanh
As compared to the sigmoid function values of tanh function lies between -1 and 1. It produces zero centred output so assisting the backpropagation process in the NN. This activation function is not providing a good result with my research datasets.

2.4. Rectified Linear Unit (ReLU)
Some text. This is the widely used activation function and provides the better result in some types of datasets. But for time series prediction it's not proving the satisfactory results.

3. Review of literature
Jianli et. al. discussed the linear and non-linear AF such as sigmoid, tanh, ReLu and various improvements of ReLu like leaky ReLu, PReLu, RReLu, and ELU. They have compared the characteristics of these AFs and also mention their pros and cons. The selection of AF is dependent on the aim and the area of our research [1]. Leonid Datta discussed the properties and problems faced by the AF. He explained the properties of widely used AFs and shared properties of AFs. He reviewed the vanishing gradient and the dead neuron problems of widely used AFs. He concluded that for the weight initialization performance of the normal initialization beats the performance of Xavier initialization. ReLU gives good performance with the normal initialization in some types of networks [2]. Heny Pratiwi et. al. used the population density dataset of Indonesia for the period 2003-2015 for prediction purposes. They have used the matlab 6 application for training and testing purposes and applied binary sigmoid activation function (logsig) in the network model. They conducted that the best architectural model can be generated using sigmoid AF in backpropagation NNs for the selected dataset [4]. Rahul et. al. provide new theoretical characterizations that support the use of the ReLU, its variants. They stated the different remarks, theorem, proposition and examples in the framework of L-splines and showed that the choice of activation implicitly defines a NN regularizer [5]. Khaki et. al. implemented CNN-RNN model models in the most efficient manner to forecast corn and soybean yields across the entire Corn Belt 13 states in the United States for 3 years 2016 to 2018 using environmental data and management practices. They concluded that the model predicted good results and it could be used in future yield prediction tasks [6]. da S. Gomes [7] and the co-author suggested the new activation function namely log-log, probit and loglog, cloglog, logsigm in NN for time series forecasting. They used the Matlab platform for conducting the different experiments. The experiment

contains a MLP algorithm with one hidden layer, variation in number of nodes. The evaluation was carried out by the two learning algorithms namely LM and CFG and the best model was chosen using values of MAPE for forecasting. Author recommends the new AFs for smaller network structure and financial time series data sets. Raju Prasad Paswan [8] applied the thnh, log sigmoid and linear transfer function in MLP neural network algorithm on maize and rice crop datasets to forecast area and productivity using the Matlab environment. The result of different AFs shows that the predicted values are within ±20% error.

4. Materials and Methods

The dataset for the result study was collected from Directorate of Agriculture, Gandhinagar and from Agro-meteorology department, Anand Agricultural University, Anand. WEKA "Waikato Environment for Knowledge Analysis" JAVA libraries were used for the implementation of various AFs. Dataset includes the weather and yield data. Weather parameters such as basic sun shine hours (BSS), maximum temperature (MAXT), minimum temperature (MINT), relative humidity morning (RH1), relative humidity afternoon (RH2), vapour pressure morning (VP1) and vapour pressure afternoon (VP2) were used for the prediction the wheat yield at regional level. Daily weather data were converted as per Standard Meteorological Week (SMW) to the weekly data for the research study. There are eight agro-climatic zones in Gujarat state. A total of six districts from different agro-climatic zones are selected namely Anand, Sabarkantha, Banaskantha, Bhavnagar, Junagadh and Ahmedabad. Data from 1990–1991 to 2016–2017 were used for the time series forecasting.

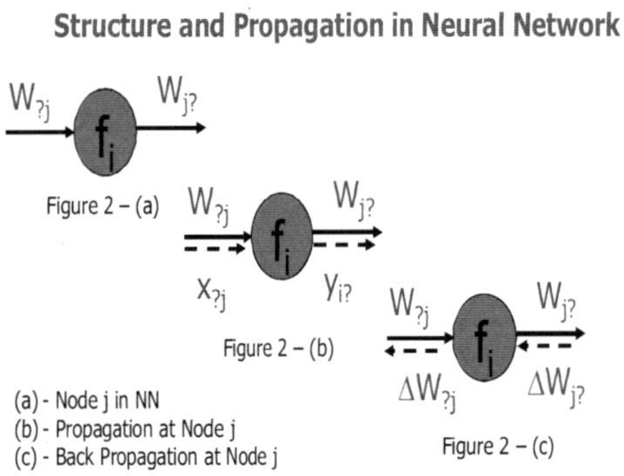

Figure 1: Structure and propagation in NN

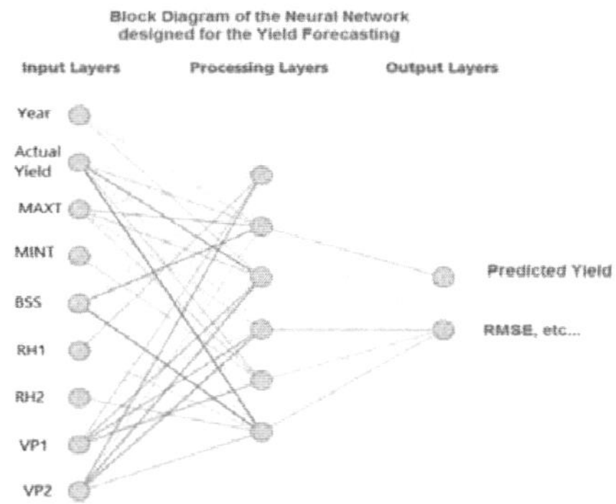

Figure 2: NN with 3 layers

Processing the data from input layer to the output layer is called propagation. The values from input nodes are sent to the subsequent nodes using some AFs and reach to the output nodes. Figure 4 shows the propagation in the neural network.

Neural network consists of processing units, neurons, and directed, weighted connections between those neurons. In NN there are many neurons that work in correspondence of weights and bias. Weight initialization method plays an important role in training and performance of a neural network AF at a hidden layer is very much important; if it is absent then outputs will be the linear combination of the input parameters. Another name of AF is Transfer Function. Traditional mathematical model-based algorithms have reasonably high clarity in how a model and the input data lead to output decisions. Conventional NN typically have a fixed, bounded continuous nonlinear AF at each neuron, which is the key to the overall nonlinear behaviour of NN[3]. Research work is carried out by WEKA data mining tool and forecasting package. MLP Classification algorithm was used for the wheat yield prediction at district level. Research work is carried out by multiple trials and uses the different values combination. However it was noticed in this study that one hidden layer with 0.3 learning rate and 0.2

momentum generates the better results. In this research it is observed that the number of nodes, lags, and instances are also dependent on each other.

5. Result and Discussion

Experimental results obtained using MLP algorithm and various AFs applied for wheat yield prediction. Various AFs listed below are used for the performance analysis.

Table1: List of Activation Functions used in this study

AF1	Sigmoid	AF5	loglog with 15 node
AF2	cloglog	AF6	loglog with 16 node
AF3	loglog	AF7	Relu
AF4	logsigm	AF8	tanh

Results of actual and predicted yield for selected districts were compared and a conclusion was drawn. After comparing the result it was analyzed that when data applied to the ReLu function the error rate of actual and predicted yield was very much higher for most of the districts. Using the different AFs, predicted yields were overestimated and underestimated for selected districts. Table 2 shows the year wise maximum and minimum error rate values for various AFs.

Table 2: Minimum and Maximum Error Rate of Actual and Predicted Yield year wise

	Error Rate of Actual and Predicted Yield						
Year		AF1	AF2	AF3	AF4	AF5	AF6
2015-16	MIN	-0.33	1.48	1.01	-3.90	1.47	-3.61
	MAX	21.34	32.50	27.25	26.96	23.74	29.99
2016-17	MIN	10.19	0.32	8.04	6.15	-2.10	-3.26
	MIX	27.23	42.49	26.63	-44.42	29.13	-34.02
2017-18	MIN	0.20	-1.77	0.42	1.89	2.36	1.09
	MAX	35.36	51.40	43.44	48.66	53.16	53.30

By analyzing table 2 we can observe that error rate values were much higher. Default sigmoid AF values were also lying between 0.20 and 35.36 which is not acceptable. Same way for each above listed AF the values are higher. Therefore it is not recommended to try and use another AF which provides better performance for the agriculture dataset. Hence, there is a need to propose some new AF and recommend amending the MLP algorithm for crop yield forecasting. .

Figure 3(a) to 3(f) shows the actual and predicted yield using the different AFs or selected districts.

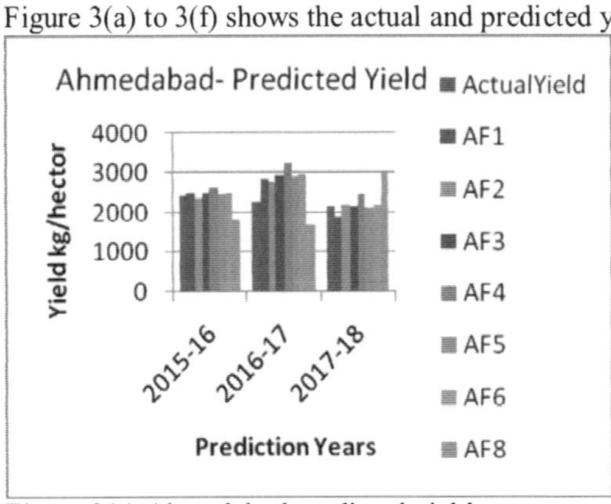

Figure 3(a) Ahmedabad predicted yield

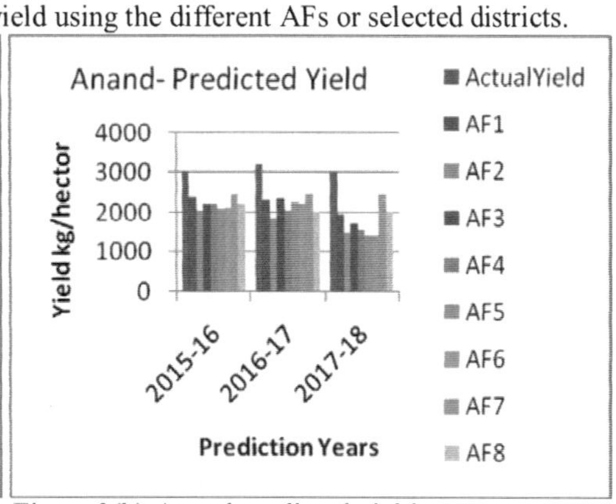

Figure 3(b) Anand predicted yield

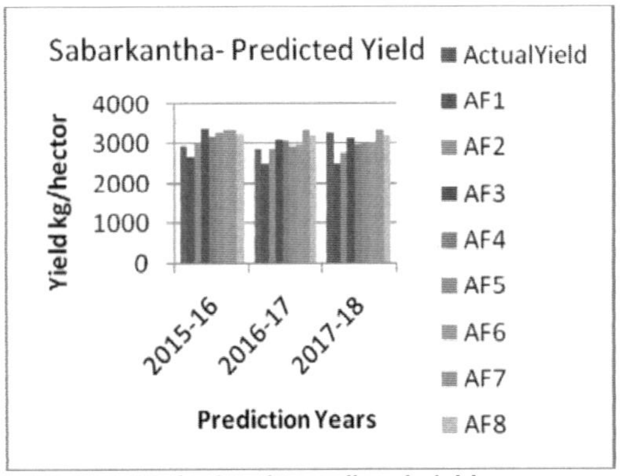

Figure 3(c) Sabarkantha predicted yield

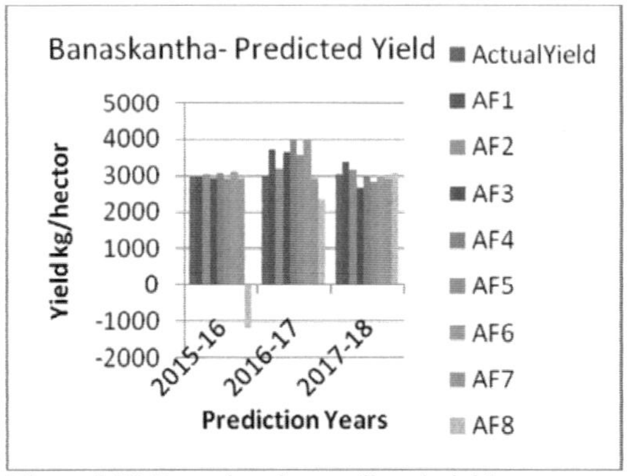

Figure 3(d) Banaskantha predicted yield

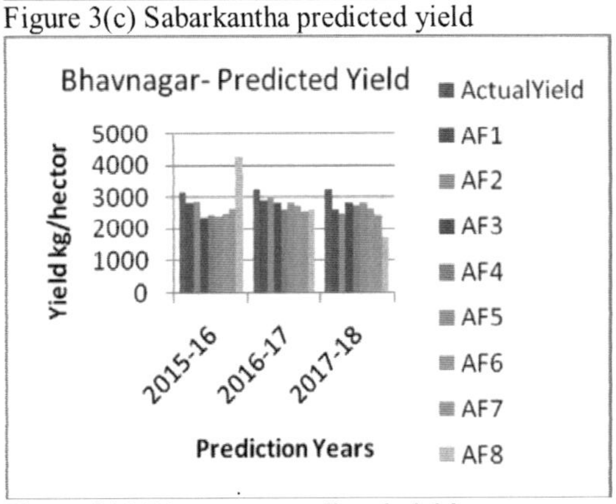

Figure 3(e) Bhavnagar predicted yield

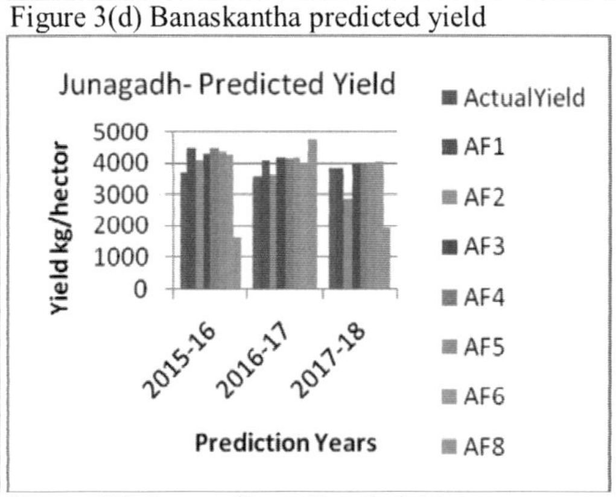

Figure 3(f) Junagadh predicted yield

The performance evaluation was carried out by the MLP algorithm using various AFs, and observed the values of mean absolute percentage error (MAPE) for wheat yield prediction. Figure 4 shows the MAPE of selected districts. As we discussed, the error rate is much higher for each district although the resulting MAPE values were quite satisfactory. For ReLu and tanh the values were in thousands and lakhs for most of the districts so its result was not included in this study. Table 3 shows the year wise maximum and minimum MAPE values for various AFs.

Table 3: Minimum and Maximum MAPE year wise

	Mean Absolute Percentage Error (MAPE)						
Year		**AF1**	**AF2**	**AF3**	**AF4**	**AF5**	**AF6**
2015-16	MIN	0.0008	0.004	0.0003	0.0003	0.0001	0.0005
	MAX	0.5432	0.8135	0.1919	0.1719	0.179	0.5323
2016-17	MIN	0.0006	0.0031	0.0003	0.0003	0.0001	0.0004
	MAX	0.4477	0.751	0.1673	0.1489	0.1821	0.5327
2017-18	MIN	0.0007	0.0032	0.0003	0.0003	0.0001	0.0004
	MAX	0.4473	0.798	0.2065	0.1515	0.1874	0.5556

MAPE is a measure of forecast accuracy percentage which plays an important role to find out the best-fitted model in statistics. It is also noticed that the value of learning rate and momentum is not affecting the result for the selected datasets. Research work is carried out by taking the confidence interval at a level of 95%. Fugure 4(a) 4(b) and 4(c) shows the MAPE for the year 2015-16, 2016-17 and 2017-18 respectively.

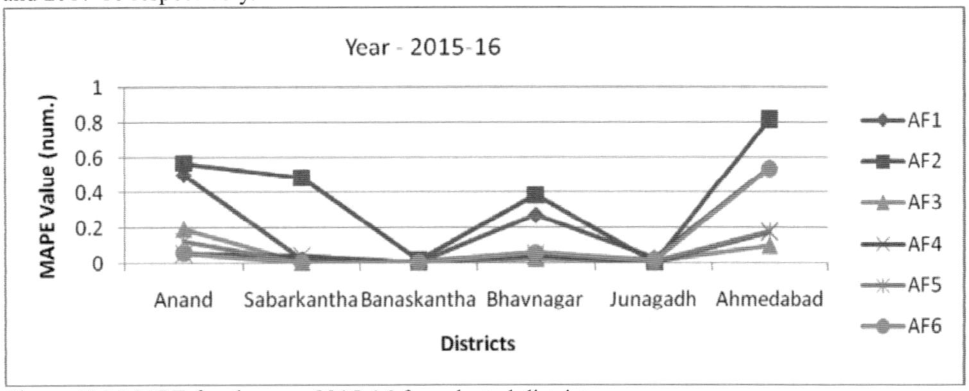

Figure 4(a) MAPE for the year 2015-16 for selected districts

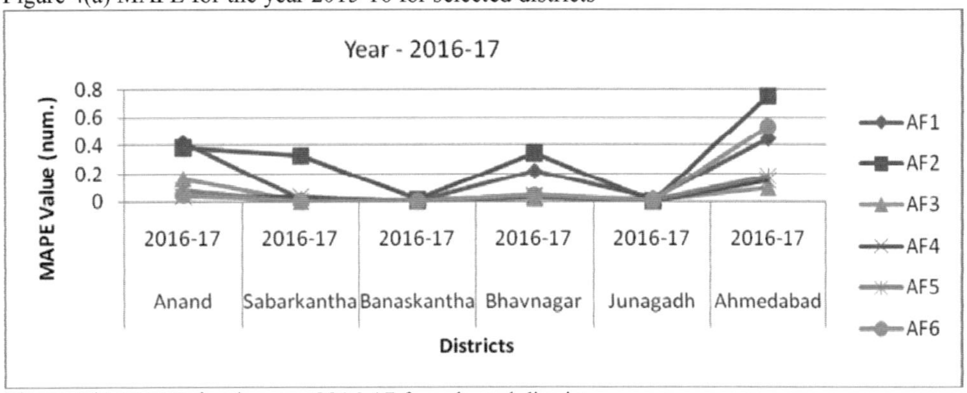

Figure 4(b) MAPE for the year 2016-17 for selected districts

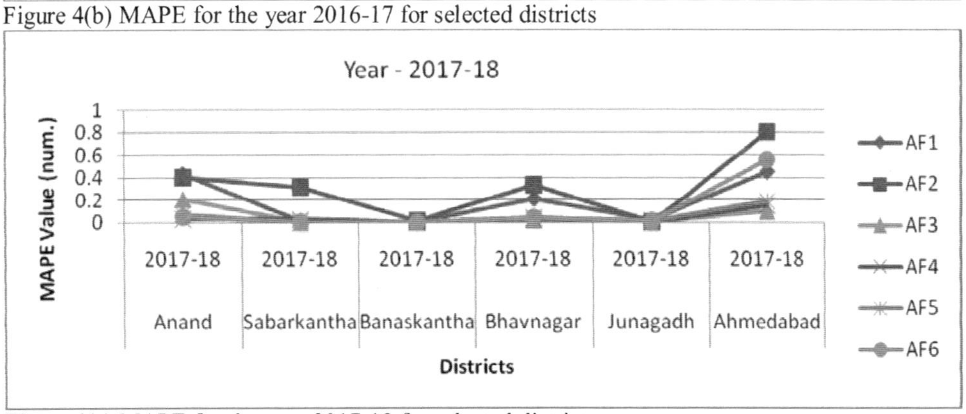

Figure 4(c) MAPE for the year 2017-18 for selected districts

6. Conclusions

In this study, we have discussed the result of various AFs applied for wheat crop yield prediction at district level. MLP provides the error rate of actual and predicted yield between 0.2 and 35, which is not acceptable. During research study it was noticed that the model was not fitting while applying my dataset to MLP algorithm using the ReLu and Tanh activation function as MAPE was much higher, reaching to thousands and lakhs. For sigmoid, loglog and logsigm the model is fitted but with higher

MAPE. So it is concluded that it's quite difficult to choose the best AF and it's mainly dependent on the aim of our research and the type of datasets.

Acknowledgments
I am grateful to the Directorate of Agriculture, Gandhinagar, and Agro-meteorology Department, Anand Agricultural University, Anand, Gujarat, for providing the yield and weather datasets respectively for my research work.

References
[1] Jianli Feng, Shengnan Lu 2019 Performance Analysis of Various Activation Functions in Artificial Neural Networks IOP Conf. Series: Journal of Physics: Conf. Series 1237 (2019) 022030
[2] Leonid Datta 2020 A Survey on Activation Functions and their relation with Xavier and He Normal Initialization arXiv:2004.06632v1
[3] Shital H. Bhojani, Dr. NiravV. Bhatt Wheat crop yield prediction using new activation functions in neural network Neural Comput&Applic
[4] Heny Pratiwi, Agus Perdana Windarto, S. Susliansyah, Ririn Restu Aria, Susi Susilowati, Luci Kanti Rahayu, Yuni Fitriani, Agustiena Merdekawati, Indra Riyana Rahadjeng 2020 Sigmoid Activation Function in Selecting the Best Model of Artificial Neural Networks IOP Conf. Series: Journal of Physics: Conf. Series 1471 (2020) 012010
[5] Rahul Parhi and Robert D. Nowak 2020 The Role of Neural Network Activation Functions arXiv:1910.02333v2
[6] Khaki Saeed, Wang Lizhi, Archontoulis Sotirios V. 2020 A CNN-RNN Framework for Crop Yield Prediction Frontiers in Plant Science Volume 10
[7] da S. Gomes, G.S., Ludermir, T.B. & Lima, L.M.M.R., 2011 Comparison of new activation functions in neural network for forecasting financial time series. Neural Comput&Applic
[8] Raju Prasad Paswan1, ShahinAra Begum and BorshaNeog 2018 Comparison of Abilities of Different Activation Functions of Artificial Neural Network to Predict Crop Area and Crop Production. Int. J. Pure App. Biosci. 6 (6): 212-220

Abbreviations

DM	Data mining
NN	Neural network
AF	Activation function
SMW	Standard Meteorological Week
BSS	Basic sun shine hours
MAXT	Maximum temperature
MINT	Minimum temperature
RH1	Morning relative humidity (%)
RH2	Afternoon relative humidity (%)
VP1	Morning vapour pressure
VP2	Afternoon vapour pressure
MAPE	Mean absolute percentage error

Regression Model for Prediction of Epidemic Outbreaks

Punna Sai Teja Shashank
Computer Science and Engineering, Hyderabad
Shashankpunna21@gmail.com

Akkaladevi Vaibhavi
Computer Science and Engineering, Hyderabad
vaibhavi.hyd@gmail.com

Jenega Vaishnavi
Computer Science and Engineering, Hyderabad.
vaishu9121@gmail.com

Dr M.A Jabbar
Professor
Computer Science and Engineering
Vardhaman College of Engineering, Hyderabad
jabbar.meerja@gmail.com

ABSTRACT:

Epidemic diseases are contagious diseases that can spread into the entire nation if the contagion measurement had reached the outbreak level and there is a chance to wipe out the entire population. So, there must be a way to track these epidemics to take counter measurements to restrict its spreading. There are many ways to track these epidemics. Nowadays, machine learning is applied in almost all the fields since data is generating rapidly from various sources. The data which is useful for tracking epidemics is also very high around us.Variousmachine learning techniques can be implemented to find out the status of epidemics. Regression is a machine learning algorithm that is used to find the relationship between a dependent variable and an independent variable. For fitting a regression line, the most common method is the method of least-squares. In this method, the best-fitting line is calculated for the observed data such that the sum of the squares of the vertical deviations from each data point to the line is minimum. In this paper, we proposed a Regression-based prediction system that is used to predict the status of an epidemic in a region. Linear regression is used for this purpose to predict the number of deaths in a region on a specific date. Various libraries of python like NumPy, pandas, matplotlib, and sci-kit-learn are used to develop the model.

KEYWORDS:
Epidemic Prediction, Electronic Health Record, Linear Regression, Covid-19, Machine Learning, Influenza

1.INTRODUCTION:

Epidemic Outbreaks are one of the major threats in the 21st century. They occur when there is a sudden increase in the number of cases of a disease. They spread very rapidly. The time period of this epidemic is usually unknown. It can last up to days, weeks, months, or years. Scientists are still finding out a way to know the time period of these epidemics. It kills almost half of the population if the contagion measurement had reached the outbreak level and expands into the entire nation. Epidemic Outbreaks have been confirmed to create severe world health issues and have caused many life losses and wiped out 41% of the population globally. Examples of epidemic outbreak seen earlier are Cholera, Dengue, Influenza, Bird flu, Pneumonia, Covid-19, etc. In recent years, there have been advances in collecting health care data. Epidemics are usually caused by direct contact from one person to another and contaminated air. By considering the causes and features of epidemics, there should be a way to monitor or track these epidemics to take counter measurements to control it.

For this reason, we proposed a predictive model, to predict the status of the epidemic in a region using social media data. Since machine learning is used in many fields to extract valuable information which requires a huge amount of data and social media is generating a huge amount of data every minute. So, we applied machine learning on social media data for prediction. We collected the dataset from the data.CDC.gov website. We imported different types of modules and libraries like NumPy, pandas, and sci-kit learn which helps in Pre-processing, regression, and data cleaning. Data cleaning helps to discard unnecessary information in the dataset. We implemented this prediction system by applying certain machine learning algorithms like linear regression. It is a commonly used technique to build a relationship between two sets of variables. The result is a linear regression equation that can be used to predict the disease outbreak. The early detection and prediction of the spread of epidemics can be very useful and faster response can take place. We also used Data Visualization for a better understanding of the data. The prediction model on the resultant data gives us information about the number of deaths in a given region or location on a certain day, week, or month. This paper deals with the methodology to predict the status of the epidemic with the application of Linear Regression. Electronic Health data is chosen for simulation.

The remaining part of the paper is ordered in this way. Section 2 is assigned to related work. Section 3 describes the methodology and Section 4 contains about our proposed work. The experimental result is discussed in Section 5. We conclude our remarks and gives direction in Section 6.

2.RELATED WORKS:

Machine Learning in the healthcare field can be used for effective disease prediction, diagnosis, and treatments.

This section discusses related works on epidemic outbreak prediction by mining social media data.

Predicting Outbreak Detection in Public Health Surveillance: Quantitative Analysis to Enable Evidence-Based Method Selection is proposed by David L Buckeridge, Anya Okhmatovskaia, Samson W Tu, Martin Joseph O'Connor[1]. They developed and validated quantitative models for predicting the ability of commonly used surveillance algorithms which are used to detect different types of outbreaks using simulated outbreak data. They used the BioSTORM software test-bed developed by their group to encode the C-algorithms and to evaluate their ability. They recorded characteristics of each outbreak signal and the parameters of the algorithms along with the results for detection.

Detecting and Tracking Disease Outbreaks by Mining Social Media Data is proposed by Yusheng Xie1,2 Zhengzhang Chen1,2 Yu-Cheng Kunpeng Zhang Kathy Lee Ankit Agrawal Wei-Keng Liao Alok Choudhary2 [2]. They stated the real-time challenge in detecting and tracking epidemic outbreaks., They aimed to provide a platform for researchers in the domain to work on the real-time challenge by releasing relevant datasets from social media. They made four distinct social media datasets publicly available:

1.Health-related blogs (BL)
2.Facebook public walls of medical brands and organizations (FBW)
3. Facebook public posts mentioning health-related terms (FBP)
4.Media uploads from Instagram (INST).

Using online social networks to track a pandemic: A systematic review is proposed by Muhammad SadiqKhan KasturiDewiVarathan GhulamMujtaba AbdelkodoseM.Al-Kabsi[3]. They conducted a systematic review of the use of Online Social Network to track a pandemic and presented challenges and future implications of using them. The electronic literature search is conducted for eligible English articles published between 2004 and 2015 and concluded that the outcomes of online social network surveillance systems have demonstrated high correlations with the findings of official surveillance systems with some limitations.

Comparing Social media and Google to detect and predict severe epidemics is proposed by Loukas Samaras, Elena García-Barriocanal& Miguel-Angel Sicilia [4]. The authors aimed at gathering evidence on which kind of data sources lead to better results. Data on influenza in Greece are collected from Google and Twitter and compared to influenza data from the official authority of Europe. The data were analyzed by using two models: the ARIMA model and a customized approximate model. Google data showed a high Pearson correlation and a relatively low Mean Absolute Percentage Error (R = 0.933, $MAPE$ = 21.358)whereas Twitter results gave (R = 0.943, $MAPE$ = 18.742).

3.METHODOLOGY:

This section will discuss the methodology adopted by our proposed methodology. As stated earlier, this research paper aims to develop a predictive model to predict the status of an epidemic in a region. We used the Deaths involving coronavirus disease 2019 (COVID-19), pneumonia, and influenza data set given by the Centers for Disease Control and Prevention (Data.CDC.gov), which is partitioned into training and test set.

3.1 DATASET:

Our dataset contains the details of the number of deaths by various epidemics like Covid-19, Influenza, and Pneumonia in different regions. Our dataset contains 1782 rows and 14 columns before pre-processing, which is 33 rows for a location. After removing the uninterested columns and rows containing missing values the resulting dataset contains 720 rows and 11 columns. To predict the status of the epidemic in a location, the rows of only that location are extracted to build the model. Here the independent variable is the date attribute and the dependent variable is the Covid-19 deaths. The dataset used in our methodology is shown in figure 1.

1	Data as of	Start week	End Week	Group	State	Indicator	COVID-19	Total Deaths	Percent of Expected Deaths	Pneumonia Deaths	Pneumonia and COVID-19 Deaths	Influenza Deaths	Pneumonia, Influenza, or COVID-19 Death:	Footnote
2	09/18/2020	02-01-2020	02-01-2020	By week	Unitec	Week-end	0	58,576	0.99	3,796	0	479	4,275	
3	09/18/2020	02-08-2020	02-08-2020	By week	Unitec	Week-end	1	59,296	0.99	3,799	0	520	4,320	
4	09/18/2020	02/15/2020	02/15/2020	By week	Unitec	Week-end	0	58,697	1	3,824	0	558	4,382	
5	09/18/2020	02/22/2020	02/22/2020	By week	Unitec	Week-end	5	58,734	1.01	3,699	1	564	4,267	
6	09/18/2020	02/29/2020	02/29/2020	By week	Unitec	Week-end	9	59,167	1.03	3,822	5	654	4,480	
7	09/18/2020	03-07-2020	03-07-2020	By week	Unitec	Week-end	35	59,557	1.03	3,957	17	632	4,606	
8	09/18/2020	03/14/2020	03/14/2020	By week	Unitec	Week-end	53	58,506	1.02	3,960	28	625	4,609	
9	09/18/2020	03/21/2020	03/21/2020	By week	Unitec	Week-end	571	59,052	1.04	4,534	252	560	5,406	
10	09/18/2020	03/28/2020	03/28/2020	By week	Unitec	Week-end	3,152	62,883	1.13	6,167	1,424	442	8,287	
11	09/18/2020	04-04-2020	04-04-2020	By week	Unitec	Week-end	10,011	72,134	1.29	9,905	4,750	479	15,409	
12	09/18/2020	04-11-2020	04-11-2020	By week	Unitec	Week-end	16,172	78,869	1.42	11,984	7,223	474	21,067	
13	09/18/2020	04/18/2020	04/18/2020	By week	Unitec	Week-end	17,052	76,536	1.41	11,378	7,298	266	21,209	
14	09/18/2020	04/25/2020	04/25/2020	By week	Unitec	Week-end	15,415	73,612	1.37	10,352	6,592	144	19,222	
15	09/18/2020	05-02-2020	05-02-2020	By week	Unitec	Week-end	13,131	69,007	1.28	8,921	5,503	65	16,594	
16	09/18/2020	05-09-2020	05-09-2020	By week	Unitec	Week-end	11,154	66,570	1.26	7,818	4,702	47	14,307	
17	09/18/2020	05/16/2020	05/16/2020	By week	Unitec	Week-end	9,152	64,215	1.22	6,741	3,767	21	12,144	
18	09/18/2020	05/23/2020	05/23/2020	By week	Unitec	Week-end	7,145	61,323	1.18	5,849	2,949	24	10,064	
19	09/18/2020	05/30/2020	05/30/2020	By week	Unitec	Week-end	6,096	59,365	1.15	5,243	2,470	12	8,881	
20	09/18/2020	06-06-2020	06-06-2020	By week	Unitec	Week-end	4,976	58,527	1.12	4,860	2,132	11	7,715	
21	09/18/2020	06/13/2020	06/13/2020	By week	Unitec	Week-end	4,179	57,624	1.11	4,498	1,850	11	6,835	
22	09/18/2020	06/20/2020	06/20/2020	By week	Unitec	Week-end	3,787	57,497	1.11	4,329	1,590	7	6,532	
23	09/18/2020	06/27/2020	06/27/2020	By week	Unitec	Week-end	3,750	57,959	1.13	4,210	1,541	10	6,428	
24	09/18/2020	07-04-2020	07-04-2020	By week	Unitec	Week-end	4,419	59,094	1.14	4,480	1,959	4	6,941	
25	09/18/2020	07-11-2020	07-11-2020	By week	Unitec	Week-end	5,610	60,794	1.19	5,421	2,657	9	8,383	
26	09/18/2020	07/18/2020	07/18/2020	By week	Unitec	Week-end	6,955	61,648	1.21	6,007	3,360	12	9,611	
27	09/18/2020	07/25/2020	07/25/2020	By week	Unitec	Week-end	7,897	62,069	1.23	6,520	3,833	10	10,589	
28	09/18/2020	08-01-2020	08-01-2020	By week	Unitec	Week-end	7,839	61,035	1.2	6,471	3,851	13	10,465	

Figure 1: Dataset used for our model

3.2 LINEAR REGRESSION:

To find the relationship between a dependent variable i.e. COVID-19 deaths and an independent variable i.e. date, the best and simple regression algorithm is linear regression. There are several variations of linear regression. If there is only one explanatory variable, then it is called simple linear regression and if there is more than one explanatory variable then it is called multiple linear regression [9].

For fitting a regression line, the least square method is used. The regression line is to be drawn for the observed data. This line is drawn such that the sum of the squares of the vertical deviation from each point to the line is minimum [9].

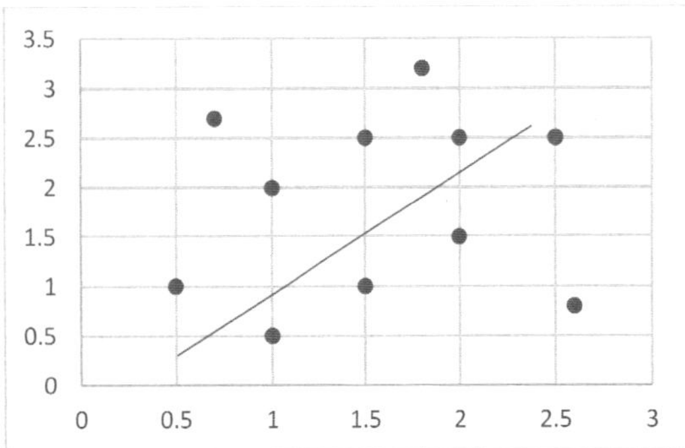

Figure 2. Best Fit line

To decide the best fit line, we use the following formula: $\sum(Y_0-Y_p)^2$

3.3 Evaluation Measures:

To evaluate the accuracy of our model we used the following evaluation measures:

- Explained variance score
- Max error
- Mean absolute error
- Mean squared error
- Mean squared logarithmic error
- Median absolute error
- R^2 error, the coefficient of determination

4.PROPOSED APPROACH:

As stated earlier, this paper aims to propose a predictive model to predict the status of an epidemic in a region. A dataset containing details of the number of deaths by various epidemics like Covid-19, Influenza, Pneumonia in different regions on different days is collected. The dataset is preprocessed to remove noise and redundancy. Further, the dataset is divided into two parts called train dataset and test dataset in the ratio of 70:30, and the linear regression algorithm is applied to the training dataset to predict the number of deaths in a region on a specific date. Lastly, our prediction model is tested using various performance measures.

Steps in our proposed method are as follows:
Step 1: Load the Dataset
Step 2: Apply preprocessing to the data set
Step 3: Visualize the data by various plots
Step 4: Build the predictive model using Linear Regression
Step 5: Predict the number of deaths by an epidemic in a region on a given date.

Algorithm: Predictive Model to predict the status of epidemic in a region

Loading is done in step 1 with the help of python's library called pandas. The dataset is loaded into a datatype called data frame which is like a two-dimensional array with indexes on both columns and rows with the help of which we can perform effective operations on datasets.

Pre-processing is done in step 2 to remove the uninterested columns and rows containing missing and misleading values with the help of python's libraries namely NumPy and pandas.

Since in our dataset, the date column is having dates of different formats i.e. dd/mm/yyyy and dd-mm-yyyy. We made all these dates to a single format i.e. dd-mm-yyyy. And also, the column COVID-19 deaths had commas in the data. So, we removed those commas and converted the string type to integer type.

Data Visualization is done in step 3 to observe the patterns, trends, and outliers in the data using the python's library called matplotlib. We observed the relation between date and number of COVID-19 deaths by plotting various graphs like bar, line, and scatter. The outputs of bar graphs and scatter graphs are shown in figure 2 and figure 3.

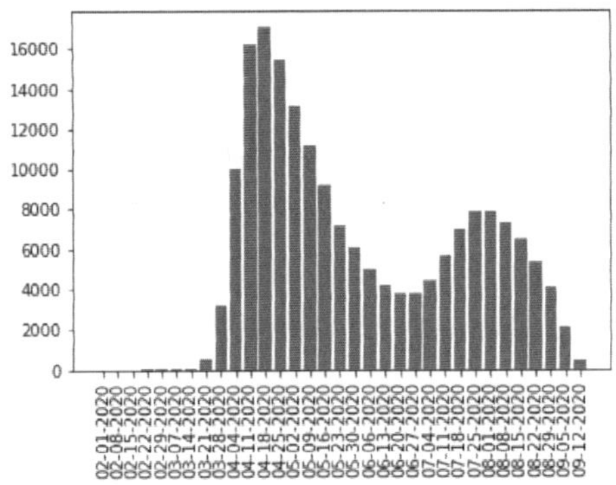

Figure 3: Bar graph

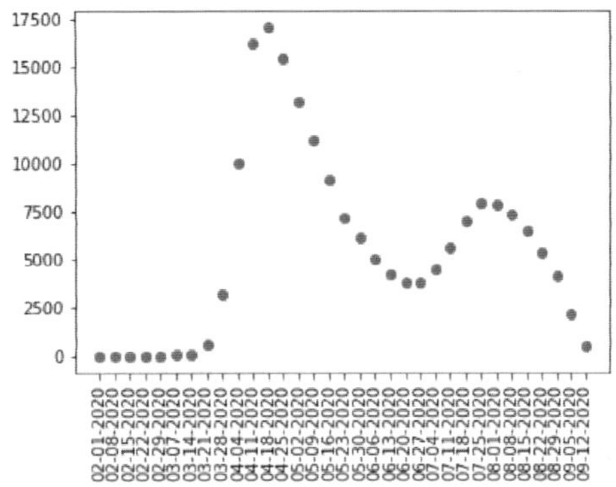

Figure 4: Scatter plot

Dataset is divided into training and testing datasets in the ratio of 70:30 and the Linear regression algorithm is applied to the training dataset in step 4 to build our prediction model. Further, we estimated the accuracy of our prediction model with the help of various performance measures in step 5.

The data flow in our proposed approach is pictorially shown in figure 4.

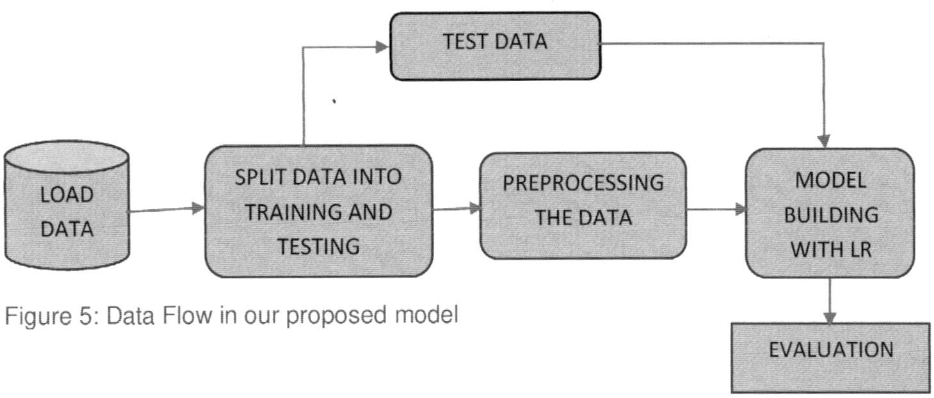

Figure 5: Data Flow in our proposed model

5.Experimental Results:

Experiments for the proposed model is conducted on the Deaths involving coronavirus disease 2019 (COVID-19), pneumonia, and influenza data set given by the Centers for Disease Control and Prevention (Data.CDC.gov). The dataset is partitioned into training and testing datasets. The training dataset is used to build the predictive model and the testing dataset is used to test the instances. The proposed model is evaluated using the following metrics:

1.R-squared (R^2)=(variance explained by the model)/Total variance

2.Explained variance score= $1-\frac{Var\{y-y^{\wedge}\}}{Var\{y\}}$

3.Max Error = max($|y_i$-y_i^$|$)

4. Mean absolute error = $\frac{1}{n_{samples}}\sum_{i=0}^{n_{samples}-1}|y_i - y_i{}^{\wedge}|$

5.Mean Squared error=$\frac{1}{n_{samples}}\sum_{i=0}^{n_{samples}-1}(y_i - \hat{y_i})^2$

6.Mean squared logarithmic error=$\frac{1}{n_{samples}}\sum_{i=0}^{n_{samples}-1}(log_{e(1+y_i)} - log_{e(1-y_{i^{\wedge}})})^2$

7.Median absolute error = median($|y_i - \hat{y_i}|,..,y_n - \hat{y_n}|$).

The results obtained by our prediction model on the test dataset is shown in table 1.

Table 1: Outputs from our model

X_TEST	Y_TEST	PREDICTED OUTPUT
737491	13131	5300.36
737456	3787	5623.94
737484	571	5023.02
737561	0	4791.89
737575	5610	5762.61
737463	5313	6039.96
737477	0	4699.44
737680	7301	5947.51
737470	17052	5207.91
737554	6478	5993.74

The results of different evaluation metrics for our model is shown in table 2.

Table 2: Output from different accuracy measures

ACCURACY MEASURE	SCORE
Explained variance score	0.03
Max Error	6618.01
Mean absolute Error	4075.24
Mean squared Error	21670427.64
Mean squared Logarithmic Error	15.03
Median absolute Error	3889.99
R-square Score	-1.80

6.CONCLUSION:

In this paper, we proposed a model for predicting the status of an epidemic in a specific area, region, or location by predicting the number of deaths occurring in a specific location. The use of conventional methods is not sufficient enough for faster and accurate results. So, we use this new approach in finding out the epidemic in earlier stages. Data cleaning, pre-processing, and machine learning algorithms like linear regression are used for this model. We use the Least Square Method for fitting the regression line. Data Visualization is used to understand the data and relationships between dependent and independent variables in a better way. We have also tested our predicted model using various kinds of performance measures. We evaluated the accuracy of linear regression by using R-squared(R2), Root-mean squared error, Residual Standard Error, Mean Absolute Error. By implementing the steps of our approach, we have proved it to be successful in predicting the epidemic.

REFERENCES:

[1]. Predicting Outbreak Detection in Public Health Surveillance: Quantitative Analysis to Enable Evidence-Based Method Selection David L. Buckeridge, MD, PhD1, Anna Okhmatovskaia PhD1, Samson Tu MS2, Martin O'Connor MS2, CsongorNyulas MS2 , Mark A. Musen, MD PhD2 1 McGill University, Montreal, QC; 2 Stanford University, Palo Alto, CA

[2]. Detecting and Tracking Disease Outbreaks by Mining Social Media Data Yusheng Xie1,2 Zhengzhang Chen1,2 Yu Cheng Kunpeng Zhang Kathy Lee Ankit Agrawal Wei-keng Liao Alok Choudhary2

[3]. Using online social networks to track a pandemic: A systematic reviewMohammed Ali Al-garadi, Muhammad Sadiq Khan, Kasturi DewiVarathan, Ghulam Mujtaba, AbdelkodoseM.Al-Kabsi . Journal of Biomedical InformaticsVolume 62, August 2016, Pages 1-11

[4]. Comparing Social media and Google to detect and predict severe epidemics.Loukas Samaras, Elena García-Barriocanal &Miguel-Angel Sicilia. Available at: https://www.nature.com/articles/s41598-020-61686-9

[5]. A review of influenza detection and prediction through social networking sites Ali Alessa and MiadFaezipour. Available at: https://pubmed.ncbi.nlm.nih.gov/29386017/

[6]. TRENDING NOW:Using Social Media to Predict and Track Disease Outbreaks. Charles W. Schmidt, MS. Available at: https://www.ncbi.nlm.nih.gov/pmc/articles/PMC3261963/

[7]. How social media can be used to track disease outbreaks. Article by healthcareglobal.com. Available at: https://www.healthcareglobal.com/technology-and-ai-3/how-social-media-can-be-used-track-disease-outbreaks.

[8]. Can Social Media Predict DiseaseOutbreaks? An Article in healthadministrationdegrees.com. Available at: https://www.healthcareglobal.com/technology-and-ai-3/how-social-media-can-be-used-track-disease-outbreaks

[9]. Linear Regression Article in wikipedia. Available at: https://en.wikipedia.org/wiki/Linear_regression.

[10]. Forecasting Influenza Levels using Real-Time Social Media Streams Kathy Lee, Ankit Agrawal, Alok Choudhary EECS Department, Northwestern University, Evanston, IL {Kathy lee,ankitag,choudhar}@eecs.northwestern.edu

[11]. The Real-World Benefits of Machine Learning in Healthcare by Ed CorbettMedical Officer. Article in healthcatalyst.com. Available at: https://www.healthcatalyst.com/clinical-applications-of-machine-learning-in-healthcare

[12].12+ Machine Learning Applications Enhancing Healthcare Sector 2020 by Prashant Kathuria. Blog in upgrade.com. Available at:https://www.upgrad.com/blog/machine-learning-applications-in-healthcare/

[13]. Numpy documentation in numpy.org. Available at:https://numpy.org/doc/

[14]. Pandas documentation in pandas.pydata.org. Available at: https://pandas.pydata.org/docs/

[15]. Matplotlib documentation in matplotlib.org. Available at: https://matplotlib.org/3.3.2/contents.html

[16]. Scikit-Learn documentation in scikit-learn documentation. Available at:https://scikit-learn.org/stable/
[17]. Regression Model Accuracy Metrics: R-square, AIC, BIC, Cp and more. By kassambara. Article in STHDA.com. Available at:http://www.sthda.com/english/articles/38-regression-model-validation/158-regression-model-accuracy-metrics-r-square-aic-bic-cp-and-more/

Multiple Face Detection Using Haar - AdaBoosting, LBP -AdaBoosting and Neural Networks

B.Thirumaleshwari Devi [1], S.Shitharth[2]

[1,2]Department of Computer Science and Engineering, Vardhaman college of Engineering And Technology

devi04091997@gmail.com

Abstract. Multiple Face Detection may be a process of identifying or recognizing quite one face on a picture. Face detections have recently attracted increasing interests thanks to the multitude of applications that end in format. There are numerous methods for identifying the face on the image but here we are using 'Haar-AdaBoost', 'LBP-AdaBoost', and 'Neural Networks' for identifying the faces on the image. And also we are comparing each of the methods to urge which method is giving highly accurate results and which method is giving results very quickly. In humans, by seeing the positions of Eyes, Nose, and Mouth another person can identify the face. The neural mechanism within the brain controls these all and provides the result. We are implementing an equivalent concept on the machine so that the machine can identify the faces by using the positions of the Eyes, Nose, and Mouth. The machine will identify quite one face or multiple faces within the image by using this, and that we also can compare which algorithm will provide a good percentage of accurate results.

Keywords: *Face Detection, Haar-AdaBoosting, LBP-AdaBoosting, and Neural network.*

1. Introduction:

In humans, by seeing the positions of Eyes, Nose, and Mouth another person can identify the face. The neural mechanism within the brain controls these all and provides the result. We are implementing an equivalent concept on the machine so that the machine can identify the faces by using the positions of the Eyes, Nose, and Mouth. The machine can identify quite one face or multiple faces within the image by using this, and that we also can compare which algorithm will provide a good percentage of accurate results.[1] consistent with this context, we've established a comparative study between three methods (Haar-AdaBoost, LBP-AdaBoost, and Neural Network). These techniques vary according to the tactic during which they extract the knowledge and thus the adopted learning algorithms. Here LBP-Ada boost and Haar Ada Boost methods are Boosting algorithms which can be used for learning and selection of strong classifier. And therefore the last one is that the "Neural Network" uses the Gabor filter to extract the characteristics. [4]

The main objective of this specification is to supply a transparent view of the wants for Multiple Face Detection. This project aims to develop a system for users to upload and detect the multiple faces within the image. The system will identify the faces using Harr-AdaBoost, LBP- AdaBoost, and GF-NN and build a graph determining the face detection time and efficiency for every method.

Identifying human faces during digital images by using a device is named as Face detection. Face detection is one of the required instances of object class detection. during this model, a major importance is given to the highest face detection. In all the possible regions human eye areas are firstly detected and evaluated in this model. It aims to develop a system for users to upload and detect the multiple faces within the image. The system will identify the faces using Harr-AdaBoost, LBP- AdaBoost, and Neural Network and build a graph determining the face detection time and efficiency for every method. During this project we start with an introduction, then we present

within the first section, the detection methods supported Haar, LBP, and Gabor extraction techniques. Then within the second section, we expose the approaches of automatic learning Boosting and Neural Networks. [4] Then we are showing the results a comparison between the three methods (Haar-AdaBoost, LBP-AdaBoost, and Neural Network) consistent with the time interval of the test images and therefore the detection. Then the conclusion is that the last step.

We are implementing this idea on the windows OS using Python as a programming language and Pycharm as a tool for developing or editing the code. For identifying the faces we are using the Cascading files using a picture as an input.

2. Feature Extraction

2.1 Haar features

A. Haar method:

In real-time, efficiently detecting objects in an image viola and Jones method is the most important method. This method is particularly invented for face detection. Two or three rectangles of Haar features are composed and candidates' faces are searched and scanned for the current stage of Haar features By learning the algorithm the generated weights are constant. The value of each Haar feature is calculated by the area of multiplying each rectangle by their particular weights. For haar feature an example is given fig (1). [1, 8]

Figure1. Haar features an example.

B. Integral image:

By calculating the values efficiently and quickly of a grid of rectangle subsets are known as Integral image or Summed area table which is an algorithm for random computation of Haar features the viola and Jones method is applied to the integral method. Following is the integral image construction. [1, 4, 12, 19]

2.2. LBP:

LBP is popularly known as a local binary pattern. It is a texture descriptor that has different applications such as environmental modeling, remote assessment, image recovery, motion analysis and face image analysis, medical image analysis, and face image analysis. The process implies dividing a facial image into collective regions where LBP exhibits concatenated and extracted to features vector which is used for a further facial descriptor. [2, 4 , 20]

A. LBP principle:

Ojala and al are the first users of the original LBP operators. LBP works on using a 3X3 window of a pixel in a neighborhood from the image and it will extract LBP code. It has three parameters on which LBP works the central pixel value used as a threshold. Value '1' is assigned to a pixel when the neighboring pixel has a gray color when the central pixel has less value or it is assigned with '0'. The concatenations of the eight 1or0 counterclockwise or clockwise to a binary code which is produced by LBP code for the central pixel. The process of LBP is explained in fig (2) by labeling method. [6, 10, 11, 17, 18]

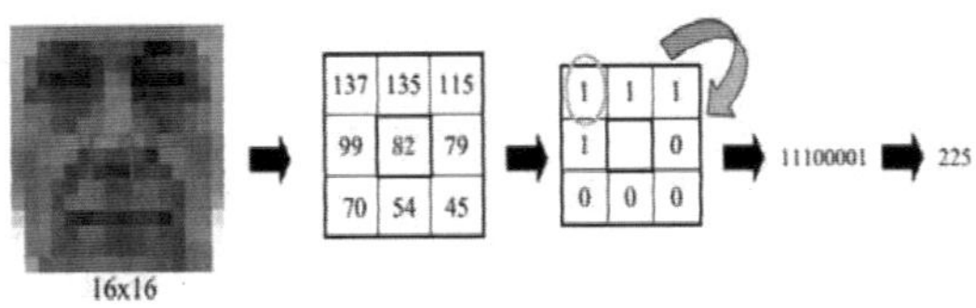

Figure2 LBP Labeling.

Let (x_c, y_c) be pixel positions, 8-bit word resulting in a decimal value that can be expressed as follows:

$$\text{LBP}\ (x_c\ ,y_c) = \sum_{n=0}^{7} S(I_n - I_c)2^n \qquad (1)$$

Where

l_c=gray value for central pixel (x_c, y_c).

l_n=gray values of the 8 neighboring pixels.

S (K) ={1 $if\ K \geq 0$, 0 if K<0.

B. Facial Representation of LBP:

Micro-pattern compositions that are effectively identified by the LBP operators are considered from an individual face of the image. To represent facial recognition Ahonen and al introduced a face-based LBP representation. Shape information of the faces is probed by dividing the faces in the image into M and regions (R0, R1… RM) small non-overlapping. Histogram of the single spatially enhances into sub-regions. [3, 4, 10, 16]

$$H_{i,j} = \sum_{x,y} I(f_l(x,y) = i) I\left((x,y) \in R_j\right) \qquad (2)$$

With

i=0… L-1,

j=0… M-1. The overall shape and the local texture are described by the histogram of the extracted features of the face images.

2.3. Neural Network

Neural networks use the Eigen faces through a classification of the coefficients calculated. The database of the network is competent in the pictures first, later it will recognize the given picture faces. Eigen's faces represent the facial characteristics that are used in persons face recognition. In this technique, a set of images are tested and trained to determine the eigenvectors of the covariance matrix. Eigenvectors do not correspond directly to face features like mouth, eyes, and nose as they are Ortho normal to each other. The ghostly face is displayed when the eigenvectors are shown and Eigen's faces are termed. In fig (3) Network Architecture is explained clearly. [5, 7, 13]

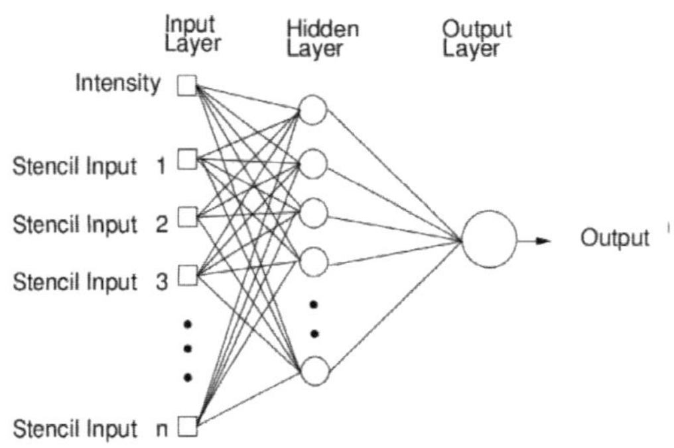

Figure3 Network Architecture

2.3.1Classification of the neural network:

The face recognizer is planned for recognition after the training data is trained. The face detection unit from the image sequence detects the face and Eigen face space is calculated on its projection. In the neural network unit output signals are given by the new face descriptor Ω, it produces the number 'T' produces the index number of the training set of the person by the following equation.[1,5,14]

$$\text{Index Number} = \text{Log10}\ (T/\text{Log10}\ (2.0)). \quad (3)$$

Confirmation of the capable name is confirmed by the back propagation error if the index number is from the person list. If the comparison method is finished the name of the person is generated if identified.

3. Classifications

3.1. AdaBoost Learning:

AdaBoost on learning algorithm of the Adaboost of a machine learning algorithms also known as a weak learner by using the output of weighted classification algorithms. The only better way to be randomly is convenient for the right orientation is a weak learner. In the same way, strong learners are also represented to a set of classification algorithms that have more correlated with the perfect orientation. So, that poor learners can change to minimize errors in subjects, but they are not classified. By doing this way other images except human images are detected. In the formulation process, the AdaBoost learning algorithm can be explained in the following way. [1, 3]

Let us consider n images in a set (x_1..........x_n) and associated with labels (y_1........y_n). If x_i is a negative example then y_i=0 and if x_i is an object to detect example then y_i=1.T is the number of boosting algorithms contained, t is iteration, j is characteristics, h_j is constructed as a weak classifier and h is sample classifier.

$$y_i = h(x_i)\forall_i \in \{1, \dots \dots, n\} \quad (4)$$

When the error is produced and the classifier is not found then in practice it is given by:

$$E_j = \sum_{i=1}^{n} w_i |h_j(x_i) - y_i| \quad (5)$$

3.2. Cascading Classifier:

The impact of the cascade structure will also have on the training process of an image. Face detection has a detection task rarely. The detection algorithm of the organization is an important idea of the method to reduce the cost in a cascade of classifiers. Consequently, the classifiers will decide on acceptance. The example is passed to the next classifier when the window contains the object, or it may be rejected, when the window does not contain an object and for this case this process is clearly explained in fig (4), the example is commonly not appeared. [9]

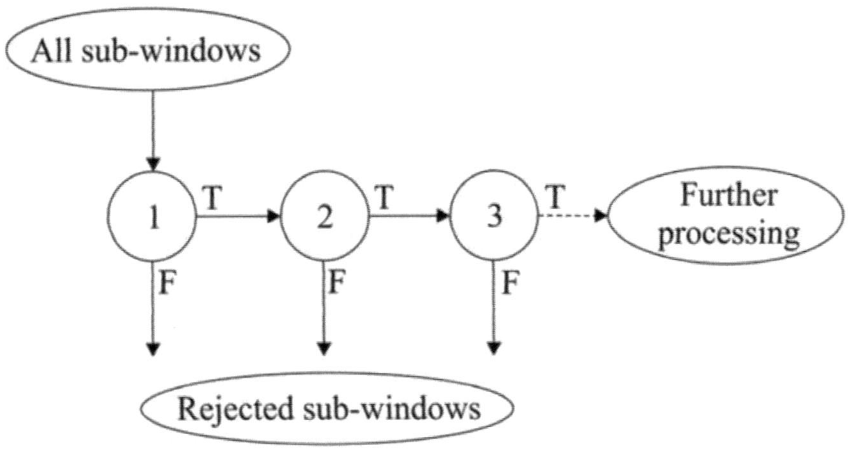

Figure4 cascade architecture

4. The Testing Results:

In the testing process, the results obtained by giving an image to these three methods are different based on the type of the algorithm. The accuracy and time are taken by each method are calculated.

4.1 Haar-AdaBoosting Method: By using this method 11 faces are identified in the given fig (5).

Figure5 Finding multiple faces by using the Haar-AdaBoosting method.

4.2.LBP-AdaBoosting Method: By using this method 9 faces are found in a given fig(6).

Figure 6 Finding multiple faces by using the LBP-AdaBoosting method.

4.3 Neural Network Method: By using the neural network method the 13faces are identified in the given image fig (7).

Figure7 Finding multiple faces by using the Neural Networks method.

4.4 .Accuracy and Time Graph images: Based on the faces identified in the image by each method the graph is plotted using results obtained from each method. And the time graph is drawn by the time taken for each algorithm to identify all faces in the image are explained in fig (8) and fig (9).

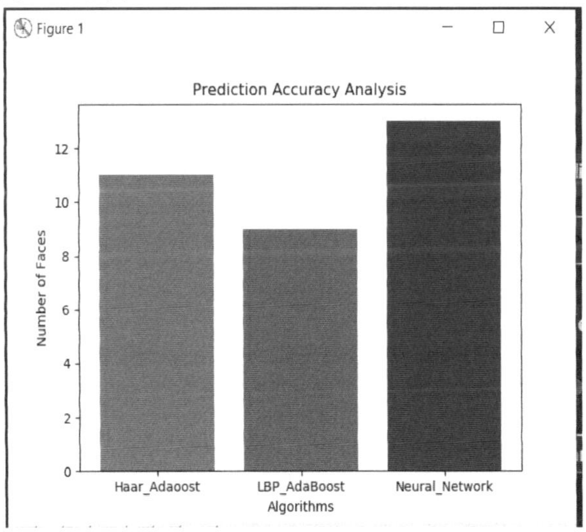

Figure 8. Accuracy Graph Image

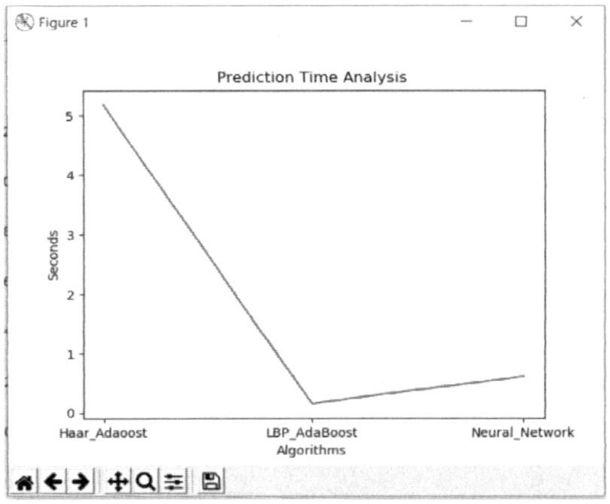

Figure 9. Time Graph Image

5. Conclusion

In this paper, we find that Face detection is currently an active area of research. Recent years have shown great advances in algorithms handling complex environments. a number of the simplest algorithms are still too expensive in terms of calculation to be applicable in real-time, but this is often likely to vary with upcoming improvements in hardware. during this project, we present in-depth research on face detection methods, all methods have their advantages and drawbacks, for instance, the characteristics of Haar utilized in the trouble by Viola and Jones are very simple and effective for the detection frontal of the face, but they're less ideal for random faces. the foremost direct direction to return is to further improve the algorithms and learning features. It's interesting, to ascertain face detection techniques increasingly utilized in real applications. for instance, most digital cameras today have built-in face sensors, which may help the camera to raised focus or autofocus and auto exposure, which is additionally a crucial technique for interfaces. Man-machine, to permit a more natural interaction between a person's and a computer.

References

[1] Jamil, N., Iqbal, S., & Iqbal, N. (n.d.). Face recognition using neural networks. Proceedings. IEEE International Multi-Topic Conference, 2001. IEEE INMIC 2001. Technology for the 21st Century. DOI:10.1109/inmic.2001.995351

[2] M.Gopi Krishna, A. Srinivasulu, Face Detection System On AdaBoost Algorithm Using Haar Classifiers, International Journal of Modern Engineering Research (IJMER) www.ijmer.com Vol. 2, Issue. 5, Sep.-Oct. 2012 pp-3556-3560 ISSN: 2249-6645.

[3] Arezou Chehrehgoshal*, Mehran Emadi2, Face Detection using Fusion of LBP and Adaboost, Article in Journal of Soft Computing and Applications · June 2016 DOI: 10.5899/2016/jsca-00064.

[4] B.Thirumaleshwari Devi, S.Shitharth, 'An Appraisal over Intrusion Detection systems in cloud computing security attacks', 2nd International Conference on Innovative Mechanisms for Industry applications, ICIMIA -2020, Conference Proceedings, pp. 122.

[5] Zhang and Zhengyou Zhang, A Survey of Recent Advances in Face DetectionCha, June2010Technical ReportMSR-TR-2010-66Microsoft Research.

[6] Shitharth, D.Prince Winston, "A Novel IDS technique to detect DDoS and Sniffers in Smart Grid", IEEE

WCTFTR – Proceedings of 2016 World Conference on Futuristic Trends in Research and Innovation for Social Welfare, 2016, DOI:7583897,2016

[7] Nazish Jamil, Samreen Iqbal, Naveela Iqbal, Face Recognition Using Neural Networks.

[8] Shitharth, D.Prince Winston, "An Appraisal on Security Challenges and Countermeasures in Smart Grid", International Journal of Applied Engineering Research, Vol.10, No.20, 2015, pp. 16591-16597

[9] T. Ahonen, A. Hadid, and M. Pietik¨ainen. Face recognition with local binary patterns. In Proc. of ECCV, 2004.

[10] S.Shitharth, Sangeetha, Praveen Kumar, 'Integrated Probability relevancy classification (IPRC) for IDS in SCADA', Design Framework for wireless network, Lecture notes in network and systems, Springer , vol. 82, Issue 1, 2019, pp. 41-64

[11] M. M. Campos and G. A. Carpenter. S-tree: Self-organizing trees for data clustering and online vector quantization.Neural Networks, 14(4–5):505–525, 2001.

[12] C. Huang, H. Ai, Y. Li, and S. Lao. Vector boosting for rotation invariant multi-view face detection. In Proc. of ICCV,2005.

[13] Shitharth, & D.Prince Winston, D 2017, 'An Enhanced Optimization algorithm for Intrusion Detection in SCADA Network', Journal of Computers and Security, Elsevier, vol. 70, pp. 16-26.

[14] Shitharth, D.Prince Winston, "A New Probabilistic Relevancy Classification (PRC) based Intrusion Detection System (IDS) for SCADA network", Journal of Electrical Engineering, Vol.16, No.3, 2016, pp. 278-288.

[15] Y. Ban, S. K. Kim, S. Kim, K. Toh, S. Lee, Face detection based on skin color likelihood, Pattern Recognition, 47 (2014) 1573-1585. *http://dx.doi.org/10.1016/j.patcog.2013.11.005*.

[16] K. B. Ge, J. Wen, B. Fang, Adaboost algorithm based on MB-LBP features with skin color segmentation for face detection, International Conference on Wavelet Analysis and Pattern Recognition (ICWAPR), 6 (2011) 40-43.

[17] S.Shitharth, Masood Shaik, Sirajudeen, Sangeetha, 'Mining of intrusion attack in SCADA network using clustering and genetically seeded flora based optimal classification algorithm', Information Security, IET, vol. 14, Issue 1, 2019, pp. 1-11

[18] H. D. Cheng, X. Jiang, A. Sun, J. Wang, Color image segmentation: advances and prospects, Pattern Recognition, 34 (2001) 22-59.

[19] Shitharth, T.Manikandan, "Removal of Selective Black Hole Attack in MANET by AODV protocol" in International Journal Of Innovative Research In Science, Engineering And Technology , Vol.3, Special Issue 3, March 2014, pp. 2372-2377.

[20] Senthil Kumar, C.Sebastian Albina, S.Shitharth, T.Manikandan, "Modified TSR protocol to Support Trust in MANET using fuzzy" in International Journal Of Innovative Research In Science, Engineering And Technology , Vol.3, Special Issue 3, March 2014, pp. 2551-2555.

IOP Conf. Series: Materials Science and Engineering 1042 (2021) 012018 doi:10.1088/1757-899X/1042/1/012018

WSN Smart Irrigation System and Weather Report System

J. Akhila Srinidhi[#1], A. Aasish[#2], N. Kishore Kumar[#3], T. Ramakrishnaiah[#4]
[#]Vardhaman College of Engineering, Hyderabad, India

E-mail: [1]akhilasri.j@gmail.com
[2]aasisharumilli99@gmail.com
[3]kishoreyadavu108@gmail.com
[4]t.ramakrishnaiah@vardhaman.org

Abstract: This paper proposes a cost-effective system that can be used by farmers easily. In these modern times also, we find farmers going to their fields to check the water level and to see that no cattle are present near the field. Even if there are cattle he can't do anything at that moment. So, we proposed a system where he can be notified if there is any animal or if the water content is low on his registered mobile number. In this system, an irrigation-based Wireless Sensor Network (WSN) using Raspberry Pi and the solar panel is implemented, Raspberry Pi being used as Coordinator node and Arduino as an end node, nRF24L01 as a transceiver. The sensor input values such as soil moisture level, PIR; the temperature is collected by the Arduino and is wirelessly sent to the Raspberry Pi. The water pump can be turned on based on sensor values using the Raspberry Pi web server and those values are also posted on the webpage. For temperature and humidity detection, we used the DHT11 sensor and the solar panel is used as an additional source of power. A PIR sensor is used to check if there are any animals present near the field.

Keywords: Irrigation, Wireless sensor network, Python, Raspberry Pi, Solar Panel.

1. Introduction

In India, most people depend on agriculture as their livelihood, in ancient times groundwater level was high so crop yield was high. With the regular irrigation system, the field has to be monitored by the farmer continuously which is a huge challenge. In modern times, technology has improved a lot, and internet connectivity even available at cheaper rates. So, farmers can use WSN based irrigation system to control the crops through their smartphones. Farmers can know the temperature, humidity, Water level of their crops in the smartphone through which they can efficiently secure their crop yield. It is also a major reason for the reduction in crop yield and cattle. If we have Irrigation and atmospheric condition systems we can solve these problems and improve the crop yield.

2. Literature Review

Many efforts have been made to propose by Agrawal and S. Singhal [1], to design an energy-efficient and cost-effective home automation system for ready to use. Internet of Things and WSN based field security which was implemented by Tanmay Baranwal, Nitika et al [2] to detect rodents and insects in the field and inform farmers immediately using Cloud Computing and Big Data. Gutierrez J, Villa-Media, J.F, et al [3] proposed a system of automated irrigation for effective utilization of water for crops. This system checks with sensors the temperature and soil moisture at the root regions of the plant using a distributed wireless network. Besides, the gateway unit which handles sensor results and activates actuators is used for a web application. R. Hussain, J. Sehgal, et al [4] discussed different techniques of soil fertility measure because of checking productivity of the crop. Chuang Yu, Yi Wu, et al [5] proposed a system that controls the output pressure of channel pipe with an intelligent meter by adopting a fuzzy control algorithm. R.Karthikeyan, P.Mahalakshmi, et al [6] proposed a hybrid solar energy and wind power in rural India for pumping water effectively. A. Satya. J. Kishore et al [7] proposed an irrigation system that controls water flow and also monitors the soil moisture status of the field. An SMS is delivered to the farmer about the amount of time water is to be supplied to the crops. Kamalesh and Usha Rani [8] developed web-based irrigation using ZigBee which monitors the soil moisture and motor status. K. K Namala Krishna Kanth Prabhu et al [10] proposed an intelligent Irrigation system, which could be used to control water flow for flower plants. So human involvement is reduced for watering plants.

3. Hardware Description

Figure1 below shows the Wireless Sensor Network based automatic irrigation system using a solar panel as a backup power source and using Raspberry Pi for security. The embedded system includes various components that communicate with each other and making it much easier to monitor the field parameters. The system includes an end node and the main node. The main node is a Raspberry Pi module which collects various field parameters from the end node and updates those values in the web server so it is easy for farmers to monitor the status of the field in real-time. Arduino Uno board is the end node that is interfaced with components like temperature sensor, soil moisture sensor, PIR sensor, and buzzer.

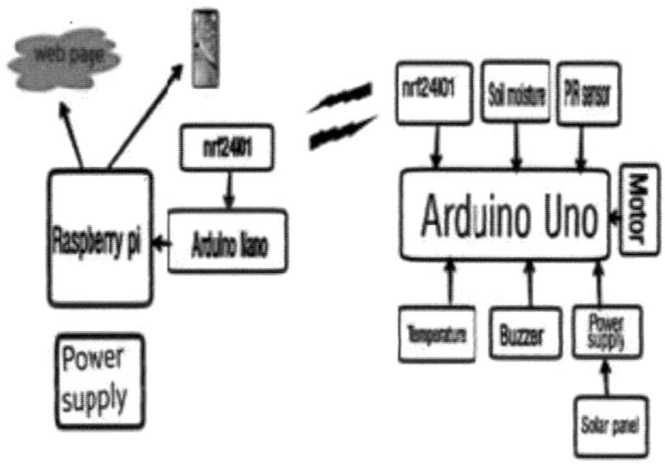

Figure1. Wireless Sensor Network Block Diagram

An nRF24L01 transceiver module was used for communicating with the main node. The temperature and soil moisture sensor at the end node senses the moisture and temperature parameters and sends these values to the main node. The PIR sensor is used for cattle intrusion detection if any cattle are entering the field the PIR status becomes high and the buzzer will turn on and drive away, the cattle from the fields. The values from various end nodes are sent to the main node then it compares the threshold value been set and does the functions like turn on or turn off the pump accordingly. The status of sensor parameters of each pump node is displayed on a web page which is to be observed by the farmers with their mobile device or a device that is facilitated with the internet through the web page.

3.1. Raspberry Pi3 Model B

In this paper, Raspberry Pi is the main node and it has the following specifications. The system comprises 1.2 GHz Quad-Core, 1GB RAM, Broadcom BCM2837, a 64 bit CPU. It has 40-pin GPIO and 4 serial USB ports. For loading and storing an operating system and program files, a Micro-SD card is used. In this paper wireless communication is used to gather the sensor values with Raspberry Pi via nRF24L01.

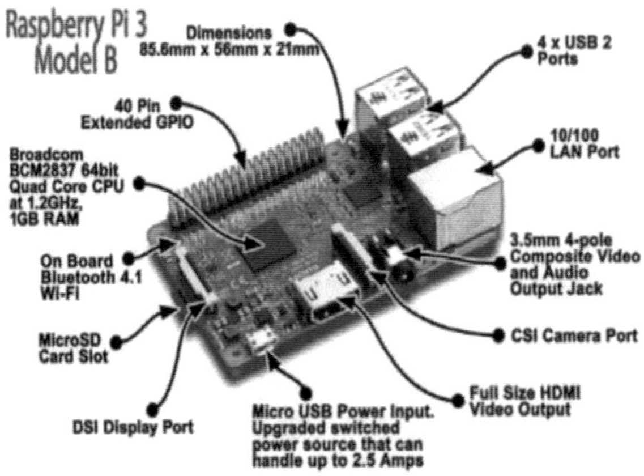

Figure2. Model B Raspberry Pi3

3.2. Arduino Uno

It is an ATmega328P general-purpose microcontroller. It has a total of 14 input/output digital pins, 16MHz ceramic resonator, 6 analog pins, USB connecter, a power jack, reset button and finally ICSP header.

Figure3. Arduino Uno

3.3. nRF24L01 Transceiver

It is a wireless transceiver that operates with 2.4Ghz frequency and a supply of 1.9-3.6V. The information was sent at a data rate of up to 2Mbps. In this paper, nRF24L01 transceiver pairs are used to link Raspberry Pi and Arduino. The Raspberry Pi receives sensor values from the nRF24L01 device. Here in this paper transceiver is used as it can handle detected data concurrently from multiple nodes.

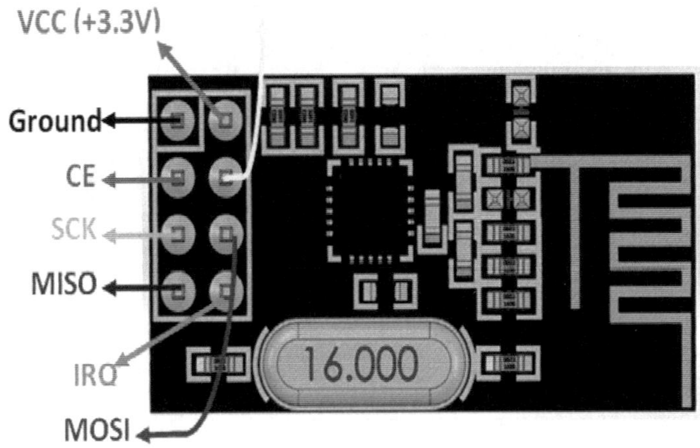

Figure4. nRF24L01

3.4. PIR Sensor

A passive Infra-Red sensor is used to detect the motion by sensing changes in infrared levels. The output is in the digital format and it can be either 0 or 1. The range it can detect humans and cattle is around 20ft.

Figure5. Passive Infra-Red Sensor

3.5. Humidity and Temperature Sensor

Both humidity and temperature can be measured using a DHT11 sensor. The sensor has an inbuilt NTC which measures temperature and to send the values of temperature and humidity it uses an 8-bit microcontroller that sends the data in serial form. The temperature can be measured from 0°C to 50°C and 20% to 90% of humidity range using Humidity and Temperature sensor. This sensor can be used where the end node exists in this range of temperatures.

Figure6. Temperature and Humidity Sensor

3.6. Soil Moisture Sensor

In this paper, a sensor for sensing moisture consists of LM 393 comparator circuit along with two probes. The LM393 comparator was used to compare soil moisture levels with a predefined threshold value. When the soil moisture is more than the predefined threshold value the sensor output becomes low and vice versa. Therefore these sensors are used by farmers to monitor field moisture status in turn they can manage the water flow to the fields, thereby it saves water and results in increased crop yield.

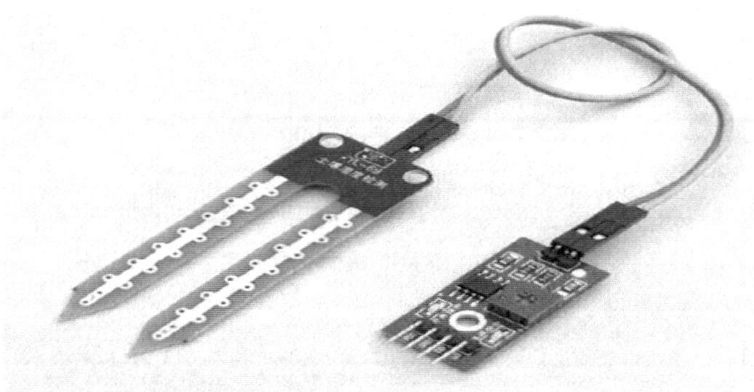

Fig. 7 Soil Moisture Sensor

4. Software Tools

To develop a Wireless Sensor Network based Irrigation System along with security and backup power as a solar panel, different types of software are used like Arduino IDE, or Python and Adafruit IO which is a Raspberry Pi server for IoT applications.

4.1. Arduino IDE

IDE stands for Integrated Development Environment. With this IDE the written programs are debugged and uploaded to Arduino and the remaining modules are interfaced with Arduino. The program file is saved with the extension of .ino.

4.2. Adafruit IO

Adafruit IO is a cloud service platform; here we can monitor various parameters of the field like humidity, the moisture of the soil, atmospheric temperature, and motor on/off control. Every Raspberry Pi has a specific IP address that is unique by this IP we can retrieve the information send by that Raspberry Pi. This IP is stored in the python code as soon as the program is executed the field parameters from the end node collected by the Raspberry PI is sent to the Adafruit IO server.

4.3. Python Language

It is a high-level scripting language. The specific tasks are done by pre-coded libraries. It has various inbuilt libraries for doing a specific task in the embedded system domain. The extension .py is used to save the program files.

5. Software Flowchart

Figure 8 and Figure 9 shown below described is the flow of events that occurred at the coordinator node and end node device.

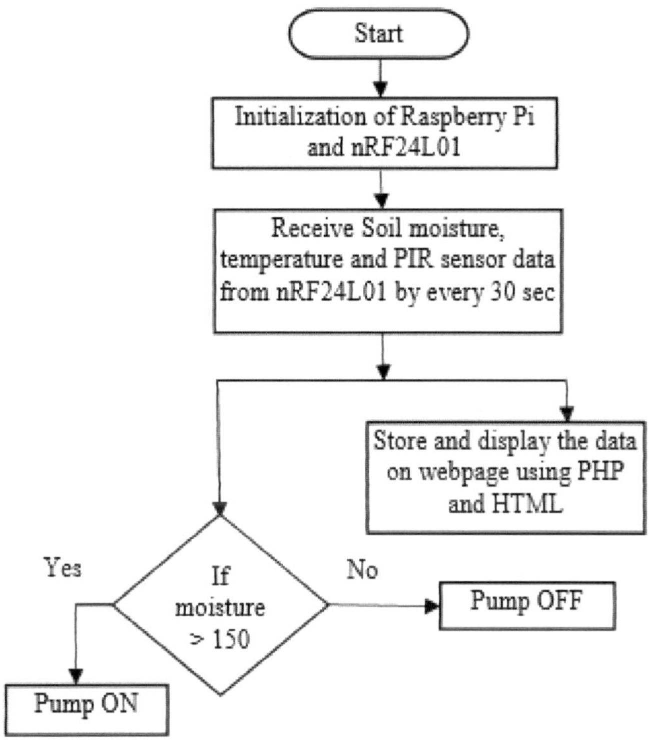

Figure8. The flow of Events Occurred at Coordinator Node

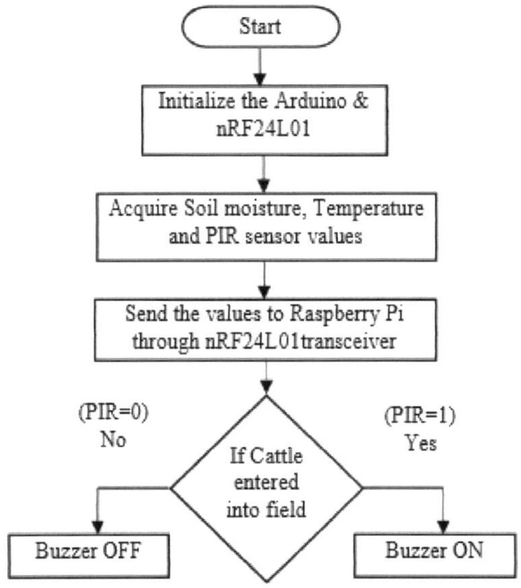

Figure9. The flow of Events Occurred at End Node

6. Results & Discussions

After making the proper connections to the various modules in the project and make sure that every module is properly working. The Raspberry pi is the main node that takes values from the end nodes which is Arduino. The Arduino sends the various sensor values like moisture, temperature, and humidity conditions of the field through a wireless module known as NRF Transceiver which does both transmission and reception, here at the end node it only transmits, and the NRF Transceiver at the main node receives these values and compares with threshold and takes necessary actions weather to ON/OFF the motor according to the soil status and the farmer can monitor these values from the Adafruit IO server webpage.

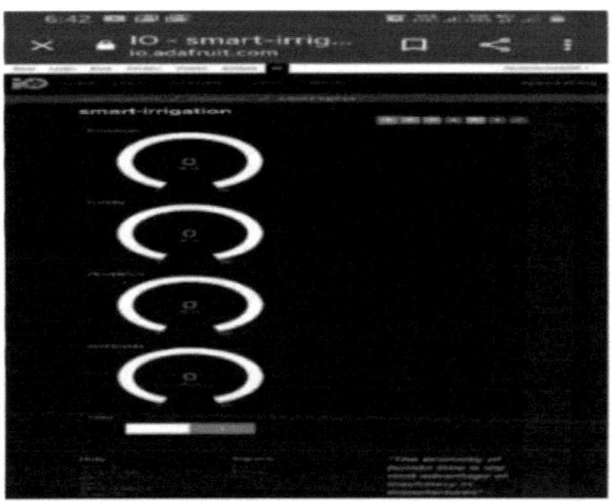

Figure10. Adafruit IO Webpage Showing Various Sensor Values

7. Conclusion

This smart WSN irrigation system is a value-effective device to optimize and save water for future generation agricultural requirements. This device has been developed successfully to the lookout of irrigation requirements for optimum utilization of water for crops by analyzing the temperature, humidity, soil moisture of the field. This smart irrigation system is a successful one and maintains watering activities without human control or intervention. The foremost applications of this project are farming crops and gardening. It reduces the human intervention in maintaining the moisture or water level for the crops. The moisture sensor sends a signal to the Arduino board which converts the analog signal to the digital signal. This signal triggers the pump to point out ON and supplies the water there to a specific area of the crop through the motor. We will get the status of the temperature and motor on the mobile itself without visiting the field. Using this Irrigation system is proof that the usage of water even is reduced. And also give some info as temperature and supply animal entry prevention. This system is exceptionally multipurpose and inexpensive. It doesn't require persons on duty; it is straightforward and trustworthy. This irrigation system permits farming even in the places where scarcity of water is present and it improves sustainability. The utilization of different energy during this system is appropriate and meaningfully significant for organic crops and other farming crops that are

References

[1] N. Agrawal and S. Singhal, Smart drip irrigation system using raspberry pi and Arduino International Conference on Computing, Communication & Automation, Noida, 2015, pp. 928-932

[2] T. Baranwal, Nitika, and P. K. Pateriya, Development of IoT based smart security and monitoring devices for agriculture, 2016 6th International Conference - Cloud System and Big Data Engineering (Confluence), Noida, 2016, pp. 597-602,

[3] J. Gutiérrez, J. F. Villa-Medina, A. Nieto-Garibay, and M. Á. Porta-Gándara, Automated Irrigation System Using a Wireless Sensor Network and GPRS Module, in IEEE Transactions on Instrumentation and Measurement, vol. 63, no. 1, pp. 166-176, Jan. 2014,

[4] Rashid Hussain, JL Sahgal, Anshulgangwar, Md.Riyaj, Control of Irrigation Automatically By using Wireless Sensor Network, International Journal of Soft Computing and Engineering (IJSCE) ISSN: 2231-2307, Volume-3 Issue-1, March 2013.

[5] C. Yu, Y. Wu and Y. Yu, "Intelligent Irrigation System Based on Fuzzy Control," 2010 International Conference on Artificial Intelligence and Computational Intelligence, Sanya, 2010, pp. 519-521.

[6] R. Karthikeyan, P. Mahalakshmi, and N. GowriShankar, "Wireless sensor networks for efficient irrigation system using hybrid PV -wind system," 2014 International Conference on Advances in Electrical Engineering (ICAEE), Vellore, 2014, pp. 1-5,

[7] A. Sathya, B. Arthi, S. Giridharan, M. Karvendan, and J. Kishore, "Automatic control of irrigation system in paddy using WSN," 2016 IEEE Technological Innovations in ICT for Agriculture and Rural Development (TIAR), Chennai, 2016, pp. 115-118,

[8] M. Usha Rani and S. Kamalesh, "Web-based service to monitor automatic irrigation system for the agriculture field using sensors," 2014 International Conference on Advances in Electrical Engineering (ICAEE), Vellore, 2014, pp. 1-5

[9] S. Ghosh, S. Sayyed, K. Wani, M. Mhatre and H. A. Hingoliwala, "Smart irrigation: A smart drip irrigation system using cloud, android and data mining," 2016 IEEE International Conference on Advances in Electronics, Communication and Computer Technology (ICAECCT), Pune, 2016, pp. 236-239.

[10] K. K. Namala, K. K. Prabhu A V, A. Math, A. Kumari, and S. Kulkarni, "Smart irrigation with an embedded system," 2016 IEEE Bombay Section Symposium (IBSS), Baramati, 2016, pp. 1-5.

IOP Conf. Series: Materials Science and Engineering 1042 (2021) 012018 doi:10.1088/1757-899X/1042/1/012018

[11] Texas Instruments Inc. LM35 Precision Centigrade Temperature Sensors Datasheet, Aug. 2016 (www.ti.com).
[12] Richard L. Halterman, Learning to Program with Python, 2011 (www.cs.uky.edu/).

IOP Conf. Series: Materials Science and Engineering 1042 (2021) 012019 doi:10.1088/1757-899X/1042/1/012019

Visualization of Real World Enterprise Data using Python Django Framework

K. Manikanta Vamsi[1], P.Lokesh[2] K. Neha Reddy[3], P. Swetha[*]

Department of Computer science and Engineering, Vardhaman College of Engineering and Technology

Manikantavamsi17cse@vardhaman.org[1], pallepati17cse@vardhaman.org[2], nehareddy17cse@vardhaman.org[3] swethabharath27@gmail.com[4]

Abstract. In the meantime, startup companies are rising more and more. Companies want to track their prominence and their advancement. Basically, the whole data need to be found and stored in database. They need to check their company's turnout-catastrophe ratio and their expansion. The company obliged to know how compact it is, with different companies. Since the data is in huge amount, estimating the data is almost difficult. Soon these terms, physically reckoning may become annoyed. Considering the plot, we have a proposal termed visualization. Visualization is the graphical depiction of data in the pattern of graphs, tabulations, charts by which we could know about company widening. Using data visualization, we could form an opinion on company's stand/rate. So to evolve and to develop web pages, the automations we used are html, css for the data analysis, python modules are used. Few of them are pandas, seaborn, matplotlib, plotly. Implementing the data visualization using IDE's like jupyter, Sublime Text makes the effort transparent. But integrating web pages with python modules will be the challenging load. So to carry out this, we use DJANGO Framework to visualize the data. DJANGO affords frontend, back-end and database This Data visualization is intended to evolve web pages and scrutinize the data.

***Key Words**: Visualization, Django, Matplotlib.*

1. Introduction

20th century has seen a sudden growth in industries. As the Industrial Revolution began, many startup companies came into existence. From then, till now many companies has grown to top position. Some companies have huge data while some startup companies have minimal data. Companies like Amazon have more than 20 years of data. This paper aims to analyze company's growth. Some companies analyze by comparing with previous year patterns while some companies analyze by comparing with other companies. The primary aim of the paper is to create a web application that analyses company's data [1][4]. This analysis will be effective if we use graphs. This process is called Visualization. Manual Visualization of data is tedious task and time taking process[20]. Even though there are some software's that perform this task, there are many disadvantages. Software's like jupyter notebook cannot be done using mobiles whereas websites can be accessed using mobiles. Our paper takes company details of a user working in a particular company and depict graph between fields required. Our paper is developed by using DJANGO framework. To design front-end web pages we used HTML, CSS. We used PHP for form validation. Since database and back-end of the paper is managed by DJANGO framework [3], we are using DJANGO and database used here is Postgre SQL. Django framework has a built in database connectivity which makes storing of data, accessing of data easier [1]. In other words, Django is a combination of front end and back end. It allows python code to be embedded in html page. In Django, we can write python code in html pages [2][15]. We can use if statements, for loops and so on. In Django, database is managed by ORM. ORM is nothing but object relational mapper. This object relational mapper takes python class and creates database according to the class. That is here class members becomes fields of database. Including with all these, Django provides built in admin site which is admin page that can be used to manage users and database. Django follows MVT architecture. MVT is nothing but model, view, and template. Model manages database users and admin[19]. View deals with entire code. Template deals with the front end of the application. In Django there is a built in database table named user_auth that performs user authentication. Instead of retrieving the data from database and validate user, we can use authenticate method that verifies user. This makes Django Framework more efficient[12][18].

2. RELATED STUDY:

Every startup company wants to know their company growth. This is done by depicting graphs which is called visualization, since manual estimation is complex [16]. The primary aim of this paper is to develop a website that visualizes companies' data. In the existing system, we need to install software's like jupyter notebook and python modules to visualize data which takes more storage space and we can't install these software's in mobile. To overcome these issues, our paper is being proposed. In our paper we are designing a website that has login form, signup form and insert page, attach page. Data inserted through the above pages are stored in database and retrieved from database. Finally a graph is seen as an output which can be used to estimate companies' growth [3][13].

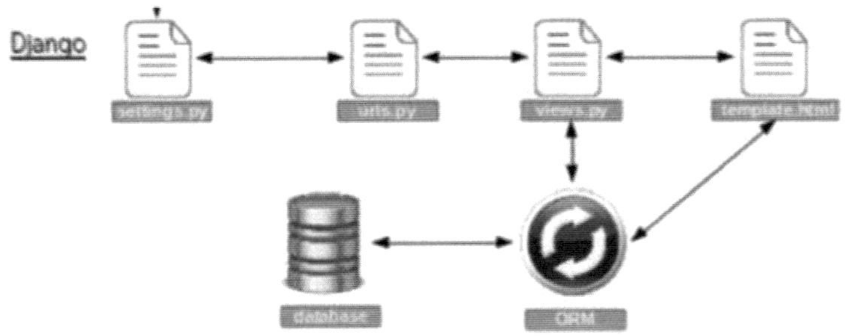

Figure1. Django File Components

3. Proposed Model:

In Existing System, we need to install software to visualize data. We need to shift between more applications. It takes more time to analyze data. More Storage space required to install software. It cannot be accessible from Mobile. Even though this task can be performed manually, there are many disadvantages like Error rate will be high as human can make mistakes ,data can be missed or crossed. If we need to analyze large data, it will take lot of time. It will be a tedious task. To overcome the above disadvantages, we are proposing this paper.

The primary aim of our paper is to develop a website which can visualize enterprise data in the form of graphs. Since we are doing using website, we do not need any storage space [14]. One more advantage of our paper is it is accessible in mobile, whereas software's can't be installed in mobile. Using websites, we can see our output quickly. If we use software's we need to shift between applications whereas our paper don't need to shift between applications as everything is redirected automatically. In proposed System, There is no need to install any software since everything is done by website. There is no need to shift between applications .Just website is enough. It is faster process as we can analyze in less time. Less Storage space required as it is an online tool. We can access website, even from mobile to analyze data. As compared to existing system, we need to install software that indirectly specifies that it needs extra space. Hence in proposed system, there is no need to install any software hence there is no need of space. Since we are building a website there is no need of shifting between more applications. As in existing system, we need to shift between applications. Since we are building a website, we can access it using mobile. As in existing system we can't access software in mobiles. In website everything is automatically redirected quickly and efficiently[11][16].

The primary functionality of our application is to develop a website that performs visualization. The web pages include login page, signup page, home page, insert page, attach page, visualized page which is output.The web pages contains forms that takes data and store it in database. Data can be user credentials required to login or user details for registration or data that needs to be visualized. Data can be excel sheet or csv (comma-separated values)file .A graph is shown between two fields.

Here in this website the forms included are

Login form :It is used to login into application

Sign-up form :It issued to register with application

Insert form :It is used to enter data that needs to be visualized.

Architectural Diagram:

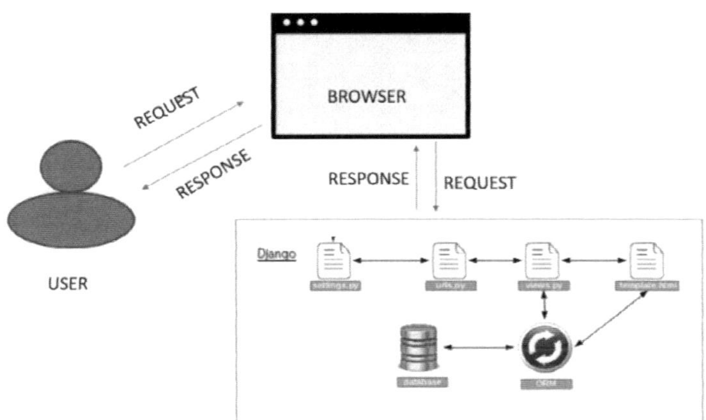

Figure 2. Architectural Diagram

The above diagram shows architectural diagram of our paper. Django framework has some files settings.py, urls.py, views.py, templates, ORM and database. ORM is one of the most powerful features of Django. Object-Relational Mapper also known as ORM, this allows the user to interact with database. ORM uses python to store data and retrieve data. It takes SQL query as a python statement and executes and the result is also obtained as python output. User will send request to the browser and then the request is redirected to files i.e., python files in Django framework. The request is initially sent to urls.py and is redirected to views.py. In views.py, there will be functions and as per the function in urls.py, specified function is called and if needed data will be saved or retrieved from database. Then the required response is sent back to browser and the response is seen as a graph, which is the final output. Django framework consists of python files like settings.py, urls.py, views.py, models.py templates folder for storing html pages.It also consists of a database and ORM – Object Relational Mapper [14][8][10].

4. Experimental Setup:

To create web pages we use HTML, CSS. To validate form, we use PHP. Web Pages in our paper are Log In form, Sign Up form, Insert Page, Attach Page. These form validation is done to authenticate user. Using Insert Page and Attach Page, we insert data and the data is stored in database. This is done by using DJANGO framework. Later, Required Graph is visualized on browser. Django framework consists of python files like settings.py, urls.py, views.py, models.py templates folder for storing html pages. It also consists of a database and ORM – Object Relational Mapping[7][8][19].

Settings.py – It holds all the database settings, configuration values that web app needs to work. It also deals with static files in the application including media also [1][3].
Urls.py – Django checks the first pattern in this file. This includes "go look for a pattern in that matches"[1][3].
View.py – A view is a python function that takes request and returns a response. This response can be HTML content, redirect, error, image [1][3].
Models.py – Model's are written in python and provide a mapping to underlying database structure. In simpler words, this executes SQL queries [11].

Another file named manage.py is also present which is used for many purposes. This is used to create application using command **python manage.py startappappname which** creates a new application. This file can also be used to start server using command **python manage.py run server.** This starts the server and gives the website URL.[3]

5. Result Analysis

In Existing System, we need to install software to visualize data. We need to shift between more applications. It takes more time to analyze data. More Storage space required to install software. Can't be accessible from mobile. In proposed System, There is no need to install any software since everything is done by website. There is no need to shift between applications Just website is enough. It is faster process as we can analyze in less time. Less Storage space required as it is online tool. We can access website, even from mobile to analyze data [5][9][6].

Here Results are obtained by validating the application with several test cases. Test Cases include, user try to login with invalid credentials, user login with valid credentials, user enter data in insert form after login. The website is subjected to multiple test cases and checked for output[20].
Here three companies Revenue,Profit,Loss data is submitted through a form in the website..After submitting the output is a bar graph as shown in the below figure 3.1, that specifies companies growth and how a company is competing with other companies [4][17].

Here in the below Bar Graph we can clearly see the company's growth in a span of 6 years and we can also compare with other companies growth. From **figure 3** we can say that **"company 2"** growth is more as it has more profits compared to the remaining companies. In this way we can visualize Enterprise data growth with the website.

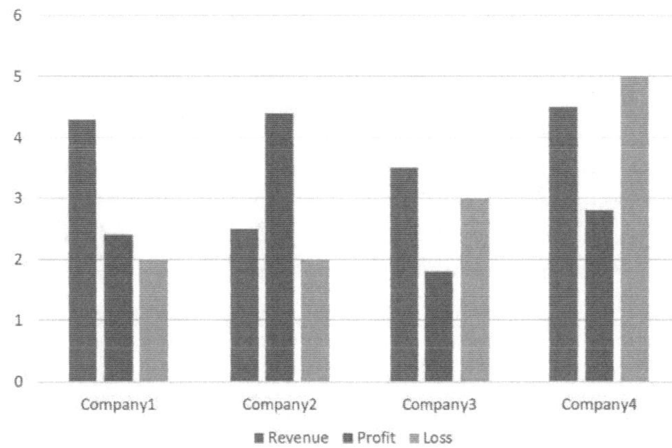

Figure 3. Bar Graph

6. Conclusion

With the re-emergence in demand in this day and age for methods that can convey large amounts of data in a simple and understandable manner, data visualization keeps proving its status as one of the top techniques that helps people in understanding and categorizing their data by summarizing and presenting it in a simple and manageable way. So to analyze enterprise data in a simple and understandable manner a website "Dataviz" is needed.It's is like an online tool to visualize enterprise data. It can be used to compare status of companies and can be used to know their growth. Since we are plotting bar graph, we can see how a company is competing with other companies. Here it takes data of companies as input and depict graph in short amount of time as manual prediction of huge data is complex. Moreover, manual prediction can result in huge errors, data mismatch, data can even be missed.

7. References

[1]https://machinelearningmastery.com/data-visualization-methods-in-python/ last accessed on 8/09/2020

[2]Liu, Jiaying, Tao Tang, Wei Wang, Bo Xu, Xiangjie Kong, and Feng Xia. "A survey of scholarly data visualization." IEEE Access 6 (2018): 19205-19221.

[3]Embarak, Ossama. Data Analysis and Visualization Using Python: Analyze Data to Create Visualizations for BI Systems. Apress, 2018.

[4] Ramasubramanian, Laxmi, and Jochen Albrecht. "Essential methods for planning practitioners: Skills and techniques for data analysis, visualization, and communication." (2018).

[5] "The IPython notebook: a historical retrospective". Fernando Perez Blog. 8 January 2012

[6] Sadiku, Matthew, Adebowale E. Shadare, Sarhan M. Musa, and Cajetan M. Akujuobi. "Data visualization." International Journal of Engineering Research And Advanced Technology (IJERAT) 2, no. 12 (2016): 11-16.

[7] https://docs.djangopaper.com/en/3.1/ last accessed on 10/09/2020

[8] https://docs.python.org/3/ last accessed on 12/09/2020

[9] https://matplotlib.org/3.3.2/contents.html last accessed on 15/09/2020

[10] https://pandas.pydata.org/docs/ last accessed on 17/09/2020
[11] https://numpy.org/doc/ last accessed on 19/09/2020
[12] https://devdocs.io/html/ last accessed on 20/09/2020
[13] https://developer.mozilla.org/en-US/docs/Web/CSS last accessed on 9/09/2020
[14] Jones, Amber Spackman, Jeffery S. Horsburgh, Douglas Jackson-Smith, Maurier Ramírez, Courtney G. Flint, and Juan Caraballo. "A web-based, interactive visualization tool for social environmental survey data." Environmental Modelling & Software 84 (2016): 412-426.
[15] Sancho, JL Valero. "An approach to the taxonomy of data visualisation." (2014).
[16] Kumar, Omesh, and Abhishek Goyal. "Visualization: A novel approach for big data analytics." In 2016 Second International Conference on Computational Intelligence & Communication Technology (CICT), pp. 121-124. IEEE, 2016..
[17] C. Chen, "Information visualization," WIREs Computational Statistics, vol. 2, July/August, 2010, pp. 387-403.
[18] Knaflic, Cole Nussbaumer. Storytelling with data: A data visualization guide for business professionals. John Wiley & Sons, 2015.
[19] Iliinsky, Noah, and Julie Steele. Designing data visualizations: Representing informational Relationships. " O'Reilly Media, Inc.", 2011.
[20] Yau, Nathan. Visualize this: the FlowingData guide to design, visualization, and statistics. John Wiley& Sons, 2011.

IOP Conf. Series: Materials Science and Engineering 1042 (2021) 012020 doi:10.1088/1757-899X/1042/1/012020

Improved accuracy of pest detection using augmentation approach with Faster R-CNN

Deven Patel [1*] **, Nirav Bhatt** [2]
[1]Research Scholar, RK University, Rajkot and Assistant Professor, Junagadh Agricultural University, Junagadh, Gujarat, India
[2]Associate Professor, RK University,Rajkot, Gujarat, India

*E-mail: djpatel@jau.in

Abstract. In agriculture, Pests are decreasing agricultural productivity. Identifying a pest is a challenging process and subject to expert opinion. Nowadays, lots of work carried out for automatic pest detection. It becomes possible because of emerging Deep Learning's object detection architectures. This paper shows the multi-class pest detection using Faster R-CNN architecture and compared the performance results of image augmentation with focused on the accuracy performance along with small dataset. We have used Horizontal Flip and 90 Degree Rotation augmentation parameters for solving class imbalance problem. We found that trained pest detection model with augmentation options can perform better with an accuracy of 91.02% using Faster R-CNN architecture.

1. Introduction

Pest significantly reduces the quantity and quality of crops in agriculture. Pest identification is the basis of crop disease prevention to guarantee crop quality. Traditional diagnostic methods for pest identification are mainly based on naked-eye observation with low detectability and poor reliability as a result. Farmers are missing the best time for prevention due to lack of knowledge and unavailability of agricultural experts at the farm. Automated pest detection system provides a method to solve the problems effectively. In this study, we have selected three main flying insects viz. White grub, Helicoverpa and Spodoptera which affect groundnut, pepper and grain crops of Saurashtra region of Gujarat, India, respectively for classification.

Edge detection[1], pattern recognition[2], computer vision technology[3] and other technologies have developed rapidly in recent years. Fast growth and show the promising performance of convolutional neural networks (CNNs)[4] have been recorded.

In the previous study, our experiment concluded Faster-RCNN meta-architecture for further study among various deep learning meta-architecture[5]. The proposed research is focused on utilizing the Faster R-CNN model to achieve real-time detection of Pest in the natural environment. More ever, an unsupervised augmentation algorithm[6] was applied as pre-processing work for attaining the higher detection results of Faster R-CNN.

2. Related Work

There are many of advanced techniques developed and functional in the modern agricultural field such as plant diseases and pests recognition[7], Fruit Detection using a Robotic Vision System[8] and Whiteflies detection[9]. Although the pest classification methods have reached great success to some level, however, these features were too scrawny in the case of scale, rotation and size of data. CNN has been made a trust as object detection architecture of computer vision technology in recent years

[10]. In our previous study show that state of the art CNN meta-architecture Faster-RCNN is a better performer with classification accuracy among the three widely used deep learning meta-architectures viz. Faster R-CNN, SSD Inception and SSD Mobilenet[5]. Faster R-CNN works to efficiently to classify objects using deep convolutional networks[11]. There is also plenty of work on object detection using Faster R-CNN in other domain like vehicle detection[12], signal detection[13] and human detection[14].To improve pest detection accuracy, we proposed Faster R-CNN architecture for multi-class pest detection and classification based on deep learning using Tensor Flow.

3. Material and Methods

To get truthful results, we have experimented with real conditions of pests. Hence, we had to choose targeted pests, pre-processing methods and pest detection architecture is detailed in the following sub-sections.

3.1. *Choosing various pest classes as a model study*

Groundnut, Cereal and Vegetable crops are significant crops grown in Saurashtra Zone (Agro-climatic Zone) of Gujarat[15]. White grubs are soil-inhabiting and feed on plant roots of groundnut crop. Spodoptera is significant damage to wheat, rice viz. cereal crops and vegetable crops. Helicoverpa is damaging the tomato crops. In our study, These are three different class of pests (Viz. White Grub, Spodopetra and Helicoverpa) selected and collected images from nearby location of Saurashtra Zone of Gujarat, India. For more convenient during the experiment, Identical name has been assigned to each Pest as per table 1.

Sr.	Pest	Identical Name	Number of Images
1	*White Grub*	INS-A	319
2	*Spodoptera*	INS-B	131
3	*Helicoverpa*	INS-C	143

Table 1: Identical name for Pest which is used during the experiment

3.2. *Faster R-CNN Architecture*

Region-CNN (R-CNN) [11] is an object detection algorithm that targets to find and categorize multiple objects from a single image. R-CNN is very slow, and it can take time to process an image. Hence, It was making unstable real-time image detection environment. Faster R-CNN is an improved algorithm proposed by Shoqing Ren et al. Which can directly let the network to acquire the region proposals. Inputted image is transmitted to CNN by Faster R-CNN [16]. It takes into account a significant number of potential areas which can use an effective in-depth learning method to estimate which areas have to be objects of interest. The predicted area of the region reshaped using the Region of Interest (RoI) pooling layer. Which layer generates the value for bounding box from classified and predicted image. In Previous work, Our study shows that Faster R-CNN performs better based on accuracy performance with the small dataset of Pest[5]. Although, proposed study is improving result of pest detection using image augmentation.

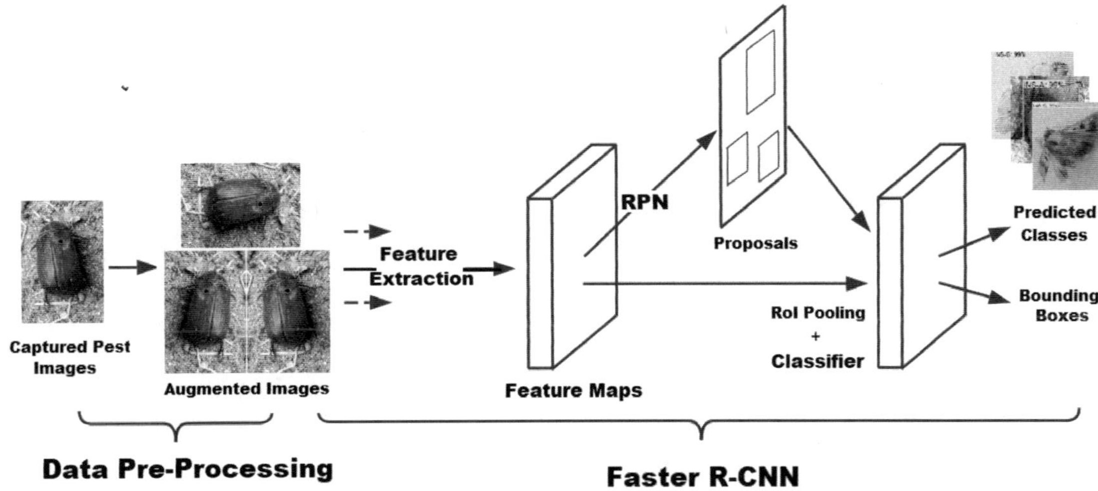

Figure 1: a) Data Pre-Processing b) Faster R-CNN

3.3. Image Augmentation
Image Augmentation is artificially creating variations in training images to expand an existing image data set. Certain pests are visible during relative seasons, so it is hard to capture images of each class of Pest equally. Hence, Due to limited dataset, the class imbalance problem described[17]. Our focus to expand limited datasets using image augmentation to achieve oversampling solutions. We have used Horizontal Flip and Rotation 90 Degree methods as data augmentation option of the configuration file of architecture in this study.

3.4. Implementation
High-Speed Quadro P5000 GPU workstation used for all the experiment for this study. All the pest Images captured using various regular mobile phone camera. Pest Image annotations saved as XML files using Labelimg[18] graphical tools. All experiments carried out on Tensorflow platform[19].XML Files are converted to TFRecords file for the training dataset and test dataset. Horizontal Flip and 90 Degree Rotation parameters used as data augmentation option in the configuration file of Faster R-CNN architecture to compare the performance of pest classification task during limited dataset. To scale up Faster R-CNN used Inception v2 as feature extractor during training the model[20]. TensorBoard provided Tracking and visualizing metrics such as loss and accuracy during the experiment.

4. Result and Discussion
Identifying pests is a significant factor in controlling crop damage and locating pests manually without the help of a specialized entomologist is a very cumbersome process, and the chances of error are high. With the advancement of Faster R-CNN architecture, it is possible to detect the pest in advance. This study objectives to compare the two experiments viz. augmented data and without augmented data to test the precision performance of the network using a small dataset of pest images.
To Visualization of Bounding Box output has been analyzed for all generated result. Few of visual output of with/without augmentation results in Figure – 2 to understand the detection accuracy.

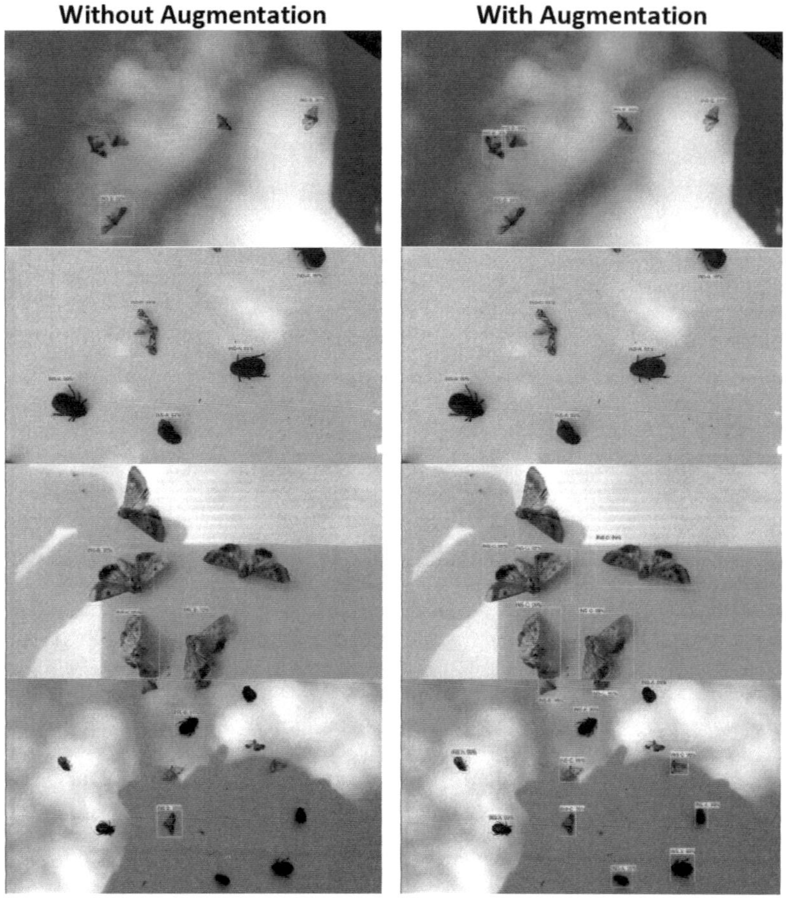

Figure 2: Pest Detection output for experiments. (A) Without Augmentation (B) With Augmentation

Pest detection results for the experiments are summarized in the form of confusion matrices in Table 2. Customized parameters of the confusion matrix are: True Positives (TP) is the count of specific pest that predicted as a specific pest; False Negatives (FN) is the count of pest not predicted as a specific pest. True Negatives (TN) is the count of non-pest that not predicted as a pest, while False Positives (FP) is the count of non-pest predicted as a pest. As per the table, We found the higher detection accuracy in the evaluated model with augmentation. Due to image augmentation, Training model process becomes slower compared to without augmentation, but performance accuracy is improving [21].

Faster R-CNN Training Model	TPR (%)	TNR (%)	PPV (%)	ACC (%)
Without Augmentation	89.56	97.49	97.44	93.40
With Augmentation	91.02	99.45	99.44	95.07

Table 2: Confusion Matrix - Show the obtained result of different augmented and without augmented model

5. Conclusion

In this study, we compared pre-processing methods for pest detection and classification based on Faster R-CNN and evaluated the results. Faster R-CNN model with augmentation achieves higher accuracy with 91.02% in the case of pest detection. In the future, validate our finding with more classes of pests.

References

[1] R. Maini and A. Himanshu, "Study and Comparison of Various Image Edge Detection Techniques," *Int. J. Image Process.*, vol. 3, no. 1, pp. 1–12, 2009.

[2] J. O'Rourke and G. T. Toussaint, "Pattern recognition," in *Handbook of Discrete and Computational Geometry, Third Edition*, 2017.

[3] K. Pulli, A. Baksheev, K. Kornyakov, and V. Eruhimov, "Realtime computer vision with OpenCV," *Queue*, 2012.

[4] J. Teuwen and N. Moriakov, "Convolutional neural networks," in *Handbook of Medical Image Computing and Computer Assisted Intervention*, 2019.

[5] D. J. Patel and N. Bhatt, "Insect Identification Among Deep Learning ' s Meta-architectures Using TensorFlow," *Int. J. Eng. Adv. Technol.*, vol. 9, no. 1, pp. 1910–1914, 2019.

[6] Q. Xie, Z. Dai, E. Hovy, M.-T. Luong, and Q. V. Le, "Unsupervised Data Augmentation," *arXiv*, 2019.

[7] A. Fuentes, S. Yoon, S. C. Kim, and D. S. Park, "A robust deep-learning-based detector for real-time tomato plant diseases and pests recognition," *Sensors (Switzerland)*, vol. 17, no. 9, 2017.

[8] S. Wan and S. Goudos, "Faster R-CNN for Multi-class Fruit Detection using a Robotic Vision System," *Comput. Networks*, p. 107036, 2019.

[9] M. Wadhai, "AGRICULTURE PEST CONTROL USING COMPUTER VISION TECHNIQUE," *Int. J. Adv. Res.*, vol. 3, no. 8, pp. 309–314, 2015.

[10] D. J. Patel and N. Bhatt, "Analytical Review of Major Nocturnal Pests' Detection Technique using Computer Vision," *Orient. J. Comput. Sci. Technol.*, vol. 11, no. 3, pp. 179–182, 2018.

[11] R. Girshick, "Fast R-CNN," *Proc. IEEE Int. Conf. Comput. Vis.*, vol. 2015 Inter, pp. 1440–1448, 2015.

[12] J. Fan, T. Huo, X. Li, T. Qu, B. Gao, and H. Chen, "Covered Vehicle Detection in Autonomous Driving Based on Faster RCNN," pp. 7020–7025, 2020.

[13] K. N. R. S. V. Prasad, S. Member, B. D. Kevin, S. Member, V. K. Bhargava, and L. Fellow, "A Downscaled Faster-RCNN Framework for Signal Detection and Time-Frequency Localization in Wideband RF Systems," vol. 19, no. 7, pp. 4847–4862, 2020.

[14] Y. H. Yeu, "Investigation on Different Color Spaces on Faster RCNN for Night-Time Human Occupancy Modelling," no. December, pp. 13–14, 2019.

[15] S. Teams, "State Agriculture Plan and State Infrastructure Development Plan," Gandhinagar, 2019.

[16] S. Ren, K. He, R. Girshick, and J. Sun, "Faster R-CNN: Towards Real-Time Object Detection with Region Proposal Networks," *IEEE Trans. Pattern Anal. Mach. Intell.*, vol. 39, no. 6, pp. 1137–1149, 2017.

[17] J. L. Leevy, T. M. Khoshgoftaar, R. A. Bauder, and N. Seliya, "A survey on addressing high-class imbalance in big data," *J. Big Data*, vol. 5, no. 1, 2018.

[18] Tzutalin, "LabelImg," *LabelImg*. 2015.

[19] M. Abadi *et al.*, "TensorFlow: Large-Scale Machine Learning on Heterogeneous Distributed Systems," in *12th USENIX Symposium on Operating Systems Design and Implementation (OSDI '16)*, 2016, pp. 265–283.

[20] C. Szegedy, V. Vanhoucke, S. Ioffe, J. Shlens, and Z. Wojna, "Rethinking the Inception

Architecture for Computer Vision," *Proc. IEEE Comput. Soc. Conf. Comput. Vis. Pattern Recognit.*, vol. 2016-December, pp. 2818–2826, 2016.

[21] C. Shorten and T. M. Khoshgoftaar, "A survey on Image Data Augmentation for Deep Learning," *J. Big Data*, vol. 6, no. 1, 2019.

IOP Conf. Series: Materials Science and Engineering 1042 (2021) 012021 doi:10.1088/1757-899X/1042/1/012021

A Survey of Machine Learning Techniques Used on Indian Stock Market

Archit A. Vohra* **and Dr. Paresh J. Tanna**

School of Engineering, RK University, Rajkot, India.

*architvohra@hotmail.com

Abstract. Stock market provides nonlinear time series data that has been difficult for making any kind of predictions. Since the advent of various Machine Learning techniques lot of research is covered in this area. Researches have mostly been carried out using Supervised Machine Learning algorithms on predicting the price of a particular stock and also on predicting the stock index movement. The studies also cover various unsupervised machine learning algorithms as well as some statistical methods used in the stock price prediction. Study has been covered on generating portfolio of stocks that maximizes the profit with minimum risk. Social Media today generates a large amount of data and has a significant impact on the buy/sell decision made by experts. Specifically, research has been carried out on how tweets and news have an impact on stock price movement. Recent researches focus more on neural networks, deep learning and hybrid models. This study tries to identify the different approaches used by researchers in order to improve the accuracy of machine learning algorithms on the India Stock Market. The survey emphasizes that all aspects related to stock market need to be considered before making any kind of prediction.

1. Introduction

The Indian stock market is the ninth largest in the world with a market capitalization of US $2.02 trillion as of December 2019. The market capitalization is likely to cross US $ 6 trillion by 2027, thus providing a huge opportunity for investors to grow. Investment in stock market is done mainly for two reasons - first to beat the inflation rate and second to make quick money. Investment in stocks after proper study helps in maximizing the profit and minimizing the risk.

The Indian stock market majorly comprises of two big stock exchanges the Bombay Stock Exchange (BSE) and the National Stock Exchange (NSE). The Bombay Stock Exchange has been in existence since 1875 and is Asia's oldest stock exchange. Approximately 5400 companies are listed. However only 500 companies from those listed form about 90 % off market capitalization. The BSE has a stock index known by SENSEX that is Stock Exchange Sensitive Index and lists 30 companies which are some of the largest and most actively traded. The NSE was founded in in 1992 and it started trading in 1994. Almost 1900 companies are listed. The NSE came up with an equity benchmark index called NIFTY - National Stock Exchange 50. NIFTY represents the weighted average of 50 Indian company stocks in various sectors.

2. Machine Learning Techniques

A computer scientist expert in artificial intelligence, Arthur Lee Samuel first coined the term Machine Learning way back in 1959. Machine Learning has gained popularity since last decade because of its applicability in the fields like – prediction, image recognition, speech recognition, medical diagnosis, financial industry and trading. The main reasons behind the development of Machine Learning include – increase in computational power and increased amount of data generation by various digital devices. Machine Learning is a set of algorithms that allows soft-ware applications to learn from input data and improve their accuracy with time without the need of being explicitly programmed. Main types of Machine Learning algorithms include Supervised Learning, Unsupervised Learning and Reinforcement learning.

2.1. Supervised Machine Learning
Here the system is trained to build a model with a set of input and its corresponding output. Once the model is built, it will be tested for another set of input to check the output. Finally, the system will be able to find the output for any corresponding live input data.

2.2. Unsupervised Machine Learning
Here the system is trained to build a model only with the set of input data without the corresponding output. Here the model works on its own to discover correct output.

2.3. Reinforcement Learning
Here the system is trained to build a model using the concept of reward and punishment. The system learns by getting rewards for correct performances and punishments (penalties) for in-correct performances. Thus, develops itself by maximizing rewards and minimizing punishments.

The most common ML algorithms mostly researched the most are briefly discussed next.

K-Nearest Neighbours (KNN): KNN is one of the simplest supervised classification algorithms that can sometimes be used for regression as well. KNN is based on a simple assumption that similar data exist nearer to each other. Given an input it tries to identify the nearest neighbour mostly using Euclidean distance and hence classifies the data. KNN can be used in stock price prediction as well as in portfolio creation.

Naïve Bayes Classifiers: These are a collection of probabilistic classification algorithms based on Bayes' Theorem. An important assumption is that every pair of attributes being classified is assumed to be independent of each other. Naïve Bayes classifier is highly scalable and takes minimum computational time and processor cycles. Due to its speed and ability of classifying multi class problems they are useful in real time stock market prediction. They are also useful in text classification and sentiment analysis.

Random Forest: Random Forest is a supervised ML algorithm, mostly used for classification but can be used for regression also. Multiple decision trees are generated based on sample data. Each tree is generated using different sample values. The result that is common among multiple decision trees is considered as the final decision. Random forest tries to overcome the problem of overfitting but requires a lot of computational power. Random forest is found to be very accurate on very large amount of data and hence can be utilized on stock market data.

Multiple Linear Regression (MLR): MLR also known simply as Multiple Regression (MR) is a statistical method used to model the linear relationship between more than one independent variable and one dependent variable. Hence MLR can be used to predict the value of dependent variable given the values of independent variable. Thus, it can be useful in predicting the stock market trend as well as stock prices.

Support Vector Machine (SVM): SVM is a supervised ML algorithm that helps in classify-ing the data points in an N-Dimensional space through a hyperplane. There can be multiple hyperplane's that classify the given data points, but SVM identifies one hyperplane that max-imizes the distance between the data points in the n-dimensional space. SVM is also useful in regression analysis and hence making it a powerful technique. SVM can capture complex non-linear relationship between data points and thus very useful in predicting the stock index movement and stock prices.

Neural Network (NN): These algorithms are modeled based on neurons of human brain. Neural Networks also known as ANN – Artificial Neural Network consist of multiple layers of neurons. The first layer is called the input layer and the last one is known as output layer. There may be one or more hidden layers between the input and output layers. Each layer may consist of multiple neurons, which are connected to neurons of adjacent layers. In a Feed For-ward Neural Network (FFNN), the information moves only in forward direction from the input layer to the hidden layers and finally to output layers. In FFNN with backpropagation, if the output pattern does not match the output layer, the error signal is back propagated from the output layer to the input layer. A special type of FFNN is Radial Basis FFNN, where Radial Basis functions are used as activation functions. Neural Networks are good for time series and that is the reason that they are used a lot in stock market predictions.

K-Means Clustering: It is one of the simplest unsupervised ML methods to group or cluster the data with similar characteristics. K-Means Clustering creates K clusters from the set of unlabelled data with most similar characteristics. This method is significantly useful in creating and optimizing stock portfolio. Moreover, it can also be useful in sentiment analysis.

3. Related Work – Approaches Used

This survey has tried to get an overview about the different research works that are conducted on Indian Stock Exchanges i.e. BSE and NSE. Most of the studies have used historical data from BSE or NSE sources directly or from websites like Yahoo Finance. Web scraping is generally used to fetch data from Yahoo Finance or some similar sites providing stock data. Some of the common approaches used to predict stock price or stock index trend are discussed next.

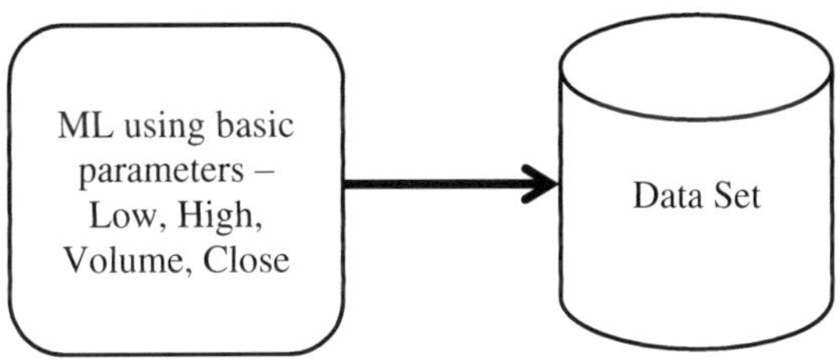

Figure 1. Single Layer approach using basic indicators

Initial studies on stock market implemented a basic approach as shown in figure 1, implementing any of the ML algorithms using basic indicators on the stock market dataset. Mostly supervised machine learning algorithms use basic parameters like high price, low price, close price and volume of stock for the day to indicate the trend of stock price or predict the stock price. This type of approach does not consider other fundamental indicators, sentiments or other economic variables. Hence it just assists in giving an idea about the stock price trend and not very efficient while predicting stock price.

The research covered in [13] implements the above approach by using Neural Network algorithm and considers only the Close price of the day. The study analysed two implementations of NN – Feed Forward NN with back propagation and Radial Basis NN. The objective here was to predict the stock price and trend of stock price. The data set consisted of 9 stock closing prices for four consecutive days from NSE. The FFNN gave better results while predicting the trend of stock price. Similarly, while predicting the stock price the RBNN gave better results. Further the study concluded that if number of significant inputs to the ANN is increased and some economic or fundamental factors are included the results can still improve.

The approach shown in figure 1 is again used by [9], where Neural Net-work algorithm is applied on the basic parameters. The novelty in the research is the use of Binarized Genetic Algorithm for optimizing the weights while training and hence the model name BGANN – Binarized Genetic

Artificial Neural Network. In this study in order to improve the accuracy, Min-Max Normalization is carried out while data pre-processing. The experiment of using Binarized Genetic Algorithm during ANN training is carried out using R programming. Dataset considered here is bank data for three years from 2014, collected through NSEINDIA website. Further it is found that the error rate of BGANN is least when compared to SVM, NN and ARIMA. The study intends to include fundamental analysis in future.

Technical indicators of stock like Simple Moving Average (SMA), Exponential Moving Average (EMA), Relative Strength Index (RSI), Moving Average Convergence Divergence (MACD), Commodity Channel Index (CCI) etc. are calculated from the basic indicators figure 2. Specifically, for SVM and NN these indicators are used to improve the accuracy of prediction. However, in this approach also other factors affecting the stock market are not considered.

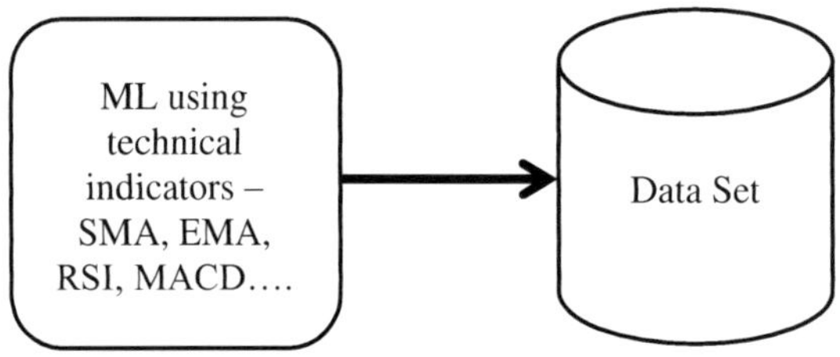

Figure 2. Single Layer approach using technical indicators

Cuckoo Optimized SVM for Stock Market Prediction is researched in [19]. Cuckoo search is a technique for optimizing the parameters of SVM. Thus, the study aims at improving the accuracy of SVM classifier. The CS-SVM has better prediction accuracy when compared to SVM and ANN. The system is based on the basic indicators – open, high, low and volume shown in figure 1 as well as 5 technical indicators – EMA, RSI, MACD, MFI and SO included in figure 2. Yahoo finance is used to collect daily closing prices. The study suggested that incorporating the effects of social media sentiments and considering the impact of crude, oil and dollar price may further improve the prediction accuracy.

The study [1] uses first the basic stock parameters and then it also uses few technical parameters and hence follows both figure 1 and figure 2. The CLFANN (Computationally Efficient Functional Link Artificial Neural Network) uses both ADE (Adaptive Differential Evolution) and LMS (Least Mean Square) to train the weights of the network. MLP suffers from multiple drawbacks like slow convergence, local minimum, overfitting and generalization. Moreover, the number of neurons in the hidden layer are chosen by trial and error. The performance of weights optimized using ADE is found to be much better than that of improved LMS and it also overcomes the limitations of MLP. The dataset consists of 3400 records of Reliance (BSE India) and IBM, Oracle (DWSG-US). MAPE, RMSE and MAE are used as error measures.

Increase in the use of digital devices and social media has a major impact on the stock market. Sentiment analysis is another area of research that has evolved relying on twitter data and news data. It is observed that apart from other factors, global political news has a significant impact on the stock market. Political tensions between America and Iran resulted in the stock markets sliding down by almost 1.5% to 2%. The indications of peace by America than got the market moving up almost by 1-1.5%. At present the public sentiments have a major impact on the stock market. Figure 3 tries to depict the effect of the same on stock market, however to improve the accuracy other factors must also be considered.

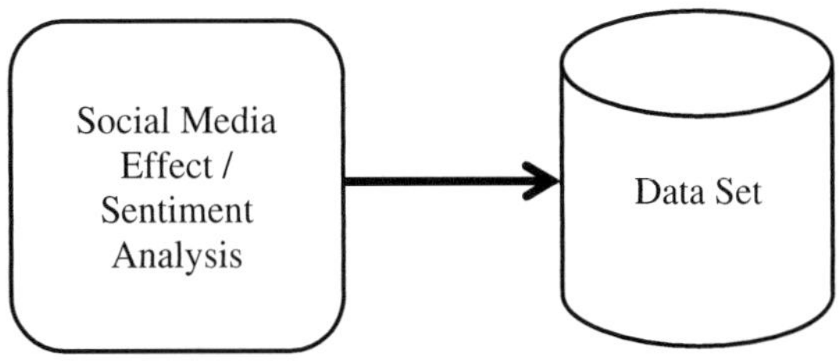

Figure 3. Single Layer approach using sentiment analysis.

The study of news sentiment effects on stock market is studied in [11]. The re-search considered pharmaceutical stocks and developed a dictionary to further develop a sentiment analysis model. A directional accuracy of 70.59% was obtained while predicting trends in stock price movement. Only news sentiments were considered excluding the effects of other parameters on the stock market. The reliable moneycontrol.com was used to retrieve news articles. To transform text into numerical vectors, python library – 'pattern' was used. The study concluded that in sentiment analysis a word weighing approach may be useful. Moreover, a hybrid approach that considers ARIMA or LSTM along with sentiment analysis may improve the directional accuracy.

Approach shown in figure 4 is a hybrid approach that considers the effect of both ML algorithms and sentiment analysis. This approach tends to better results as it considers multiple factors affecting the stock market. However fundamental analysis and many other factors that have an impact on the stock market are still missing.

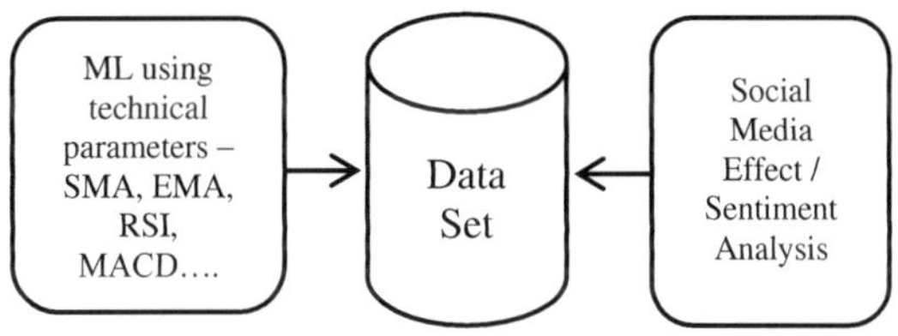

Figure 4. Hybrid approach using ML algorithms and sentiment analysis.

The research described in [16] uses the approach shown in figure 4. The study first compares the performance of three ML algorithms- ANN, SVM and MLR. The result depicts that ANN performs better than SVM and MLR. Hence ANN is combined with sentiment analysis to build a prediction model that finally gives the stock price. The stock market data is collected from Yahoo finance, news related data from economic times and reviews from twitter. The prediction model will be based on monthly prediction and daily prediction to forecast the next day open price.

The model proposed in [6] combines NN with sentiment analysis resembling the approach shown in figure 4. The paper concludes that there is significant correlation between actual stock prices and those predicted by proposed model. Here the input layer consists of 5 neurons, 4 indicating previous 4 days close price and the 5th indicating the sentiment score of previous day. The study suggested using deep multilayer neural networks with more than two hidden layers for deter-mining text sentiments. The research has collected news data from moneycontrol and historical stock data from yahoo finance.

Two different models – one for daily trend prediction and the other for monthly prediction are developed in [4]. The daily prediction model combines historical prices with sentiments, again depicting the approach shown in figure 4. Historical data is trained using supervised machine learning algorithms. Decision Boosted Tree performs better than SVM and Logistic Regression. The daily prediction model results in an accuracy of almost 70%. The monthly prediction model indicates that the trend of previous month and current month are least correlated. For sentiment analysis the data is collected from twitter and 2 different news websites while yahoo finance is used to collect stock data.

Only SVM classification method is used by the study [20] and the approach used here is as per figure 4 as technical indicators are used to predict CNX NIFTY index value. However, this study is an in-depth study and instead of doing research on multiple different ML algorithms, it has focused on four different Kernels of SVM classification model. The work has considered only 1046 data samples of CNX NIFTY and only basic parameters (open, high, low & close) are fetched from Yahoo Finance. From the 4 basic indicators, 20 other technical features are calculated and considered. Since there are many features that are considered, the SVM kernel functions can solve the problem easily with high optimization. Here 5-fold cross validation is used and hence the processing time slightly increases. Orange data mining tool is used to study all the four kernel functions i.e. Linear, Radial Basis, Polynomial and Hyperbolic Tangent. The study reveals that Polynomial Kernel function takes longer training time without good accuracy. The Linear and Radial Basis Kernel result in good accuracy with linear kernel being slightly better of the two. The research states that connections between number of features, data parts and accuracy can be further explored.

One more research [15] utilizes the approach shown in figure 4 to predict the stock price. The study uses a combination of Neural Network and Genetic Algorithm. Here a soft computing approach is used on stock market related news and social media data. This method was tested on only one stock CUB collected from NSE website. Neither the accuracy of the approach is measured, nor the method is applied to multiple different stocks.

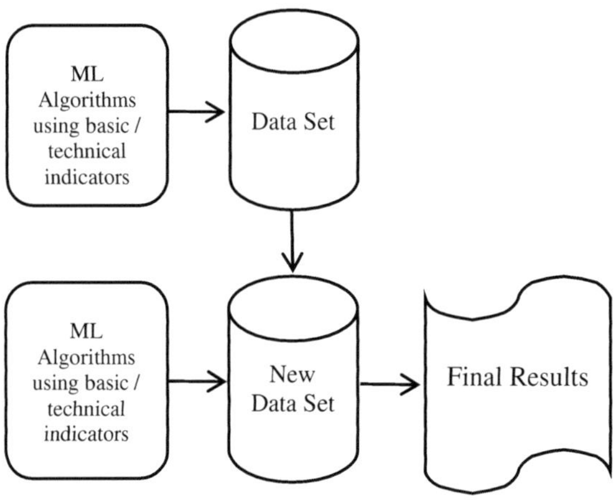

Figure 5. Multi-Layer approach using ML algorithms.

A novel Multi-Layer approach using a combination of multiple ML algorithms is depicted in figure 5. The study [8] focused on the indices S&P BSE & CNX Nifty of India. It proposed two stage fusion approach similar to the one shown in figure 5 involving SVR (Support Vector Regression) in the first stage. ANN (Artificial Neural Network), RF (Random Forest) and SVR (Support Vector Regression) are used in the second stage resulting into SVR--ANN--RF & SVR--SVR fusion model to predict stock market index. Ten technical indicators are selected as input to these models. Many of the other studies used a single stage approach. News is not included in this study. News can be categorized into various categories and their statistical weights can have an influence on the prediction.

The approach shown in figure 5 is again utilized by the paper presented in [10]. The work develops and studies three hybrid models ARIMA-SVM, ARIMA-ANN and ARIMA-RF. Thus, the linear and non-linear models are combined to capture the relationship in the time series data. The results prove that the hybrid ARIMA-SVM outperforms other models in terms of profit, volatility and risk adjusted performance. The residuals of ARIMA became the input to the SVM. The daily closing values of NIFTY was obtained from the NSE website. The study also concluded that hybrid model works well and out-performs the independent models. The proposed hybrid model can also be used by policy makers in forecasting financial and economic data.

The hybrid framework suggested by the study [3] is similar to the multi-layer approach shown in figure 5. The proposed research has considered the SVM as first layer and LSTM at second layer. The results of the hybrid model are compared with SVM, LSTM and Back Propagation. The results clearly indicate that the proposed model outperforms other algorithms both in terms of accuracy and time. The dataset of TATA Global from NSE from 2012 to 2018 was considered. The prediction was made for both one day ahead and one week ahead. Social media data was not considered in this study

The research [5] carried out not only forecasts one day ahead stock price, but also recommends a trade or no trade for a stock. The study reveals a multi-layered approach as shown in figure 5. First the technical indicators were optimized using GA technique and then used in decision tree. Moreover, two more systems one using Decision Tree – SVM and the other using Decision Tree – SVM with feature selection on the optimized technical parameters were studied. The GA optimized technical indicators decision tree SVM with feature selection gave the best results. Cipla and HUL from Bombay Stock Exchange as well as Glaxo Smithkline and Royal Bank of Scotland from London Stock Exchange were taken up for study.

A hybrid framework combining Support Vector Machine (SVM) with K-Nearest Neighbour (K-NN) is proposed in paper [14]. Initially SVM uses different kernel functions to predict the profit or loss. This output becomes the input for K-NN and computes the nearest neighbour from the training set to predict the stock index in the horizon of 1 month, 1 week and 1 day. The results also guide the user to analyse the closing price, volatility and momentum of the stock market.

The formulation of an appropriate stock selection strategy enables the development of stock market portfolio and is covered in [17]. Generally, any portfolio tries to maximize the profit while minimizing the risk. In the first step of this study, k-means clustering is applied to a heterogeneous sample of 53 stocks listed on the NSE. The monthly closing return is considered as a basis to classify the stocks in three categories (such as low stock price, medium stock price, and high stock price). The Generalized Reduced Gradient (GRG) method under the dynamic environment is used next for distributing weights to the stocks belonging to the portfolio. The study bridges the effects of unsupervised learning to supervised learning and is unique.

Stocks are classified into different clusters using data mining to optimize the stock portfolio as presented in [18]. A portfolio is built by selecting stocks from each cluster ensuring diversification and few other investment criteria. Results prove that K-Means cluster builds the most compact cluster as compared to SOM and Fuzzy C-means. The research uses timely stock returns and valuation ratios to measure the performance of the stock.

One more hybrid model [2] predicts the returns of stocks from the NSE. The model consists of one non-linear model i.e. RNN and two linear models – Exponential Smoothing and Autoregressive Moving Average. By merging the predictions from both linear and non-linear model the authors have come up with a unique Hybrid Prediction Method (HPM). In this research prediction from RNN, ES and ARMA are merged to form the HPM. Here the dataset is taken from NSE and consists of 6 stocks from different sectors, for a period of 16 weeks. Training and Test dataset surprisingly were in equal proportion i.e. 50% each. One needs to note that for training only RNN is used whereas for testing the HPM was used. The performance of the predictive models is measured using MSE and MAE. The correlation coefficient between the actual returns and predicted returns is calculated and is significantly positive. The main drawback of this research is that it does not give similar results on a different dataset. Moreover, the optimal regression order for the linear model needs to be chosen after trial and error. The computation time taken by this method is also significantly high.

A very unique and different kind of study is covered in [12] where increase or decrease in stock market performance is predicted. Normally basic indicators like (open, close, low, high etc.) or technical indicators like (MA, RSI, MACD, etc.) are used for any kind of stock prediction using machine learning. Here some of the most common macroeconomic factors such as commodity price, market history, Foreign exchange rates are used to predict BSE. In this study the four machine learning algorithms considered are SVM, RF, Gradient Boosting and Adaptive Boosting. Study includes market history data of 2 years from Jun-2015 to Dec-2016. The commodity and foreign exchange rates for 9 months are used from investing.com. The dataset is processed using Python and different proportions of training (70% & 90%) are used. The AdaBoost algorithm resulted in best accuracy. The results revealed that the performance of BSE is best correlated to only Gold price. Moreover, the research revealed a negative correlation of BSE with other factors considered and is equally useful.

The research [7] is unique in the sense it is carried out keeping in mind the fact that as the dimension of data increases, the complexity of the model also increases. Hence it tries to implement minimal variability OWA (Order Weighted Averaging) operator. The operator ensures that high dimensional data is aggregated to a single dimension. The research proves that when subtractive clustering is combined with hybrid fuzzy based network inference system based on the proposed model, the results are better compared to other models. The experiment is carried out on the Bombay Stock Exchange Index (BSE30).

4. Discussion

This survey focuses on the different types of researches carried out on Indian stock market. The survey identifies that there are few major approaches used mainly while predicting the stock price trend, stock price or the stock index trend. Almost all ML algorithms are implemented on the Indian stock market data. It is observed that the same algorithm say for example neural network may be used by multiple researchers where the input parameters are different and hence the output. Moreover, the performances of various researches are all measured using different methods. Hence it becomes extremely difficult to compare any two researches.

Some of the researches focus only on a small subset of the stocks, making it difficult to imagine as to how it will perform when a different set of stock data is used. The stock market is extremely volatile; hence the performance of one particular research may vary from time to time. However, in general it is observed that fine-tuned SVM and ANN seem to give better results compared to other algorithms. In order to improve the accuracy most of the current researches employ a hybrid or multi-layered approach as shown in figure 4 or figure 5.

5. Conclusion

Many of the researches are single dimensional i.e. only one major parameter is used in the prediction process. There are many hybrid or multilayer researches, but again they fail to cover all the parameters that affect the stock price / stock price index. Most of the work have not considered the fundamental parameters and macro-economic variables that significantly affect the stocks. One factor that cannot be ignored at present is the impact of social media data on stock market. Clearly the research that tries to cover all the parameters is missing. Moreover, when all parameters are considered, it needs to be checked whether a multilayer or hybrid approach is better. If possible based on historical evidence weights can be given to parameters and it needs to be evaluated from time to time. If possible, the research should focus on customized set of parameters and weights for each different stock to improve the accuracy.

References

[1] Rout A K, Biswal B, and Dash P K 2014 A hybrid FLANN and adaptive differential evolution model for forecasting of stock market indices *International Journal of Knowledge-Based and Intelligent Engineering Systems* **18**(1) 23–41

[2] Rather A M, Agarwal A, and Sastry V N 2015 Recurrent neural network and a hybrid model for prediction of stock returns *Expert Systems with Applications* **42**(6) 3234–3241

[3] Mundra A, Mundra S, Verma V K, and Srivastava J S 2020 A deep learning based hybrid framework for stock price prediction *Journal of Intelligent and Fuzzy Systems* **38**(5) 5949–5956

[4] Nayak A, Pai M M M, and Pai R M 2016 Prediction Models for Indian Stock Market *Procedia Computer Science* **89** 441–449
[5] Nair B B, and Mohandas V P 2015 An intelligent recommender system for stock trading *Intelligent Decision Technologies* **9**(3) 243–269
[6] Vanipriya C H, and Thammi Reddy K 2014 Indian Stock Market Predictor System *Advances in Intelligent Systems and Computing* **249** 17–26
[7] Kaur G, Dhar J, and Guha R K 2014 Stock market forecasting using ANFIS with OWA operator *Int. J. Artif. Intell.* **12**(2) 102-114
[8] Patel J, Shah S, Thakkar P, and Kotecha K 2015 Predicting stock market index using fusion of machine learning techniques *Expert Systems with Applications* **42**(4) 2162–2172
[9] Srinivas K, Srinivasa Rao V, and Sreemalli M 2018 Binarized genetic algorithm with neural network for stock market prediction *ACM International Conference Proceeding Series* 8–11
[10] Kumar M, and Thenmozhi M 2014 Forecasting stock index returns using ARIMA-SVM, ARIMA-ANN and ARIMA-random forest hybrid models *International Journal of Banking, Accounting and Finance* **5**(3) 284–308
[11] Abe N, Institute of Electrical and Electronics Engineers, & IEEE Computer Society. (n.d.). 2018 *IEEE International Conference on Big Data : proceedings : Dec 10 - Dec 13, 2018* Seattle WA USA
[12] Kohli P P S, Zargar S, Arora S, and Gupta P 2019 Stock prediction using machine learning algorithms *In Advances in Intelligent Systems and Computing* **698** 405–414
[13] Charkha P R 2008 Stock price prediction and trend prediction using neural networks *Proceedings - 1st International Conference on Emerging Trends in Engineering and Technology ICETET 2008* 592–594
[14] Nayak R K, Mishra D and Rath A K 2015 A Naïve SVM-KNN based stock market trend reversal analysis for Indian benchmark indices *Applied Soft Computing Journal* **35** 670–680
[15] Suganya R, and Sasi Kumar A 2020 Stock price prediction using tech news based soft computing approach *International Journal of Advanced Trends in Computer Science and Engineering* **9**(2) 2049–2054
[16] Nivetha R Y, and Dhaya C 2017 Developing a Prediction Model for Stock Analysis *Proceedings - 2017 International Conference on Technical Advancements in Computers and Communication ICTACC 2017* 2017-October 1–3
[17] Gupta S, Bandyopadhyay G, Biswas S and Upadhyay A 2019 A Hybrid Machine Learning and Dynamic Nonlinear Framework for Determination of Optimum Portfolio Structure In *Lecture Notes in Networks and Systems* **74** 437–448
[18] Nanda S R, Mahanty B, and Tiwari M K 2010 Clustering indian stock market data for portfolio management *Expert Systems with Applications* **37**(12) 8793–8798
[19] Nadu T, Nadu T, and Nadu T 2015 Cuckoo Optimized SVM for Stock Market Prediction
[20] Upadhyay V P, Panwar S, Panchariya R and Merugu R 2016 Forecasting stock market movements using various kernel functions in support vector machine *ACM International Conference Proceeding Series* 12-13-August-2016

Usage of time series forecasting model in Supply chain sales prediction

A Raiyani[1], A Lathigara[2], and H Mehta[3]

[1]School of Engineering, RK University, Gujarat, India
[2]School of Engineering, RK University, Gujarat, India
[3]School of Engineering, RK University, Gujarat, India

[1] **ashwin.rkcet@gmail.com**

Abstract. This paper representing a study of supply chain operation data that was used on 100 different store items from 10 stores using 5 years history of sales through open sources contest to compare the performance of time-series forecasting model mainly, decomposition, Auto-Regressive Integrated Moving Average(ARIMA) , Prophet, Box-Cox transformation. Here data is collected from 2013 to 2018 were used in real-time transaction at different store, initially model was applied on 2013 to 2017 data and based on the that predicted for 2018 then again cross checked with actual 2018 with proceed predicted data of 2018. To improve the performance and evaluation of the supply chain management system, scrutiny 3 metrices that will help to make decision on the model selection. The accuracy of the Machine learning model in forecasting future sales of supply chain store. Although the result on comparison indicates that there is no single method gives better and superior result. But present study indicates that prophet and ARIMA hybrid model gives better result compare to individual model.
Keywords: supply chain, time-series, prediction, sales,

1. Introduction

A more open and efficient supply chain has been required in terms of evolving and competitive markets, which supply chain uses, suppliers who are working for multiple customers and other suppliers in the same sector and consumer expectations. The bulk of existing metric packages with manufacturers stagnate and contribute to fire struggles. By having a part of the chain of supply transparency, current metric packages add value, and they should be used; however, the whole image is not drawn. The changing supply chain environments are not completely considered in present metric packages. With an increase in the amount of outscoring, there is a need for a new metric kit to handle larger or more dynamic supply chains and to enhance the supply chain management operations by using predictive analysis techniques. Suppliers operate with more customers than ever and their relationships also shift over time with each customer. Many suppliers have in the past been based on one client and will do anything to make that client happy. predictive and reactive analysis techniques were appropriate if a supplier relationship was built on the surviving supplier because of the customer alone.

The research has been given. The supplier's loyalty to their only customer enables consumers to exploit their suppliers so that they can trust their suppliers more with little screening, other than tracking quality goods on-time delivery. This research comparative study on the use of predictive analysis papers aims to enhance the efficiency of the supply chain in order to improve distribution performance by enhancing the risk management capabilities of the customers supply chain and increasing the accountability of their supply chain.

2. Related Work.

Sales system data are designed to facilitate the detection of abnormal buying behavior of customer. Time series models have for quite some time been of interest. The time series model allows to do prediction on sell store items based in transaction data available. There were many analysis mechanisms set on historical transactional data to have more accuracy on prediction of sales. A method like exponential smoothing and generalized regression strategies [1,2] were utilized to forecast sales historical data expectation. Addition to that, decomposition methods and multilevel time series models [3,4[were implied on better way on various sales components for prediction. ARIMA models have been rigorously utilized for future forecasting including the pattern sessional, occasion patterns and holiday trends. Machine learning sales prediction models are dependent on artificial neural networks were used to train and test the dataset.

Support vector machines (SVMs) are a new type of approach based on statistical learning theory for machine learning [5, 6]. In practical applications, they may contribute to greater potential and better performance. Typically, since of the concept of basic hazard minimization utilized in SVMs, which has more prominent speculation potential and is prevalent to the rule of experimental chance minimization grasped by customary neural systems. SVMs have been effectively connected to different time arrangement prediction issues, such as yield esteem determining within the apparatus industry[7], motor unwavering quality prediction[8], and financial time arrangement prediction[9],[10]. By utilizing SVM for time series estimating, the effective utilize of SVM in time arrangement forecast propels our research work.

The main motive of this research present paper is to have comparison on current time series methods like SVM, regression, ARIMA and exponential smoothing with reference to their real forecasting efficacy in scourge time series. This comparison may be supportive for the transmission expert to select the foremost appropriate technique in each circumstance.

3. Methodologies

3.1. Materials

Gathering of data from various sales store department for 100 product item level. Supply chain operation data that was used on 100 different store items from 10 stores using 5 years history of sales through open sources contest to compare the performance of time-series forecasting model. Again, these data were collected from 2013 to 2018 duration to measure the different parameter for decision improvement.

To improve data processing capability, the Hadoop and Spark framework is used to achieve the distribution storage and analysis work of the collected big data. Big Data analytics uses a variety of techniques such as statistics, modelling and data mining to analyse current and historical facts and make predictions for the future. [15] Big Data can fulfil correlation analysis. Correlation analysis points us whether or how two things are connected. To get better decision and improve performance, predictive analytics will be suitable. PySpark was used as python library as it supports standard data processing framework. Benefits of these library are speed, ease of use and runs everywhere. Additionally, Spark SQL and SQL used for structured data processing, MLlib for machine learning, GraphX for graph processing

3.2. Methods

3.2.1. Decomposition methods

The decomposition [11, 12] methods attempt to separate long trend pattern, a seasonal trend and residuals. To make it separate, different indices are used to indicate pattern, trend, and residual things. Decomposition time series is simple to disclose to the end user as it does not contain lots of mathematics. This is a significant bit of advantage since, in such a case that the end user has acknowledgment of how the forecast was developed, the person in question may have more confidence in its use for decision making.

So, performing the decomposition method on neural network and running the Additive Model and Multiplication Model, observed figure 1 (a) and (b) that there is not much difference on applying methods.

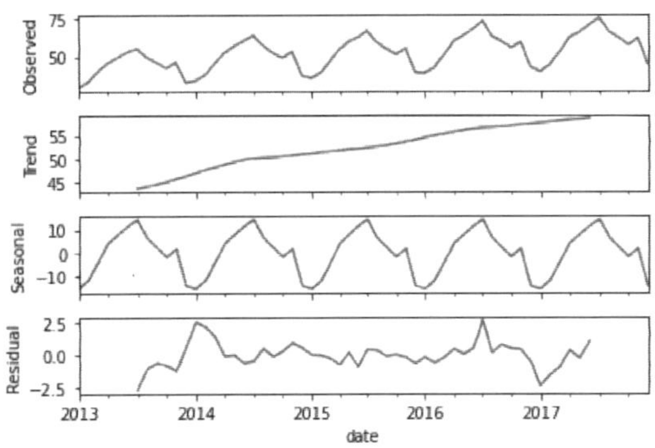

Fig. 1(a) . Decomposition – Additive Model

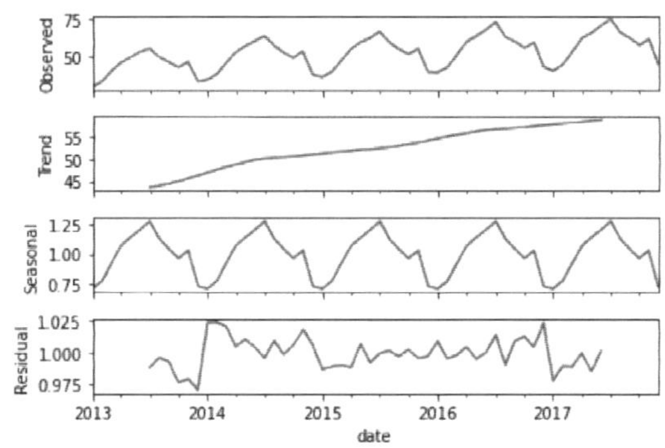

Fig1. (b). Decomposition - Multiplication

3.2.2. The Prophet Forecasting Model

Additive regression model x(t) comprising the following components:

$$x(t)=np(t)+s(t)+ho(t)+ err,$$

where np(t) is non-periodic changes, s(t) is seasonal changes, ho(t) customized holiday list with irregular schedules, err error term accounts for any unusual changes not accommodated by the model Mean Absolute Percentage Error (MAPE) as a forecast precision measurement. In this research [18], Prophet has appeared critical lower determining blunder than the other models can see from figure 2.

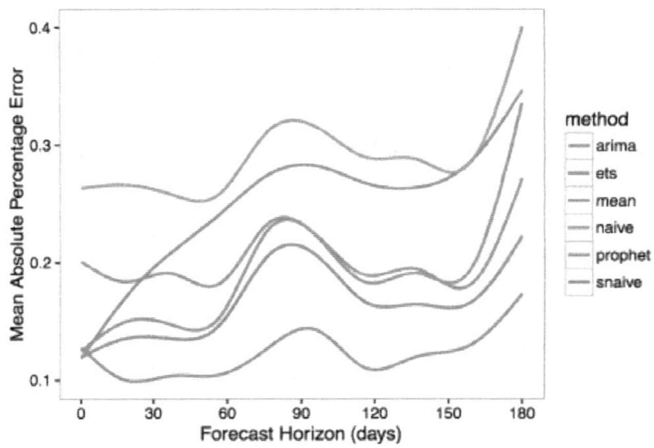

Figure 2. – prophet has lower error.

IOP Conf. Series: Materials Science and Engineering 1042 (2021) 012022 doi:10.1088/1757-899X/1042/1/012022

Again, By using Prophet.plot_components (pyspark function library) figure out and experimented on different components prophet model and taken each component like trend, yearly and weekly seasonality in separate consideration.

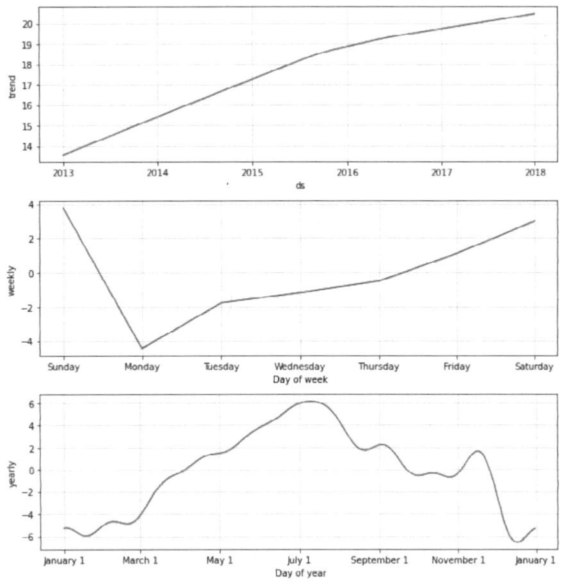

Figure 3 – prophet plot components – trend, yearly and weekly seasonality.

It seems like People go to Shopping Mostly in July, they are peaking on Sunday and Saturday. So we should add the holidays effect to make Prophet perform better. Now adding parameter types of Holidays like Less-Important public holidays and weekends. And Festive Holidays with high importance. There may occurs a high increase or decrease in sales due to these holidays e.g. New Year, Christmas etc.

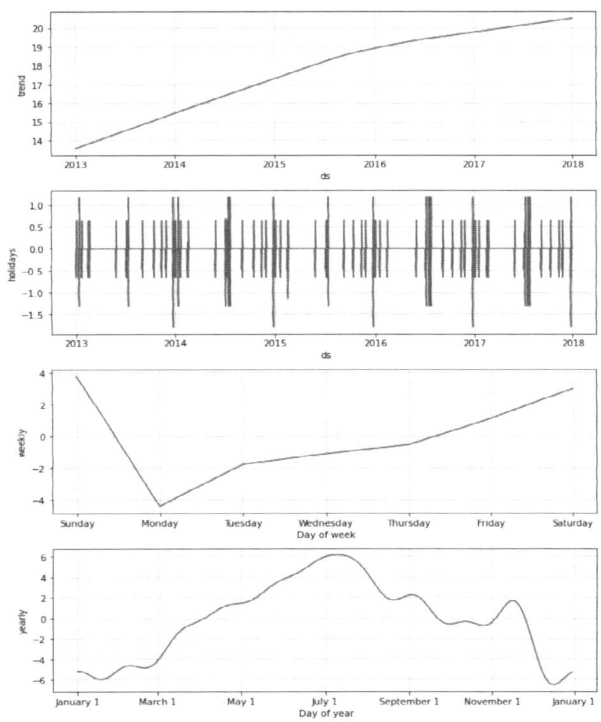

Figure 4 – with weekend and festival holiday.

Above figure 4. result shows that there is spike on the days around less important days holiday appearances, with an especially large spike for the festival holidays. With this configuration running and calculating Forecasting error.

```
def calculate_forecast_errors(df, prediction_size):
        df = df.copy()
    df['e'] = df['y'] - df['yhat']
    df['p'] = 100 * df['e'] / df['y']
    predicted_part = df[-prediction_size:]
    # Define the function that averages absolute error values over the predicted part.
    error_mean = lambda error_name: np.mean(np.abs(predicted_part[error_name]))
        # Now we can calculate MAPE and MAE and return the resulting dictionary of errors.
    return {'MAPE': error_mean('p'), 'MAE': error_mean('e')}
```

```
[77] for err_name, err_value in calculate_forecast_errors(cmp_df, prediction_size).items():
         print('Non Holiday', err_name, err_value)

Non Holiday MAPE 27.482259074941975
Non Holiday MAE 3.5469262470026868
```

```
[78] for err_name, err_value in calculate_forecast_errors(cmp_df_holi, prediction_size).items():
         print('Including Holiday', err_name, err_value)

Including Holiday MAPE 27.63731122003189
Including Holiday MAE 3.5463176952152513
```

As a result, MAPE is about 27.5%, and ME is about 3.54 predicts. It indicates that relative error is low and on an average 3.54% model goes wrong. Again, to make improvement on machine learning model, prophet combine with Box Cox Transformation model.

3.2.3. Box-Cox Transformaiton

A Box Cox transformation is a way to transform non-normal dependent variables into a normal shape. Normality is an important assumption for many statistical techniques; if data isn't normal, applying a Box-Cox means that able to run a broader number of tests.

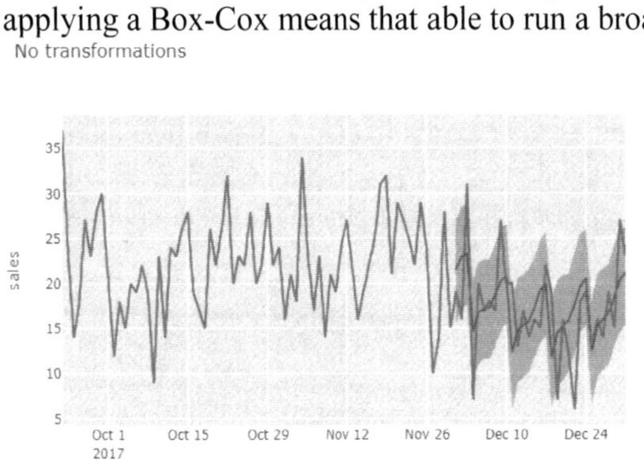

Figure 5 (a) no transformation.

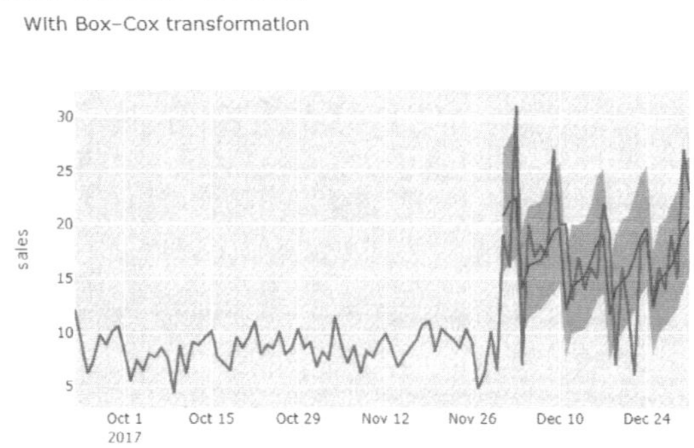

Figure 5 (b) with Box Cox transformation

In box cox transformation, as shown in above figure 5 (a) and (b) we have upper bound predict value and lower bound predict value which gives closer to our actual sales values. It will help in nearby decision making for sales store items.

In [84]:
```
cmp_df2 = make_comparison_dataframe(df, forecast2)
for err_name, err_value in calculate_forecast_errors(cmp_df2, prediction_si
ze).items():
    print('Including Holidays', err_name, err_value)
# We Get Slight Improvement Over No Transformations !!!
```

```
Including Holidays MAPE 26.26371415946375
Including Holidays MAE 3.491021444677818
```

In [85]:
```
show_forecast(cmp_df, prediction_size, 100, 'No transformations')
show_forecast(cmp_df2, prediction_size, 100, 'With Box-Cox transformation')
```

As result is not that much improved compare to no-transformation model but yes it indicates slightly positive result at the end.

3.2.4. Proposed Approached

ARIMA models are denoted with the notation ARIMA(p, d, q). These three parameters account for seasonality, trend, and noise in data.

AR: Auto-Regressive (p): AR terms are just lags of dependent variable. For example lets say p is 3, we will use x(t-1), x(t-2) and x(t-3) to predict x(t)

I: Integrated (d): These are the number of nonseasonal differences. For example, in our case we take the first order difference. So we pass that variable and put d=0

MA: Moving Averages (q): MA terms are lagged forecast errors in prediction equation.

The ARIMA [13,14] models are inclusion of 3 models, autoregressive (AR), moving average (MA), and autoregressive moving average (ARMA). The current estimation of the time arrangement is a direct capacity of its past qualities and arbitrary clamor in the AR model; though the current estimation of the time arrangement is a straight capacity of its current and past estimations of residuals in the MA model. The ARMA model is the blend of AR and MA, which takes into consideration of historical values and residuals. Stationary processes [16] are the time series required in the AR, MA, and ARMA models. This can be interpreted and implies that the mean value of result and the covariance value do not change with time. For non-stationary time series, the transformation of the series into a stationary one must be achieved first. In general, the ARIMA algorithm matches the time series data based on the ARMA model and a separation method that converts the non-stationary data into a stationary one efficiently.

By applying proposed approached forecasting model found following results:

```
==========================================================================================
Dep. Variable:                        sales   No. Observations:                       60
Model:            SARIMAX(2, 0, 1)x(2, 2, 0, 12)   Log Likelihood                   -1.683
Date:                      Fri, 07 Aug 2020   AIC                                 15.366
Time:                              13:51:08   BIC                                 17.181
Sample:                          01-01-2013   HQIC                                13.374
                               - 12-01-2017
Covariance Type:                        opg
==============================================================================
                 coef    std err          z      P>|z|      [0.025      0.975]
------------------------------------------------------------------------------
ar.L1         -0.8498      0.494     -1.722      0.085      -1.817       0.118
ar.L2         -0.4951      0.383     -1.293      0.196      -1.246       0.255
ma.L1         -1.0001   1161.623     -0.001      0.999   -2277.739    2275.739
ar.S.L12      -1.0664      0.018    -57.817      0.000      -1.103      -1.030
ar.S.L24      -0.0255      0.007     -3.437      0.001      -0.040      -0.011
sigma2         0.0686     79.636      0.001      0.999    -156.016     156.153
==================================================================================
 Including Holidays MAPE 20.563128311735583
 Including Holidays MAE 2.495684718273
```

Figure 6. Proposed model result with MAPE and MAE.

As shown in figure 6. result, the improvement over transformation

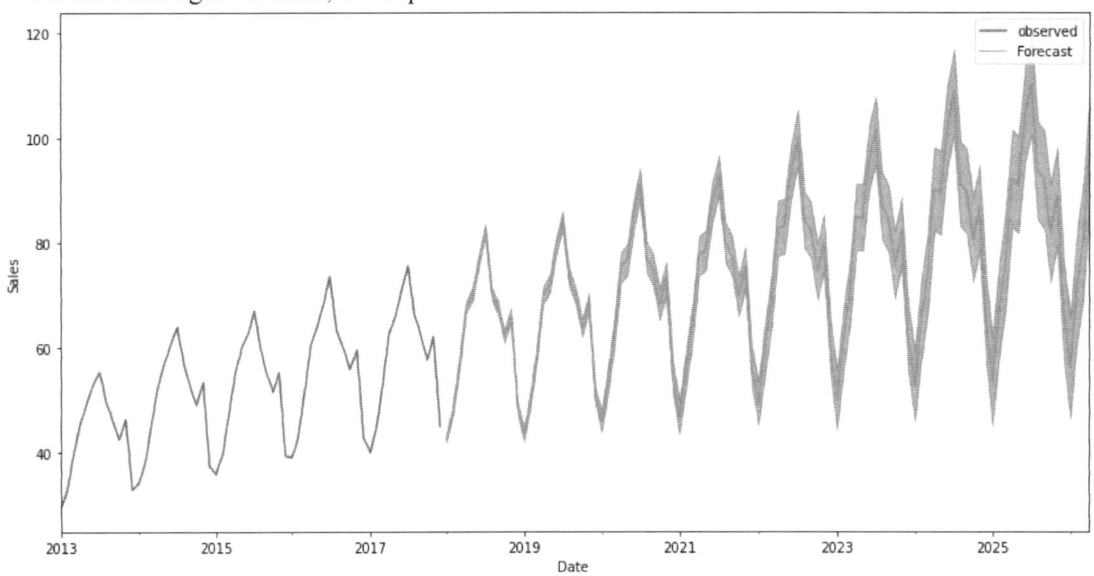

Figure 7 accuracy of proposed model for next 100 month predictions.

In figure 7, we can see overlapping lines for 2018 indicating the accuracy of model and it also indicated that how upper and lower bound predicted values of sales store items for upcoming 100 months.

4. Conclusion

Compared to current models, the new smart business architecture for market prediction is more accurate for future demand. It helps inventory managers better control their output in the supply chain by simultaneously raising days of scope and operation. Prediction using proposed require much prior knowledge or experience of forecasting time series data since it automatically finds seasonal trends beneath the data and offers a set of 'easy to understand' parameters. Proposed approach for forecasting time series dataset dependent on an additive model where non-linear patterns are fit with yearly, week by week, and everyday seasonality, in addition to holiday effects. It works best with time series that have solid seasonal impacts and a few seasons of chronicled data. No single model time-series is perfect as individual so combining technique/model it gives better outcome. In this context, the proposed model is suitable for products that are strongly linked to a particular profile or profile to a significant share of sales. In such cases, the proposed model can be further expanded to consider the migration and immigration dynamics of individuals with specific interest profiles in a region.

5. References

[1] Farrington C, Andrews N 2003 Outbreak detection: application to infectious disease surveillance. Monitoring the Health of Populations: Statistical Principles and Methods for Public Health Surveillance 203–231.

[2] Chadwick D, Arch B, Wilder-Smith A, Paton N 2006 Distinguishing dengue fever from other infections on the basis of simple clinical and laboratory features: application of logistic regression analysis. Journal of Clinical Virology **35**: 147–153.

[3] Gonzalez-Parra G, Arenas AJ, Jodar L 2009 Piecewise finite series solutions of seasonal diseases models using multistage Adomian method. Communications in Nonlinear Science and Numerical Simulation **14**: 3967–3977.

[4] Spaeder MC, Fackler JC 2012 A multi-tiered time-series modelling approach to forecasting respiratory syncytial virus incidence at the local level. Epidemiology and Infection **140**: 602–607.

[5] Chang C-C, Lin C-J 2011 LIBSVM: a library for support vector machines. ACM Transactions on Intelligent Systems and Technology (TIST) **2:** 27

[6] Thissen U, V B R, De Weijer A, Melssen W, Buydens L 2003 Using support vector machines for time series prediction. Chemometrics and intelligent laboratory systems **69**: 35–49.

[7] Pai P-F, Lin C-S 2005 Using support vector machines to forecast the production values of the machinery industry in Taiwan. The International Journal of Advanced Manufacturing Technology **27**: 205–210.

[8] Hong W-C, Pai P-F 2006 Predicting engine reliability by support vector machines. The International Journal of Advanced Manufacturing Technology **28**: 154–161.

[9] Müller K-R, Smola AJ, Rätsch G, Schölkopf B, Kohlmorgen J, et al.. 1997 Predicting time series with support vector machines. Artificial Neural Networks–ICANN'97: Springer. pp. 999–1004.

[10] Tay FE, Cao L 2002 Modified support vector machines in financial time series forecasting. Neurocomputing **48**: 847–861.

[11] Bowerman BL, O'Connell RT, Richard T 1993 Forecasting and time series: An applied approach: Belmont CA Wadsworth.

[12] Hamilton JD 1994 Time series analysis: Cambridge Univ Press.

[13] Zhang GP 2003 Time series forecasting using a hybrid ARIMA and neural network model. Neurocomputing **50**: 159–175.

[14] Pai P-F, Lin C-S 2005 A hybrid ARIMA and support vector machines model in stock price forecasting. Omega **33**: 497–505.

[15] Ashwin R, Amit L, Hardik M, 2018 Supply Chain Management and Big Data Analytics (SCMBDA): Perception to SCM Business International Journal on Future Revolution in Computer Science & Communication Engineering ISSN: 2454-4248 **Volume 4** :Issue:5

[16] Waller and Fawcett, 2013 Data Science, Predictive Analytics, and Big Data A Revolution That Will Transform Supply Chain Design and Management,

[17] Ma C, Zhang HH, Wang X. Machine learning for big data analytics in plants. Trends Plant Sci. 2014;**19(12):**798–808

[18] Taylor SJ, Letham B. 2017. Forecasting at scale. PeerJ Preprints 5:e3190v2

Health For You

Pooja Surapur , Vishnu Preetham Reddy, Munavar Hussain

Mr. C Satya Kumar Assistant Professor

Vardhaman College Of Engineering ,Shamshabad , Hyderabad, Telangana

vishnupreethamreddy17cse@vardhaman.org,hussain17cse@vardhaman.org
c.satyakumar@vardhaman.org

Abstract:

Health for you is an application which acts like interface between patients and doctors. All the patients can fill the application regarding the disease they had been suffering with. Any doctor who is interested in serving people for free can register using their license during the sign up process and give solution to the patient's disease. The doctor can examine the patient's details, write the prescription or second opinion and mail it to the patient. The problem was to create an application which efficiently does one of the most strenuous and huge tasks like connecting poor people with well experienced doctors at free of cost and gain very valuable suggestions. Finally, the app should act like a natural hospital where the treatment is available free of cost from your living room. To avoid certain doubts to our doctors we have made a bunch of clarifications to our patients and users that we validate for a unique like license of doctor and make a search in our national database we register him as a doctor after having some typical steps of registration.

Keywords:

Sorting Algorithms, Adapter Views, National Database, Regex, Hash Maps, Validation, Tree

Maps, Online medication bills, Online consultancies.

1.Introduction

Health for you is cashless portal to provide our consultancies through doctors to our patients via our mobile application. Health for you purely works based on nonchargeable environment[1]. The main aim of this application is to provide effective suggestions in a portable manner. Future is a period of many pandemic situations like the present situation. People may be ceased to come on to the streets and visit the doctors. To avoid such kind of problems we came up with an idea of home consultancy through our application. All the patients can fill an application regarding the disease they had been suffering with. Any doctor who is interested in serving people for free can register and give solutions to patient’s disease. The doctor can examine the patient’s details, write the prescription or second opinion and mail it to the patient.[2] This application efficiently does one of the most strenuous and huge tasks like connecting poor people with well experienced doctors at free of cost and gain very valuable suggestions. In existing system people are charged and they need to wait for the doctor appointment. Even online consultation are charged. It is not easily accessible to everyone. To avoid such kind of disadvantages in the existing project this mobile application is

developed in such a manner that our customers get benefits from this application. As our country is digitization in coming day's smart phones are to be a common part of every being's life so, we are developing this kind of applications to avoid huge crowded places.

2. Related Study

In existing systems as a matter of fact every online consultancy are charging a huge amount by providing more features and by upgrading to gold plan etc. We do not have any of that we clearly have come across every problem through the present online consultancies.[3] So, we have come up with a non-profitable application for us and non-chargeable for our customers. Everything is related to existing online mobile applications. In the proposed model there is no amount to charged and mainly we are using some effective algorithms and high security algorithms to validate the user or a doctor. Every doctor is validated with his unique license and it is validated against the national database of doctors to avoid fraud doctors. And then each doctor must be rated compulsorily such that we can assign customers to our best doctors and give a wonderful experience when they are using the application. The doctor can examine the patient's details, write the prescription or second opinion and mail it to the patient[4]. The objective of this application is to provide the quality suggestions to poor people regarding the treatment for their diseases from well experienced doctors at free of cost. This app can be used by anyone around the world. The patients can fill an application regarding the sickness they had been suffering with. Any doctor who is interested in serving people for free can register with us and give solutions to patient's disease. The doctor can examine the patient's details, write the prescription or second opinion and mail it to the patient. This application efficiently does one of the most difficult and huge tasks like connecting customers with well experienced doctors at free of cost and gain their valuable suggestions.

2.1 Process Model

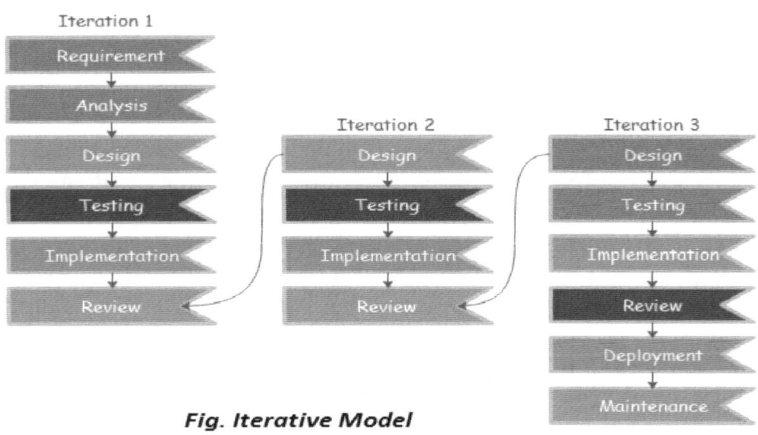

Figure 1: Iterative Model

2.2 Architectural model

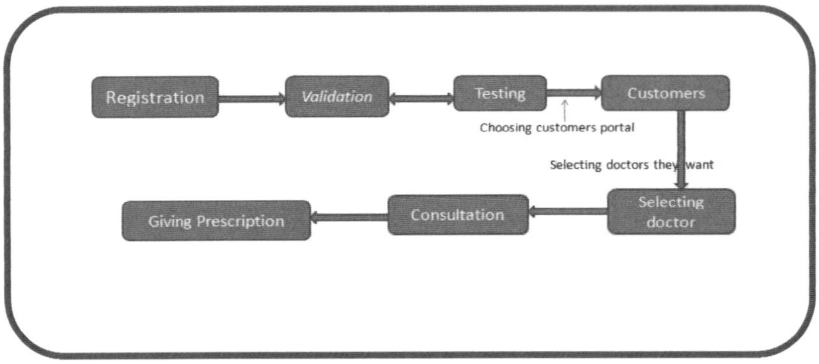

Figure 2:Architectural Model

3. Proposed Model

The doctor can examine the patient's details, write the prescription or second opinion and mail it to the patient. The objective of this application is to provide the high quality suggestions to poor people regarding the treatment for their diseases from well experienced doctors at free of cost. This app can be used by anyone around the world .The functions are divided according to the function types.

1. One user (patient) can fill, edit his details along with the scanning, X-ray etc., reports and save it.

2. Another user (doctor) who can see the patient's reports, analyze and send their suggestions to the patients through the email.

In the proposed model there is no amount to charged and mainly we are using some effective algorithms and high security algorithms to validate the user or a doctor. Every doctor is validated with his unique license and it is validated against the national database of doctors to avoid fraud doctors. And then each doctor must be rated compulsorily such that we can assign customers to our best doctors and give a wonderful experience when they are using the application[5]. The doctor can examine the patient's details, write the prescription or second opinion and mail it to the patient. The objective of this application is to provide the quality suggestions to poor people regarding the treatment for their diseases from well experienced doctors at free of cost. This app can be used by anyone around the world. The patients can fill an application regarding the sickness they had been suffering with. Any doctor who is interested in serving people for free can register with us and give solutions to patient's disease[6]. The doctor can examine the patient's details, write the prescription or second opinion and mail it to the patient. This application efficiently does one of the most difficult and huge tasks like connecting customers with well experienced doctors at free of cost and gain their valuable suggestions. We validate our doctors through a high prioritized database mode and the ER-Diagram for the same is given below

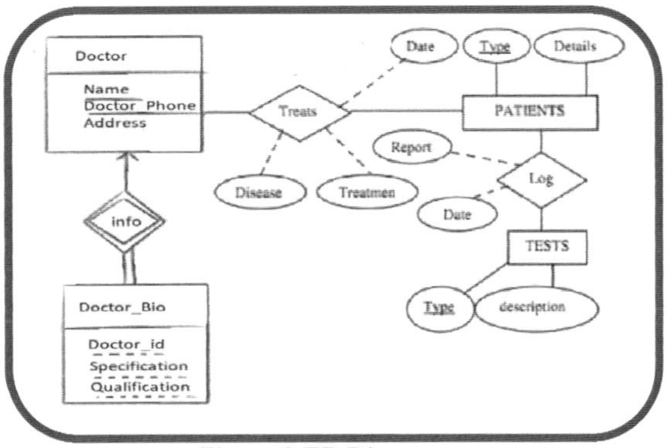

Figure 3:ER Diagram

4. Experimental Setup

Health for you experimental setup is in Android studio. This mobile application is implemented in java language. Each algorithm useful for application is coded in java language. There are various algorithms used in this application is vast. We used Tree Maps to store the best doctors based on their rating and validation of doctor is done with the browser in the form of J son format. We displayed the doctors data in the adapter view to consume less space and use less memory for the mobile device. We have two broad tabs divide in our application one for doctors and other for patients some of the experimental screen shots of our mobile application is shown below…

4.1 Layout

Figure 4.1

Figure 4.2

Figure 4.3

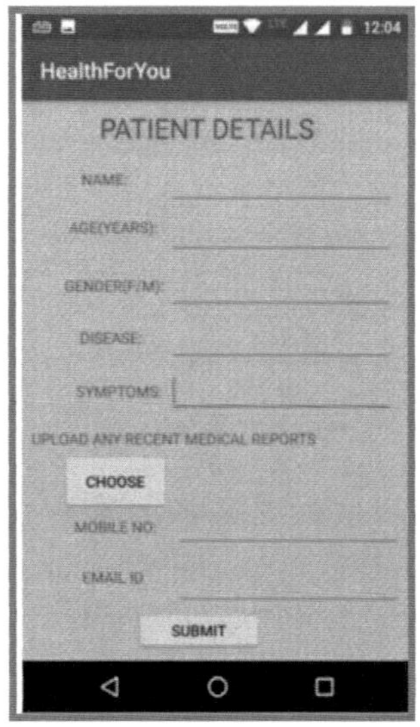

Figure 4.4

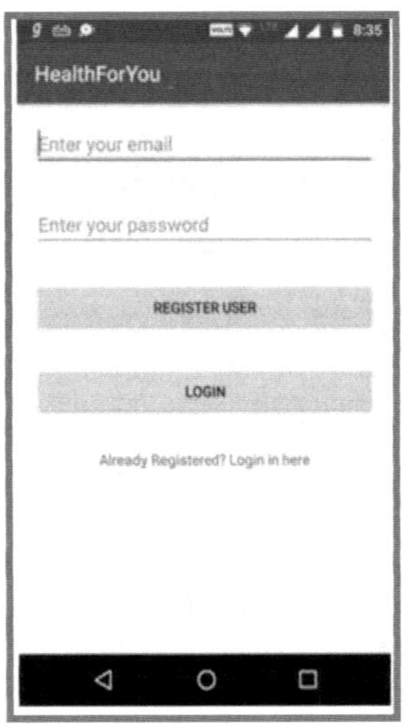

Figure 4.5

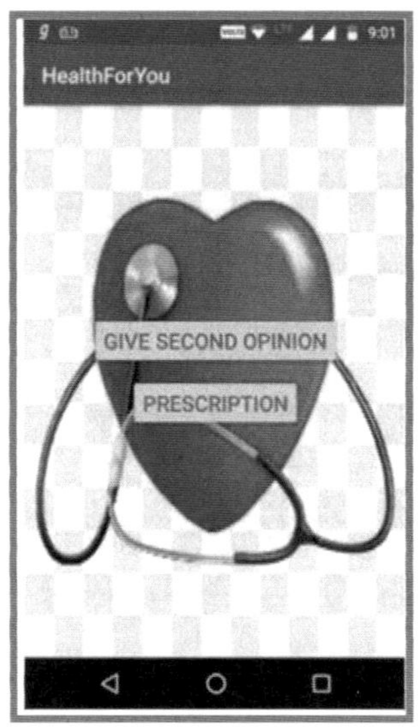

Figure 4.6

4.2 Data Flow Diagram

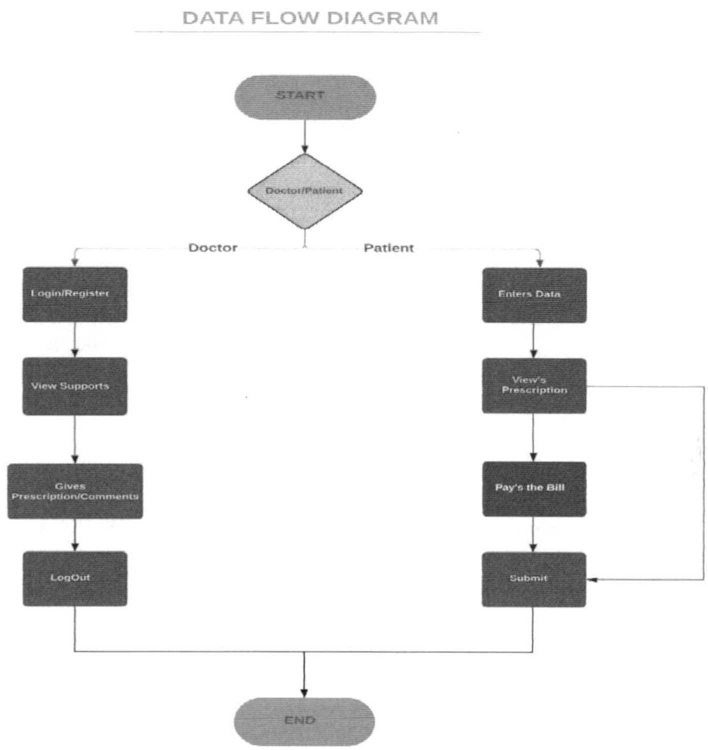

Figure 5:Data Flow Model

Data flow diagram is unlike flow chart. This diagram helps to understand the entire experimental setup of the application. In this app we start with patients and doctor's registration and after validation of doctor details it displays the patient information who want to consult him then he selects the patients to consult one after the another. The patient registration is quite different compared to the doctor's. After entering his/her data it is registered in apps database and it validates the user when he/her exit's the app. After consulting he/her can view the prescription an exit the app. There may be payment portal for the medical bill payments in the future as an update feature

5. Result Analysis

Previously there were government hospitals to provide free treatment. But now a days government hospitals are not up to the mark in the matter of keeping the hospitals clean and tidy and the main factor is most of the skilled doctors are moving their base to the private hospitals to secure their future for their families. So, the skilled doctors are less in government hospitals and patients are showing less interest towards government hospitals because the equipment provide there is not as good as the private ones.Not all people in India can afford for the doctors in private hospitals and following the disadvantages of government hospitals they are forced to visit government hospitals because they can't afford high rates in the private hospitals. As the consultancy fee is also quite high in private hospitals and including the medical bills the prices would be sky rocketing. By keeping in the view we planned this project to be helpful to all classes of people including poor.

Health for you is an application which acts like interface between patients and doctors. All the patients can fill an application regarding the disease they had been suffering with. Any doctor who is interested in serving people for free can register and give solutions to patient's disease. The doctor can examine the patient's details, write the prescription or second opinion and mail it to the patient. This application efficiently does one of the most strenuous and huge tasks like connecting poor people with well experienced doctors at free of cost and gain very valuable suggestions. Health for you provides a clean and transparent interface for interaction between doctor and the patients. No consultancy fee shall be charged. Private doctors are available for consultancy in any time. This is a mobile application so, this can be easy and simple to use. As per the current situation it is not good to step out from the houses and especially not good to visit hospitals. Our application is completely useful for the current situation because after consultancy doctor writes a prescription and medics for the patient and all this done without stepping out from your house.

6.Conclusion:

The Health for you application is now complete and patients can use it to expose their problem disease through reports while doctors can analyse and give response to it. The next step would be to increase the features of the application by including medical facilities. Over all we can conclude that our application can connect people with experienced doctors at free of cost and gain very valuable suggestions. Finally, the app should act like a natural hospital where treatment is available at free of cost.

7. Future Scope

Using this application, patients can fill the application giving the details of their disease and files like scanning, X-ray. At the same time, doctors can see the patient's reports and send their suggestions to the patients through the email. There is an option for every patient to edit their details. Tackling following concerns:

Concerns about information privacy: Concerns have been expressed about the possible implications for privacy and confidentiality when material related to patients is stored in one's personal device.

Superficial learning: Some researchers have demonstrated apprehension that these devices might enable learners to access information very rapidly. This may inhibit the "internalization of knowledge," which is a traditional part of medical education, leading to a potential for "superficial learning."

Trusting information sources: Concerns have been shown about how learners should navigate the swarm of information available from the devices, how they should find "good applications," and how they would know if the information provided was of high quality and validated.

8. References

[1] https://developer.android.com/training/implementing-navigation/nav-drawer.html

[2] https://stackoverflow.com/questions/41273503/send-receive-jsonrequest-from-android-to-firebase

[3] https://developer.android.com/design/material/index.html

[4] https://developer.android.com/training/volley/index.html

[5] https://developer.android.com/training/picasso/index.html

[6] https://developer.android.com/reference/android/os/OkHttp.html

Tyre Pressure Monitoring System

Amogha Abbi[#1], T. Ramakrishnaiah[#2],
[#]Vardhaman College of Engineering, Hyderabad, India
[1]aamoghaabbi@gmail.com
[2]t.ramakrishnaiah@vardhaman.org

Abstract: The Tyre Pressure Monitoring System which is used today indicates only whether the air pressure in any of the tyres is below a particular level of safety. This avoidable signal has made it possible for the statistical death rate to come up to 650 deaths per year. This project displays values through continuous monitoring. Therefore, this research proposed a method to cumulate the various factors of the tyre and to display it to the driver for a safer experience. The proposed TPMS has an electronic device unit that is attached to the tyre and transmits the values of the pressure, temperature, and the wheel alignment to the unit to be placed inside the vehicle, observable to the driver. This unit includes pressure sensors, microcontrollers, RF transmission, and batteries. An alert is introduced which indicates if the pressure exceeds the maximum or minimum safe pressure level. Several experiments have been carried out to analyze the proposed system. The integrated TPMS has proven to be an effective alternative to maintain the tyres and aims to improve the comfort and safety of the driver.

Keywords: Pressure, Temperature, Wheel alignment, Sensor, Microcontroller, Radio Transmission

1. Introduction

According to research, there is an increase in the number of accidents that take place due to the carelessness drawn towards the low air pressure in tyres. In India alone, 35-40% of accidents that have taken place, occur due to tyre bursts in the past 5 Years. Tyre blowouts are one of the most common reasons for accidents that take place. T. Xiangjun, proposed [1] the event of a sudden flat tire accounted for 49.81% of highway accidents, in 2006. N. N. Hasan, A. Arif, [2] et al proposed the current tyre pressure sensors indicate a symbol in the dashboard of the car that people usually tend to ignore. Currently, there are two basic methods to measure tire pressure; direct and indirect. The direct technique is implemented but with the use of a differential pressure sensor. This project will ensure that the driver will have the exact air pressures of all the tyres of the car displayed continuously. It also deals with the wear and tear of the tyre. One more aspect of the project is the temperature and wheel alignment of the car. When the car is driven in the most moderate conditions, there is greater longevity of the tyres while simultaneously provides a smoother and safer drive. In the currently used products, they aren't available for the wear and tear of tyres. However, through more research and technological interventions, we're looking forward to this product displaying.

- Tyre pressure is measured and displayed constantly.
- The wheel alignment that will help in parking and keeping the tyres in an optimum condition during parking.
- The temperature sensor that indicates the wear and tear of the tyre, along with the frictional changes applied to it.

2. Literature Review

Globally, we are heading towards a more secure and error-free future. Our project and the proposed methodology aim toward the same. In the interest of having a safer drive, the tyre used plays an important role and so does the maintenance of it. In the bigger picture, we are concentrating on the three most important features of a tyre. According to research, these are three of the most wide-ranged reasons that cause accidents.

Acceleration (degrees)	Pressure (psi)	Temperature (°C)	Output
< 20	>25	>10	Right Low pressure Min temperature
> -20	<32	<40	Left High pressure Max temperature
-20 ←→ 20	25 ←→ 32	10 ←→40	Straight Normal pressure Normal temperature

Figure1. Three important features handled

The unit can measure tire pressure whether the vehicle is still or in motion. The unit consists of a pressure sensor, a pressure switch, a microcontroller, an RF transmitter, and a long-lasting battery (+12volts) [1]. All units are switched ON and inserted into the tires. The alert message is displayed to the driver when the pressure goes below the set threshold value of 25 PSI. The pressure sensor has great sensitivity and takes 15 seconds to update its values.

The TPMS was first introduced in the 1980s in the European countries, generally as an additional feature since it was safe, friendly, and environmentally friendly. Some researchers and manufacturers define TPMS as a vehicle's safety device to maintain a vehicle's safety and check its condition, especially its tire pressure condition. TPMS is broadly divided into two methods, namely WSB (Wheel Speed Base) and PSB (Pressure Sensor Base).

This paper inclines towards Pressure Sensor Base, the pressure is directly proportional to the produced voltage at the output terminal of the sensor.

3. Block Diagram and Description

3.1. Transmitting side:

The block diagram below represents the transmitting part of the tyre pressure monitoring system. The proposed system can be understood in a pictorial format with the micro-controller differentiating sensors from the TLP 433MHz transmitter. Firstly, the pressure sensor (MPX10DP) senses the pressure of the tyre in PSI. Then this data reaches the Arduino through the analog pin (A0). The Arduino board has an in-built function of converting the analog value into its digital equivalent. Similarly, the accelerometer has its serial data and serial clock connected to A4 and A5 respectively.

The alignment of the car based on the pressure inside it is given by the accelerometer. Whether the car is aligned to the left or right or straight will be the three possible outcomes from the accelerometer. Along with these values the final parameter, temperature due to the friction caused in the tyre is also displayed. The temperature sensor is connected to the PIN (D2) of the transmitter at a frequency of

IOP Conf. Series: Materials Science and Engineering 1042 (2021) 012024 doi:10.1088/1757-899X/1042/1/012024

433MHz. Amplitude shift keying is the modulation technique that is utilized in this transmitter and receiver duo.

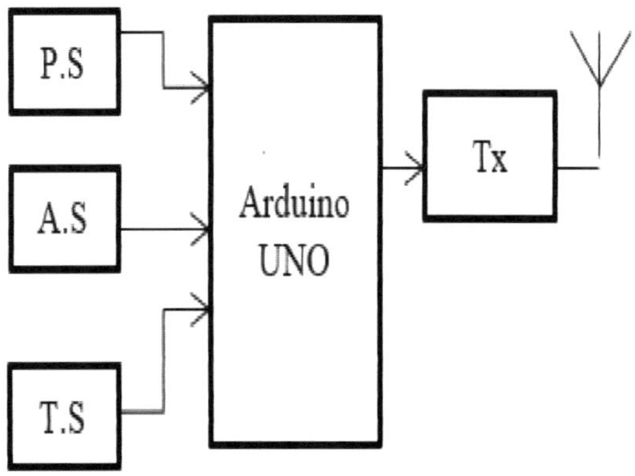

Figure 2: Block Diagram of the Transmitting unit

P.S: Pressure sensor
A.S: Acceleration sensor
T.S: Temperature sensor

3.2 Receiving side

The above block diagram represents the receiving part of the tyre pressure monitoring system. Identical to that of the transmitting part, the receiver side of the system also contains Arduino UNO as its brain. The receiving module is simpler when compared to that of the transmitter with the RF receiver and the LCD along with Arduino UNO. The RF receiver is connected to the PIN (D2) to receive the data packet sent by the transmitter. The L.C.D pins are connected to the pins ranging from (D8) –(D13). The potentiometer connected to the L.C.D functions to increase or decrease the brightness of the display.

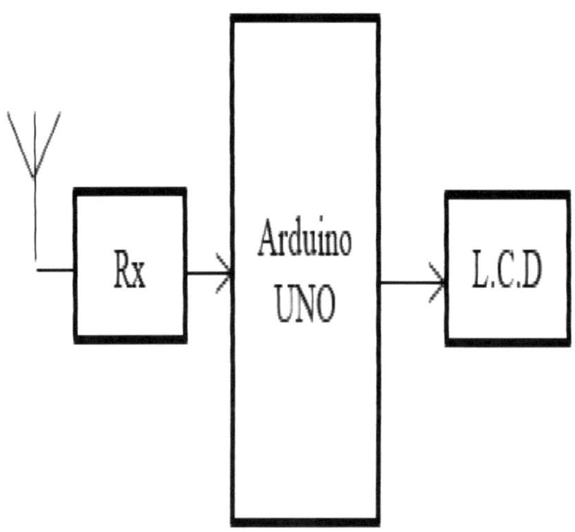

Figure 3: Block Diagram of the Receiving unit

4. Program Flow Process

Firstly, a Radio head library is installed in the Arduino IDE for enabling RF communication at 433MHz frequency. Later, the tyre pressure is sensed by the MPX10DP Pressure sensor in Pound per Square Inch (PSI). This value along with the temperature and acceleration is transmitted by the RF communication module with the help of the inbuilt functions of the library included.

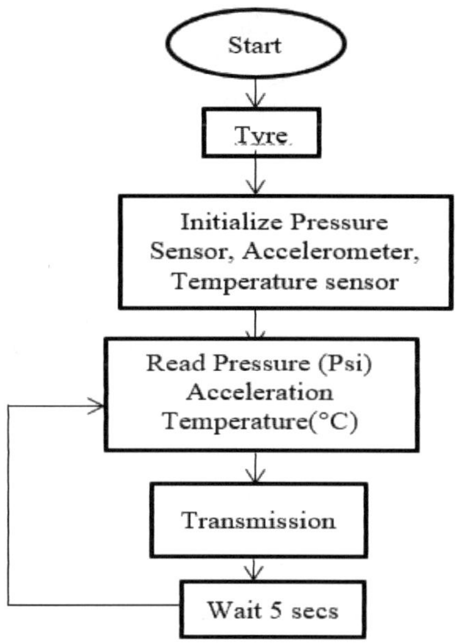

Figure 4: Flow process of transmission

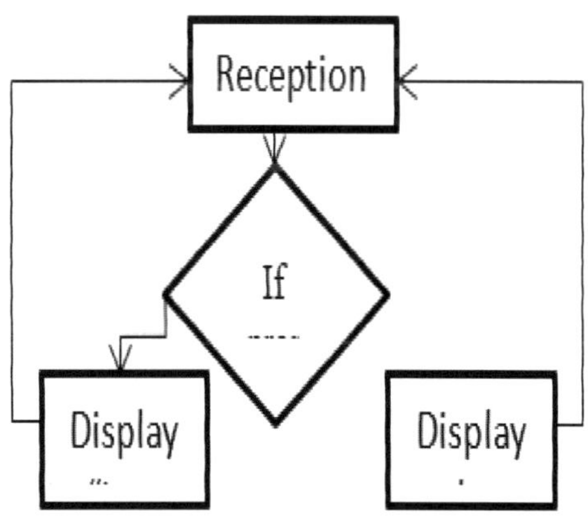

Figure 5: Flow process of reception

At the receiver end, the transmitted values are received and the pressure value is examined for a drop below 25 PSI. Once the value is below the threshold, the LCD would display an alert message to the driver.

5. Results and Discussions

The transmitting module can be seen in the below figure. This module is an interface of many sensors like temperature sensor, pressure sensor, and accelerometer. Among these, the pressure sensor attached to the tyre is the most important one. This pressure sensor is a dual-port uncompensated silicon sensor for level indicators. It is a silicon piezoresistive pressure sensor, directly proportional to the applied pressure outputs a linear and accurate voltage [3]. This standard uncompensated sensor permits manufacturers to design and add their external temperature compensation and signal conditioning networks. It is used for sensing & Instrumentation, Industrialization, Automation & Process Control.

Figure 6: Transmitting Module

Figure 7: MPX10DP Pressure sensor

The other sensors are for additional parameters like temperature and acceleration. The ADXL335, generally known as an accelerometer, is an electromechanical device that measures the force of acceleration due to gravity in g unit [4]. The analog voltage output proportional to the acceleration along x, y, and z axes is measured and given by it. The microcontroller can process these voltages by converting them to digital signals using ADC.

The DHT11 is a basic, ultra-low-cost digital temperature and humidity sensor. It uses a capacitive humidity sensor and a thermistor to measure the surrounding air and spits out a digital signal on the data pin (no analog input pins needed). Thus it is used in this application.

Figure 8: Receiving Module

Both in transmitter and receiver there are a few components that are in common and they are the microcontroller Arduino UNO and RF transmitter-receiver.

The above two figures 6 and 8 show the realized systems of transmitter and receiver modules. Below figure 9 shows the alert message that is displayed on the LCD. For the transmitting and the receiving unit, any small general-purpose microcontroller can be used. For ease, Arduino UNO was used. It is also capable of the UART protocol, which makes it suitable for RF transmission. The number of pins is sufficient enough to control the LCD at the receiver. At the same time, it also provides ADC channels at the transmitting side.

The TLP 433MHz transmitter and receiver work on the ASK modulation technique. This pair of communication devices is one of the least expensive data communication options we get in the market.

Figure 9: Alert message for low pressure

6. Conclusions and Future Work

The response obtained during the testing of the system gave stable and reliable outputs. This design is just a prototype that is capable of displaying only the pressure, acceleration, temperature of a single tyre. It is possible to extend this idea onto the other tyres of the vehicle without any confusion between them by adding an ID to each tyre. There is a chance for a size reduction, by using a PIC microcontroller instead of Arduino UNO. The reduction in size or the change in the microcontroller will not affect the working of the system. These modifications will make the system more users friendly.

At the receiver end, the transmitted values are received and the pressure value is examined for a drop below 25 PSI. Once the value is below the threshold, the LCD would display an alert message to the driver.

REFERENCES

[1] T. Xiangjun, "The Design and Research of Tyre Pressure Monitoring System," 2016 International Conference on Intelligent Transportation, Big Data & Smart City (ICITBS), Changsha, 2016, pp. 479-481.

[2] N. N. Hasan, A. Arif, M. Hassam, S. S. Ul Husnain and U. Pervez, "Implementation of tyre Pressure Monitoring System with wireless communication," 2011 International Conference on Communications, Computing and Control Applications (CCCA), Hammamet, 2011, pp. 1-4.

[3] S. Swathi, T. Ramakrishnaiah, "Driver Fatigue and Alcohol Detection System for Prevention of Accidents" 2020 International Journal of Emerging Trends in Engineering Research, 8(9), September 2020, 6275 – 6279

[4] Datasheet of MPX5700DP

https://datasheet.octopart.com/MPX5700DP-Freescale-Semiconductor-datasheet-22110.pdf

[5] Accelerometer
https://www.electronicwings.com/sensors-modules/adxl335-accelerometer-module

[6] DHT11 temperature sensor
https://components101.com/dht11-temperature-sensor#:~:text=DHT11%20Specifications%3A&text=Temperature%20Range%3A%200%C2%B0C,%C2%B0C%20and%20%C2%B11%25

[7] Stephen McClelland. "Remote tyre pressure monitoring system" U.S. Patent number: 5,963,128.

IOP Conf. Series: Materials Science and Engineering 1042 (2021) 012025 doi:10.1088/1757-899X/1042/1/012025

Survey on: Applications of Smart Wearable Technology in Health Insurance

Apeksha Shah[1], Dr. Swati Ahirrao[2], Dr. Shraddha Phansalkar[2], Dr. Ketan Kotecha[3]

[1,2]Symbiosis Institute of Technology, Symbiosis International (Deemed University)
[3]SCAAI, Symbiosis Institute of Technology, Symbiosis International (Deemed University)

*apeksha.shah.mtech2019@sitpune.edu.in

Abstract. Humanity is considered as the most intelligent species on the earth and are health reactive. Today, due to increase in sluggish behavior there is reduction in physical activity and that leads to health problems. To overcome this cause, wearable technology helps human being to get updates about physical health behavior e.g. heart rate, blood pressure, steps count, BMI, calories and many more. Wearables available in market helps to measure all kind of metrics related to health. The aim of this research is to study and analyze the research to check whether wearable technologies with augmented smart AI solutions help to predict health status and its outcome. Leveraging Artificial Intelligence (AI) and wearable technology could help health insurance to cover the risks associated with the uncertainties in the health, increased life span and high cost of treatments. In this study, we first discuss the works where wearable technology and AI can create value in the field of healthcare and importantly discuss the use case of leveraging the insurance domain with a smart health-monitoring service for the customers. Although, there are inhibitions in implementing a smart and fitness monitoring system in health-insurance sector like privacy concerns and hesitations towards sharing the data, health-insurance sector can look at this use-case as an incentivization to the fitness conscious clients.

1. Introduction

Wearable devices are influencing nowadays in medical health and health insurance industry through digital platform by continuously monitoring health attributes. As people are more concerned towards maintaining their health, healthcare practitioners, insurers are willing to adopt wearables for better health of people [5]. In market, many wearables are available which can track health related parameters such as heart rate, blood pressure, calories, step count, physical activity, etc. and is gaining popularity in the market for monitoring fitness. The wearable technology has been used for health insurance company for gaining profit as well as maintaining person's lifestyle healthier. Also, helps people in getting benefit from company such as discounts, vouchers, reduction in claim payment, etc. Hence, the data collected from wearable device helps medical company and insurance company to predict disease outcome [13].

Nowadays, technology such as Artificial Intelligence (AI) is increasing and has shown its effectiveness in the field of healthcare and health insurance industry for several application i.e. Pricing, Prevention, Prediction, Personalization applying on historical data. Generally, these are the use cases of AI which Insurance company are utilizing [22]. In some situations, when a person gets very exhausted, he/she may not be able to see symptoms of any diseases on the spot or sometimes it happens they may neglect about diseases occurring in their routine activities. To this problem, to alert people many researchers have come up with idea of predicting health and diseases risk with the static data available from hospital. But predicting huge amount of data may cost time, for this a model should be created which helps in predicting health risk [16].

Many AI technologies such as, Machine Learning, Deep Learning, Natural Language Processing, etc. are useful in making models. On applying data i.e. structured or unstructured to model, we can predict different things which can health firms in identifying future risk [25]. In this section, we will see some use cases of using wearables in healthcare and health insurance industry, which are mentioned below:

- Health Data Acquisition: Use of wearables to collect the vitals of individuals, helps in improving claim cost prediction.
- Incentivization: It will help making people lifestyle healthier and get rewards from wellness program.
- Health monitor: People gets alert message or notification from wearable to know about their health status.
- Health predictions: Several predictions can be made regarding health problems, customer retention and churn rate and fraud detection.
- Improved competitive position: Helps insurance company in strengthening their competitive position in the market.

In this paper, we would see what research has been done in health sector and health insurance using wearables in Section 2. Then, in Section 3 we will discuss and analyze what are the key findings after examining the literature. And lastly, we conclude this paper how wearables are useful for prediction purpose in Section 4.

2. Related Work

2.1 Application of Wearables in Life Style Monitoring

Patel MS et.al., gives idea for the wearable technology and devices which delivers benefits to human health behaviour, habits and well-being by using digital applications that are being used by many large technology companies. It focuses on the usage of the wearable device how it makes easy for human to adopt new technologies. Also, to adopt the new change in human behaviour and their habits these devices has to work on design strategies to make it effective for an individual. Hence, to connect with human behaviours through these wearable devices encouraging individual and feedback should be considered [1].

June Quah, gives idea about wearable is a medium through which we can monitor human health. There are several wearables available in the market but the main category of wearables in market lies to smart watches and fitness tracker. These wearables can collect human data such as physical activity in the form of step count, activity minutes, calories gain and burned, distance walked, etc. Also, many wearables nowadays can capture heart rate and sleep patterns. They have evaluated the effectiveness of human physical activity which are measured by wearable sensors helps in finding the mortality risk of a person. This was implemented on the dataset of U.S. population collected from clinical research studies. The main aim of this research is to find the relationship between person's health and lifestyle behavior that is, physical activity, sleep patterns, calories gain and burned, etc. [2].

Asthana S et.al., surveys nowadays, in market numerous wearable devices are available that shows many attributes somewhat similar to each other which helps in monitoring and measuring an individual health attribute. However, due to ample devices available it becomes difficult for a person which wearable he/she should purchase for their health risk monitoring. For this, in this paper they have prepare a model which recommend an individual which is beneficial for them. This model works as they first gather all data such as diseases, the person is at risk of; his/her attributes, age, gender, state; and medical history, smoking, any fracture in past, etc. Then these diseases are combined to the attributes we collected in order to monitor or predict diseases, this was done via textual analytics model [3].

Paul G et.al., focuses on the privacy issues related to wearable health devices. The services provided by these devices are marketed by providers to end users and all are ready to buy these devices which help in monitoring health of an individual. But, regarding the usage of these devices questions arises of user privacy. So, it investigates and compares the privacy policies of four services (Fitbit,

Jawbone, Nike+, BASIS). On comparing privacy features offered by each service, privacy score was calculated out of which Nike+ provides higher privacy than other services [4]. Dunn J et.al., depicts the wearable technology has broad range of application in the field of medicine and medical health system. As wearable is revolutionizing medical healthcare system; devices are used for general health management system to improve care of an individual when he/she is in hospital or outside the hospital. When a person is outside of the hospital wearable devices combined with sensors helps in monitoring and tracking condition related to heart, metabolic, gastrointestinal disorders, sleep, neurology, movement disorders, mental health, environmental exposures and many more. Several studies conducted shows using wearables in medical system helps person not to only monitors their activities but also helps in reducing hospital visits, hospital readmission rates, improving an individual health and costly medical events. Wearables can also inform person about any ongoing symptoms such as any side effects. Thus, wearable data and sensors helps in collecting information and transferring fast to hospital or any medical service in case of an emergency [5].

Jin CY., summarizes how AI technologies are applied on wearable devices and various types of application are inspired by it. It shows different types of wearable gadgets available in market, which collects data and helps in model training and its application. They have categorized different types of wearable device such as Smart phones, watches, wristbands, Smart glasses, clothes and socks, shoes, earphones. In addition to that, these gadgets can collect various data such as Sport data, Physiological data, Environmental data and Communication data. For processing these collected data from wearables different models have been proposed which are effective to use that are Machine Learning, Convolutional Neural Network, Long Short-term Memory and Ensemble Learning. Also, it notes different application where machine learning and wearables can be used. Still there are some implication which need further exploration [6].

Wright R et.al., surveys the wearable devices and technology are merged into clothes and accessories that are worn easily on the human body. These devices can be any of these: glasses, jewelry, headbands, and watches. These devices have great influence in the fields of healthcare, medical and body fitness. In the field of healthcare, several devices are developed such as Fitbit, wrist bands, and clip-ons which can help to track steps, activity, calorie burnt, distance and other physical activity. Many healthcare firms have developed smart glasses in the form of eyewear includes video glasses/goggles which is used for gaming, watching television or movies. Medical community has planned to use these Google glass to look up on patient records which was based on application. Where physician can wear these glasses and before entering to patient room, he can scan the QR code through glass and find out the details of patient. But using this application it provides only limited information, does not allow to add/modify information with any gestures or voice commands. Along with that battery only last about two which was insufficient for shifts in the department [7].

2.2 Wearables-Artificial Intelligence (AI) in Health Sector

In this paper, Yu KH et.al., describes Artificial Intelligence which is also said as machine intelligence has application in various fields such as Education, Healthcare, Business, Autonomous vehicles and many more. AI in Healthcare has played a vital role in medical diagnosis which takes information, process it and gives result to user. These analyses of data are done by applying AI technology that include machine learning technique for deep learning, structured data and natural language processing (NLP) for an unstructured data. Generally, AI in healthcare helps to analyze the relationship between the prevention/ treatment techniques and patient outcomes. Several medical institutions and companies are developing AI algorithms and predictive analysis for their betterment in improving business operations. Using AI in healthcare gives accurate results and better prediction in patient's treatment. Future scope is to develop Brain Computer Interaction that help people who are facing difficulty in moving, speaking and spinal cord injury [8].

Bayro-Kaiser E et.al., surveys as number of wearable devices are growing high in market of U.S but there are some consequences to know if population of U.S are willing to adopt the wearable technology for health insurance wellness program and the incentives through which they may adopt it. For this, they have conducted a survey on U.S residents who were willing to accept wearables and was conducted in six different use cases where it was mandatory to access wearable device and give all the metrics gathered within an insurance company. Those six use cases were, promoting health, early detecting of health diseases, predicting health risks, finding certain diseases disorder which may happen in coming future, adherence tracking activity, personalizing products and services and automated traditional process i.e. underwriting. On the basis of economic, privacy and technical circumstances, it says that two out of the three U.S population were willing to adopt wearable device for personal benefits related to health prevention and promotion [9]. Angelides MC et.al., shows how wearables can help an individual to improve their health lifestyle by bringing change in their daily routine and habit which may lead to a healthier lifestyle. They have developed a mobile app that supports data analytics and machine learning approach to help with inference of wearable data and recommending an individual's personal lifestyle improve. They have shown a workflow where configuration and goal are set up in smart watch then with the goal set watch tracks an individual's activity, those data are synchronized via Bluetooth, the raw data are analyzed and generates recommendation and forwards for visualizing the change or any improvement in habit or routine and then decision is taken and again after finding any change, goal is synchronized and whole cycle is repeated. Here, the analyses are merged with machine learning techniques to visualize patterns and correlation between activity such as sleeping and exercising which was very low. Thus, wearables require many sensors to monitor personal fitness during in case of any emergency. This would have high impact in health sector, insurance industry for better health and fitness [10].

Prabhu et.al., describes prediction of health risk at an early stage using machine learning technique. Many people in America have more chronic diseases and many of them have spent fee for the treatment of chronic diseases. This crisis has intended towards preventative measures where the major and primary concern is recognizing and taking actions at the early stage possible. These can be done on a person's lifestyle, habits and medical records. They have proposed a Collaborative Assessment and Recommendation which depends on person medical history. For this they have built a new algorithm using Convolutional Neural Network which is based on Multimodal Diseases Risk Prediction using both types of data: structured and unstructured. This model was implemented on dataset available from the hospital. Compared their work with existing work, they have focused on both types of data in data analytics. With these, the prediction accuracy of their proposed algorithm was faster. Hence, this performs better at predicting future diseases risk using machine learning algorithm [11].

Shinde SA et.al., gives idea on machine learning techniques and its applications how they are used in predicting health risk and health care system. The main purpose of this papers is categorized in four folds. They are about providing guidelines for the researchers who want to explore in machine learning field, providing survey on machine learning, its application in predicting health risk and showing further direction to be implemented in coming future into health prediction system. Also, it shows machine learning methods in predicting health risk is gaining popularity in research area as well as for the future direction [12].

Burnham JP et.al., reviews and analyzes the literature to check whether using wearables we can determine health outcomes or not. For this, they researched various studies that using wearable technology for predicting health outcome. Also, it involves human studies where wearable technology tracks human health parameter and modelled that data for a prediction of mortality, readmission and emergency department visits. They study resulted out of eight, six studies developed a model for readmission and two studies modelled for mortality. But each of studies, have captured data from a wearable device that were predictive. Hence, wearable technology has high influence in predicting clinical outcomes, but still there is more to find in health care system for future aspects [13].

Repaka AN et.al., focuses on predicting heart diseases using data mining technique by examining and collecting data which can be useful in finding patterns for making business related decisions. They have used Naïve Bayesian classification technique to predict heart related diseases. They have collected data from various sources which includes patient particulars such as person's span, blood pressure, sex, cholesterol, blood sugar, etc. Based on collected attributes it acts as an input for the Naïve Bayesian classification method. It is based on application platform where user has to login into its web portal and provides data and data are then transfer by Advanced Encryption Standard which are secured. Then the output is produced. The output reveals that the proposed system works systematically and effectively assists in predicting risk related to heart diseases [14].

Ambekar S et.al., surveys in healthcare data plays an important role. It has large amount of data to handle which can be used for health diseases prediction and patient care. Due to some missing data values, accuracy of the model decreases. To overcome the problem of missing values, they have performed method of data cleaning and transforms incomplete data into complete data. For this, they have applied data mining technique called Naïve Bayes classification and KNN algorithm applied on the dataset. To predict risk, they have used structured data and method CNN based unimodal risk of disease prediction algorithm. Overall, the accuracy of CNN-UDRP method have high accuracy in predicting risk [15]. Amin P et.al., describes nowadays identifying diseases and diagnosis the diseases are the forefront of machine learning in healthcare. Today, there is increase in urbanization and decrease in physical activity which leads to health problems. To overcome this issue, wearables are available in the market which helps in monitoring physical data and notifying user about their health condition. But not all wearables are accurate in measuring metrics and predicting health risk. For that, they have proposed a machine learning model that captures data collected from wearable device and process it and helps in prediction whether there is risk of cardiovascular diseases. Method such as Logistic Regression gives good accuracy in predicting health risk [16].

Massaro A et.al., surveys on a proposed work which states decision support system based on AI algorithm and the wearable device helps in predicting health report. The data captured by wearables are stored on Big Data and processed by Support Vector Machine algorithm that able to predict health status. Another algorithm, Long Short-Term Memory was applied on the dataset, that proves the function of physiological data prediction. Hence, data mining techniques applied in healthcare industry can increase prediction analytics which offers advantages [17].

Salamon J et.al., have experimented sentiment analysis and heart rate analysis through four different data sources. At first, they measured heart rate through wearable device that is Fitbit Charge HR and collect data regarding their heart rate, stress, excitement, tiredness, as well as all physical activity, BMI, blood pressure, etc. Also, they measured heart rate by the Basis Peak and the ECG and collect data. Then, they compared both ECG heart rate data and HR heart rate data, where the precision of HR heart rate data was good enough than ECG heart rate data. Secondly, for sentiment analysis they conducted a survey where people have to expressed their sentiments in tweets by writing short texts and Twitter as a medium was used as a social recording medium and the result was evaluated using positive and negative hashtags. These data were merged together to find the mutual dependency of human activity: physical and emotional [18].

The literature discussed the potential of AI leveraged wearable devices in health data acquisition, processing and prediction for effective health monitoring.

2.3 Wearables-Artificial Intelligence (AI) in Health Insurance Sector

Rao et.al., surveys wearable devices that are been used by insurance companies which helps in increasing their business profits, sales opportunities and reduce cost of customer retention and acquisition. For this, Big data and Wearables in Life Insurance offers them to face today's challenge to more targeted approach to attract customer and retaining them. Many companies have adopted wearable technology to attract customer to use this device which captures customers personal data such as all types of physical activities, blood pressure, BMI, heart rate, etc. Through these they set target for customer to achieve in given period which helps customer in maintaining their health lifestyle and getting better life and also helps customers to get benefits from insurance company. Also, Big Data helps insurance companies to identify new opportunities in business and helps in expanding relationship with customer for greater profitability [19].

Ravi Malhotra et.al., depicts data plays an important role in an insurance industry. The data are collected from various places such as wearable device connected with sensors, social media activity, voice analysis and many more. Many insurance companies are examining how they can take advantage of these data and merge it to Artificial Intelligence (AI) and Machine Learning (ML) to solve business challenges in an insurance industry which includes underwriting, claim handling, fraud detection, prevention, pricing, sales and customer experience. Also, they are finding ways how to explore ways to use predictive modelling and machine learning to enhance customer satisfaction, increase business efficiency and operations. It states graphically, trend of global AI market for coming years where it predicts in coming future market will overtake Artificial Intelligence (AI) technology. Also, adopting this technology some of the challenges may face by insurers related to data in an insurance company about security which should be solved [20].

Daniel Thyssen et.al., surveys how wearables are helpful and used in an insurance sector using predictive analysis. It shows how business can be processed better using Predictive modelling, Handwritten Recognition, Text Analytics and Data Capture. These can be done with the help of new technology such as wearables device, Internet of Things (IoT), Blockchain, Cognitive Technology and many more. Using Predictive modelling, it can help insurance company for claim processing, dynamic pricing, better underwriting and reserving decision. These predictive modelling can be done with the help of wearable devices which are integrated with sensors that collects data from human body and provides several benefits such as managing diseases at early stage, reducing claim in an insurance company by achieving goal set by an insurer, better rehabilitation, etc. [21].

Samad Masood et.al., shows nowadays, Artificial Intelligence (AI) is changing Insurance company and the way it engages customers. From many years, AI technique has been useful in detecting fraud but now insurers are adopting these technologies for better uses such as to speed up their operational efficiency. To improve operational efficiency and customer experiences insurers are merging Internet of Things with AI. Many companies started adopting AI that helps in claiming insurance on an ad hoc basis. Other than that, there are several uses of AI in Insurance Sector such as for Prediction, Pricing, Processing, Personalization and for Prevention techniques. For these, several companies are using wearable gadgets providing customer to achieve some physical activity by setting goal and based on results, it helps customer by offering different claims, discounts, wellness benefits with insurance. Hence, still this was challenging as data plays an important role, insurance company need to develop a data strategy for securing data [22].

Joanne Buckle et.al., focus on the electronic device that is wearable devices how it is implemented and what are the practical use cases of wearables in the private medical insurance in the market. It also shows how insurers are willing to use these wearables devices for their own benefit for business growth and profit. These wearables devices help in capturing different metrics of an individual health data. In their research, they have conducted a survey in market to understand customers view points and their interaction levels with wearable device. From this, insurer may determine the premiums of customer. Also, with the help of wearables medical insurance company may get to know about health risk assessment as a part of scoring which metrics had better predictive value for the outcome. Using wearables in medical insurance company, helps customer to develop better health and based on that they may be awarded by insurance company on completing some set target. Some of the insurance company are currently using wearable technology such as Aditya Birla Health, Aetna, Oscar United Healthcare and many more and thus providing some benefits to customers in the form of rewards, discounts, vouchers and recommendations. Also, it shows about what types of device available and what do they track and fitness activity. And, though wearables may attract customers to purchase and that helps in increasing their physical activity levels, but some key metrics are not yet captured by wearables for their health risk assessment scores [23].

Chris Falkous et.al., depicts how wearables are used in Life insurance company. There are several applications of using wearable gadgets such as tracking physical activity, improving underwriting capabilities, some value-added services for customer, also for insurance protection and main concern is to improve user health and well-being. Insurers using wearable devices need to know about the metrics and data collected from wearables in order to effectively quantify the impact on human health. Also, insurer can understand the relationship between metrics from devices and mortality rate. It helps insurer to communicate effectively and respond to evolving market [24]. Kumar N et.al., presents the use case of Artificial Intelligence in an Insurance Company to overcome issues for better customer satisfaction. For that, Insurance company built a conceptual model which calculates the relationship between use cases and AI. To verify the model, an empirical quantitative research which is Proof of Concept (POC) was used which shows uses for Insurance Sector to overcome customer troubles. Also shows various Insurance companies who have adopted AI technology in field of Insurance Company providing various benefits to customer to have a positive influence on business growth, customer satisfaction, minimizing frauds and profits in the insurance industry [25].

Balasubramanian R et.al., surveys how insurance companies will adopt AI technology in the future for their benefit. Where insurance company will need to understand some factors, which will be varied and how AI amend its claims, underwriting, distribution and pricing. Technology using CNN and other deep learning algorithms generally used for image, text and unstructured text processing is used for variety of applications. Using AI technology change from distribution to underwriting and pricing to claim will be automated. Also, for secure payment transfer blockchain technology will help to provide authority and authentication. The underwriting process will be automated is shrink to minute seconds supported by combining machine learning and deep learning models. Claims for business firm and personal lines will also be automated by reducing claim processing times into few minutes. In future, insurance worker will need to adopt these new technologies, must be creative and willing to work with mix of semiautomated and machine supported task that evolves [26]. Riikkinen M et.al., purpose is to find how chatbot in insurance support customers value creation. For developing chatbot, AI technology uses customer data for providing services in insurance sector. Chatbot designed provide interaction to customer influencing them with additional service. Here, they have built conceptual framework which plays 4 different roles: Info desk, Intelligent, Butler and Life Coach, these 4 are the chatbots that shows how customer value creation are supported. For future, chatbot can be created providing additional opportunities to customers for future research [27].

Burri RD et.al., gives idea about data are the key element in Insurance sector and how it converts machine learning methodology to insurance sector. Nowadays, insurance company are more interested in predicting future which has less financial loss for the company, optimize marketing strategy, to enhance company's income and to improve its business. The main purpose of this paper was to predict claim and helps insurance company to charge customer its premium that to be paid. They presented several machine learning algorithms such as Random Forest, Logistic Model Tree, Naïve Bayes, Multilayer Perceptron and J48 and compared each techniques performance using various metrics. On evaluating results of above-mentioned techniques, Random Forest algorithm outperforms having maximum accuracy [28].

Su J et.al., surveys, these days, the health insurance industry is facing difficulties and competition in market. One of the main concerns is customer retention which is an important factor for business growth development. In this, they have addressed the problems faced by health insurance company in maintaining right customers and for that they have developed a retention predictive model which shows scores of customer retention who are likely to retain within the company. In that model, they have also mentioned some factors that were affecting retention. In creating model, they have applied machine learning techniques such as supervised learning K-Means clustering and Hierarchical clustering. Through these they demonstrate the customer retention scores in an insurance company [29].

Guelman L et.al., focuses on customer who are more likely to retain within an insurance company. Many approaches are done to predict customer churn who are likely to switch another company. But this paper, concentrates on right customers who reacts positively to the company's persuasion. For that they have proposed an approach using method called Random Forest which was successful to predict customer retention. Also, developed algorithm of Random Forest for uplift problem. For processing a model, data set was taken from a Canadian insurer. Based on data set availability, they divided retention rates by groups that is Retained policies, Canceled policies and Retention rate. For modelling, input variables were taken as Policy, Driver and Vehicle characteristics. Using Random Forest algorithm, we can easily assume the change that a customer shift to different company when he/she approaches to company for renewal. These can help insurers also to change in their marketing skills towards customer churn [30].

Sabbeh SF., surveys, by comparing machine learning techniques for predicting scores of customer retention. As nowadays, customers are more indulged in getting Quality of service rather than quantity services. Also, organization needs to develop Customer Relationship Management system to find new customers and maintain the relationship with them and get bigger rate of customer retention for more profit gain by business. For the work to be done easily, Customer Relationship Management uses machine learning approach to build model which helps in predicting behavior and personal data of customer. This will help in predicting rates of customer retention and customer churn. Several machine learning models such as Discriminant Analysis, Decision Trees, KNN, SVM, Logistic Regression, ensemble-based machine learning algorithms were used. Overall, ensemble-based machine learning techniques outperforms all achieving higher accuracy than other methods [31].

Rawte V et.al., presents, number of frauds are increasing day by day in an insurance company which cause issues to health insurance field. To minimize these frauds, data mining techniques were applied to data available with insurance company. Before detecting fraud, it needs some preliminary data of health care system, some of fraud cases that happened in past. Here, using Data Mining technique they have divided these learning into two parts that is supervised and unsupervised learning to find fraud claims. Well both these methods have two sides: pros and cons but by combining pros of both method they have created a hybrid model for analyzing fraud claims in an insurance company. They have classified original claim and duplicate claim by clustering using Support Vector Machine algorithm (SVM). These SVM technique is used that provide scalability and usability required by data mining system [32]. Roy R et.al., surveys Insurance fraud that are happening nowadays in an insurance company. Many companies are facing this issue when person or entity makes false claim insurance to get compensation and benefits which are not owned by them. To overcome this challenge, this paper proposes a traditional approach for detecting fraud. It is based on heuristics development which indicates fraud but still there were some implication. Here, they have detected fraud in an auto or vehicle insurance company. They used machine learning techniques to indicate fraud. They have compared the accuracy of traditional method and machine learning method, where accuracy, precision and recall scores of machine learning technique was much higher than traditional algorithm [33].

Rayan N., presents how Insurance Company are experiencing fraud claim and processing. So, to analyze and detect fraudulent claims in Health Insurance sector the concepts of data mining and machine learning approach have been applied to identify it. Previously many techniques were applied to identify fraud but there was some implication with old method such as evaluating large database manually, new healthcare fraud was continuously increasing and requires regular update. To overcome these challenge, new approach was introduced for fraud detection system that is, advanced data mining, machine learning and statistical modeling method. Several factors were taken in consideration such as types of fraud happening in health insurance and healthcare data. They developed a workflow for analyzing and detecting fraud using supervised learning and unsupervised learning technique. This new approach was quite useful than old method [34].

Boodhun N et.al., surveys, in present times, companies perform underwriting method to make decision on application and to price the policies claimed by customer. Likely, there is increase in huge amount of data and it should perform underwriting method accordingly which causes some risk in assessment. So, with the help of increase in technology and data analytics, the process of underwriting method can be automated and the risk of assessing an application can be reduced. To do this, they have proposed a method using supervised learning where it helps in predicting risk. For this, machine learning algorithm namely, Random Forest, Multiple Linear Regression, REPTree and Artificial Neural Network were implemented. These methods were experimented on a dataset for predicting risk level. Among all these, Multiple Linear Regression outperforms showing higher accuracy [35].

2.3.1 Use cases of Artificial Intelligence (AI) in Health Insurance field.

The above presented literature helps us summarize the role of wearable technologies and AI in insurance domain with the help of the representation in figure 1 and figure 2.

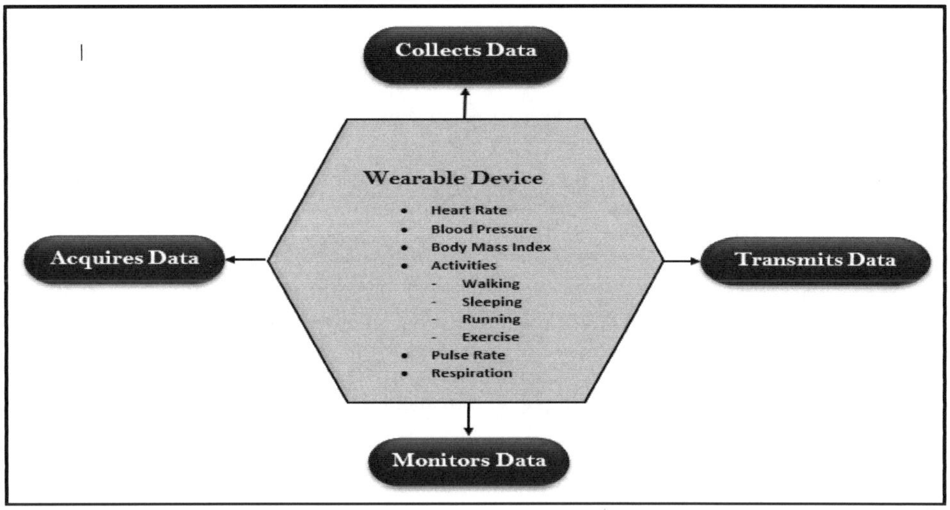

Figure 1. Representation of wearables

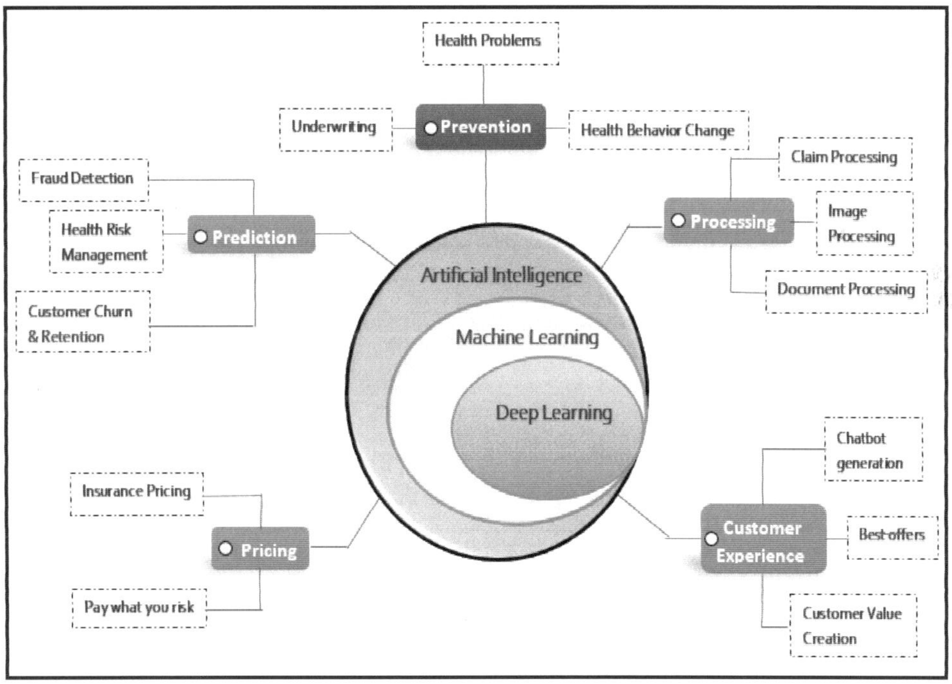

Figure 2. Representation of AI in Insurance use-case

2.4 Wearable-Artificial Intelligence (AI) in diseases prediction

We now discuss the works in the specific use case of AI and wearable in insurance. The chosen use case of disease prediction contributes as:

- Health Data Acquisition: Use of wearables to collect the vitals of individuals, helps in improving claim cost prediction.
- Incentivization: It will help making people lifestyle healthier and get rewards from wellness program.
- Health monitor: People gets alert message or notification from wearable to know about their health status.
- Health predictions: Several predictions can be made regarding health problems, customer retention and churn rate and fraud detection.
- Improved competitive position: Helps insurance company in strengthening their competitive position in the market.

2.4.1 Analysing. key findings and limitation of work done in insurance field.

Table 1. Key Findings and Limitation of work in the analysis

Author Name	**Year**	**Key Finding**	**Problems**
Alessandro Massaro	2020	Used AI algorithm- SVM and LSTM to predict health status on the data collected from wearable device	Still the goal of finding match between measured and predicted values need to be done
Poojitha Amin	2019	Using FitBit data as a wearable applied machine learning algorithm to predic heart- cardiovascular diseases	Data obtained has lower percentage o disease It only predicted heart disease
Anjan Nikhil Repaka	2019	Proposed approach was Application Based and applied Data mining technique- Naïve Bayes classification to predict hear disease	As it is application base, in overall process only patien can notice whether he/she has heart disease or not
PRABHU. T	2019	Concern was to recognize disease risk at an early stage using machine learning algorithm- CNN UDRP	Here, data was collected from Medical dataset in huge amount
Jin chun yu	2019	Summarize use of AI technology for wearable devices, its types, applications, what data are collected and models used	Problems were wearables don't give accurate measurement and due to that it effect: in prediction

Jessilyn Dunn	2018	Shows application of wearable helps in monitoring health to improve healthcare	Wearable High cos is main concern, does not measure accurate attributes

3. Discussion

3.1 Application of using wearables and AI in Healthcare and Insurance sector

In this section, we addressed how Wearables helps medical health and health insurance industry by providing solution to human health behavior. Also, we see using AI and wearables in healthcare and insurance industry has several use cases and application which helps in monitoring different attributes. AI technology merged with wearables provides solution for:

- Prediction - It helps insurance industry to offer a new way to predict real time events for their business benefit such as detecting fraud, customer retention and churn as well as also alert user for their health-related problems such as early health risk.
- Customer Experience – It helps insurance firms to know about customer learning using chatbot, providing best offers and customer value added services for better progress in company.
- Pricing – Price of Insurance policy may be reduced to an insure if they are properly motivated and convinced to used wearables otherwise, they have to pay as per the policy price.
- Prevention – Insurance company may get benefit to prevent human health fitness and illness from tracking physical activity and their daily payouts. Also, it bundles life insurance policies with change in human behavior and to reward customers with some benefits.
- Processing – Using AI in insurance sector helps insurer for faster claim processing, image processing as well as document processing. It will help in minimizing time and to underwriting policy.

Some benefits and application of using wearables:

- More than just tracking, helps doctors to monitor the patient remotely.
- Using wearable helps fitness loving people to get better healthier lifestyle by getting alert notification for health status.
- Also, incentivizes patients with reduction of paying claim cost.
- Helps insurance firms to improve business competency by providing wearables to its customer and maintain its position in the market.
- People can get benefit of getting rewards, discounts, coupons for wellness program using wearables.

The paper shows an exhaustive survey of different use case of AI and Wearable technology in insurance domain like claim processing, fraud detection, customer retention, customer churn as well as for predicting human health risk or diseases prediction using different datasets. AI techniques implemented were Convolution Neural Network, Support Vector Machine, Random Forest, Logistic Regression and many more. On analyzing and findings, we found work done in the field of healthcare for predicting health risk on dataset collected from different sources and least from wearables. The literature study shows the results were limited by the data set availability, choice of correct methods, and lesser accuracy in predicting health risk. Moreover, hesitation to share the health meter and activity tracker data are inhibitions towards the adoption. A better incentivization policy can be effective solution to the adoption of a service.

4. Conclusion

In this paper, we analyze how smart wearables with AI techniques find their applications in the field of healthcare and specifically health insurance. It helps healthcare system to get status about person's health, his exercise patterns and predict early diseases for better health in future. It demonstrates, how insurance company get benefit from wearable device to transform their relationship with customer for their growing business and profit. Our literature work opens the door for researchers, insurance domain experts, healthcare practitioners to find how the data collected from the end users can create value to the customers, insurance companies by creating a health-aware ecosystem. This will be however challenged with privacy concerns which have to be addressed. To discuss the far-reaching effects, it can also help in making the claims less fraudulent and help in quick and justified claim settlement.

References

[1] Patel MS, Asch DA, Volpp KG. Wearable devices as facilitators, not drivers, of health behavior change. Jama. 2015 Feb 3;313(5):459-60.
[2] The future is now: Wearables for insurance risk assessment [Internet]. Insureintell.com. Available from: http://www.insureintell.com/future-now-wearables-insurance-risk-assessment
[3] Asthana S, Strong R, Megahed A. Healthadvisor: recommendation system for wearable technologies enabling proactive health monitoring. arXiv preprint arXiv:1612.00800. 2016 Dec 2.
[4] Paul G, Irvine J. Privacy implications of wearable health devices. InProceedings of the 7th International Conference on Security of Information and Networks 2014 Sep 9 (pp. 117-121).
[5] Dunn J, Runge R, Snyder M. Wearables and the medical revolution. Personalized medicine. 2018 Sep;15(5):429-48.
[6] Jin CY. A review of AI Technologies for Wearable Devices. InIOP Conference Series: Materials Science and Engineering 2019 Nov (Vol. 688, No. 4, p. 044072). IOP Publishing.
[7] Wright R, Keith L. Wearable technology: If the tech fits, wear it. Journal of Electronic Resources in Medical Libraries. 2014 Oct 2;11(4):204-16.
[8] Yu KH, Beam AL, Kohane IS. Artificial intelligence in healthcare. Nature biomedical engineering. 2018 Oct;2(10):719-31.
[9] Bayro-Kaiser E, Soliño-Fermández D, Ding A, Ding EL. Willingness to adopt wearable devices with behavioral and economic incentives by health insurance wellness programs: results of a US cross-sectional survey with multiple consumer health vignettes.
[10] Angelides MC, Wilson LA, Echeverría PL. Wearable data analysis, visualisation and recommendations on the go using android middleware. Multimedia Tools and Applications. 2018 Oct 1;77(20):26397-448.
[11] Prabhu, Darshana J, Kumar D, Nazreen H. Health risk prediction by machine learning over data analytics. 2019
[12] Shinde SA, Rajeswari PR. Intelligent health risk prediction systems using machine learning: a review. International Journal of Engineering & Technology. 2018;7(3):1019-23.
[13] Burnham JP, Lu C, Yaeger LH, Bailey TC, Kollef MH. Using wearable technology to predict health outcomes: a literature review. Journal of the American Medical Informatics Association. 2018 Sep;25(9):1221-7.
[14] Repaka AN, Ravikanti SD, Franklin RG. Design And Implementing Heart Disease Prediction Using Naives Bayesian. In2019 3rd International Conference on Trends in Electronics and Informatics (ICOEI) 2019 Apr 23 (pp. 292-297). IEEE.
[15] Ambekar S, Phalnikar R. Disease risk prediction by using convolutional neural network. In2018 Fourth International Conference on Computing Communication Control and Automation (ICCUBEA) 2018 Aug 16 (pp. 1-5). IEEE.
[16] Amin P, Anikireddypally NR, Khurana S, Vadakkemadathil S, Wu W. Personalized Health Monitoring using Predictive Analytics. In2019 IEEE Fifth International Conference on Big Data Computing Service and Applications (BigDataService) 2019 Apr 4 (pp. 271-278). IEEE.
[17] Massaro A, Ricci G, Selicato S, Raminelli S, Galiano A. Decisional Support System with Artificial Intelligence oriented on Health Prediction using a Wearable Device and Big Data. In2020 IEEE International Workshop on Metrology for Industry 4.0 & IoT 2020 Jun 3 (pp. 718-723). IEEE.
[18] Salamon J, Mouček R. Heart rate and sentiment experimental data with common timeline. Data in brief. 2017 Dec 1;15:851-61.
[19] https://www.intel.com/content/dam/www/public/us/en/documents/solution-briefs/insurance-analytics-and-wearables-brief.pdf
[20] https://www.accenture.com/_acnmedia/pdf-84/accenture-machine-leaning-insurance.pdf

[21] Qeli E. Predictive analytics and wearables in insurance [Internet]. Swissre.com. Available from: https://www.swissre.com/dam/jcr:c6c46106-2965-4d83-84da-b18ec2c105bf/Presentation_Daniel_Thyssen_Emir_Qeli.pdf
[22] https://www.infosys.com/about/knowledge-institute/insights/Documents/ai-insurance.pdf
[23] The role of wearables in private medical insurance [Internet]. Milliman.com. Available from: https://us.milliman.com/en/insight/the-role-of-wearables-in-private-medical-insurance
[24] Wearable technology in life insurance [Internet]. Rgare.com. Available from: https://www.rgare.com/knowledge-center/media/research/wearable-technology-in-life-insurance
[25] Kumar N, Srivastava JD, Bisht H. Artificial Intelligence in Insurance Sector. Journal of the Gujarat Research Society. 2019 Nov 2;21(7):79-91.
[26] Balasubramanian R, Libarikian A, McElhaney D. Insurance 2030—The impact of AI on the future of insurance. McKinsey & Company. 2018 Apr.
[27] Riikkinen M, Saarijärvi H, Sarlin P, Lähteenmäki I. Using artificial intelligence to create value in insurance. International Journal of Bank Marketing. 2018 Sep 3.
[28] Burri RD, Burri R, Bojja RR, Buruga SR. Insurance claim analysis using machine learning algorithms. Int J Innov Technol Explor Eng 2019;8(6 Special Issue 4):577-582.
[29] Su J, Cooper K, Robinson T, Jordan B, BlueShield B. Customer retention predictive modeling in HealthCare insurance industry. 2009
[30] Guelman L, Guillén M, Pérez-Marín AM. Random forests for uplift modeling: an insurance customer retention case. InInternational Conference on Modeling and Simulation in Engineering, Economics and Management 2012 May 30 (pp. 123-133). Springer, Berlin, Heidelberg.
[31] Sabbeh SF. Machine-learning techniques for customer retention: A comparative study. International Journal of Advanced Computer Science and Applications. 2018;9(2).
[32] Rawte V, Anuradha G. Fraud detection in health insurance using data mining techniques. In2015 International Conference on Communication, Information & Computing Technology (ICCICT) 2015 Jan 15 (pp. 1-5). IEEE.
[33] Roy R, George KT. Detecting insurance claims fraud using machine learning techniques. In2017 International Conference on Circuit, Power and Computing Technologies (ICCPCT) 2017 Apr 20 (pp. 1-6). IEEE.
[34] Rayan N. Framework for Analysis and Detection of Fraud in Health Insurance. In2019 IEEE 6th International Conference on Cloud Computing and Intelligence Systems (CCIS) 2019 Dec 19 (pp. 47-56). IEEE.
[35] Boodhun N, Jayabalan M. Risk prediction in life insurance industry using supervised learning algorithms. Complex & Intelligent Systems. 2018 Jun 1;4(2):145-54.

Multi Variant HandWritten Telugu Character Recognition Using Transfer Learning

Tejasree Ganji [1], Muni Sekhar Velpuru [2], Raman Dugyala [3],
[1,3] Department of Computer science and Engineering,
[2] Department of Information Technology, Vardhaman College of Engineering and Technology

tejasreeganji@gmail.com
Abstract. Optical Character Recognition (OCR) has become one of the most important techniques in computer vision, given that it can easily obtain information from various images. However, existing OCR techniques cannot recognition Telugu literature characters (Handwritten Golusu Kattu writing) due to a lack of datasets and trained deep Convolutional Neural Networks (CNN). Since the Kakatiya Empire (12th to 14th century) the glorious era of Telugu literature spread across the region. Thereupon, several handwritten documents consist of ancient knowledge, health care tips, wealth information, and several land records written in Telugu Golusukattu writing. Therefore, getting that information has become a major problem because of a lack of expertise in Golusukattu writing skills in skills. In order to solve the above problem, we are proposing deep learning aided- OCR for Telugu literature.

Keywords*: Optical Character Recognition, Deep Convolutional Neural Network, Telugu handwritten recognition, Deep Learning.*

1. Introduction

In this era, there is an increasing demand for applications that can recognize handwritten characters automatically when it is feed to the computer. Now a day there is an immense demand for storing this kind of information available in either books or in any kind of documents into a computer for future use which can be retrieved later by search technique. This can be done by scanning the information which had to store and save them. But the problem with this is one can face difficulty to read the document as we cannot search the content of the image. The reason for this difficulty is those handwritten font's characters differ from the font of the characters in computer systems. Thus, the computer fails to recognize the characters while reading them.

This concept of storing the contents of paper documents in a computer storage place and then reading and searching the content is called document processing. Sometimes in this document processing, we need to process the information that is related to languages other than the English in the world. For this document processing, we need a software system called the character recognition system. This process is also called Document Image Analysis (DIA). Thus, we aim to build a character recognition system that can perform DIA which automatically transforms input documents into electronic format. There are distinct techniques to carry out this process, among these techniques we had chosen the Optical Character Recognition technique.

The goal of Optical Character Recognition (OCR) is used to classify optical patterns into corresponding alphanumeric or other characters. It is a type of document image analysis where a scanned digital image contains a machine-printed document or handwritten script is input to an OCR software engine and translate it into an editable machine-readable digital text format (like ASCII text).

Optical character recognition is the most important technique in computer vision. It can easily obtain information from various images, but it cannot recognize Telugu handwritten characters due to a lack of data sets and trained deep convolutional neural networks. Optical character recognition is the process of converting scanned images of handwritten, typed, or printed text into machine-encoded text. OCR technology deals, with the problem of recognizing all kinds of different characters. Handwritten and printed characters can be recognized and converted into a machine-readable or digital data format. By using OCR we can transform serial numbers or code consisting of numbers and letters into a digital output.

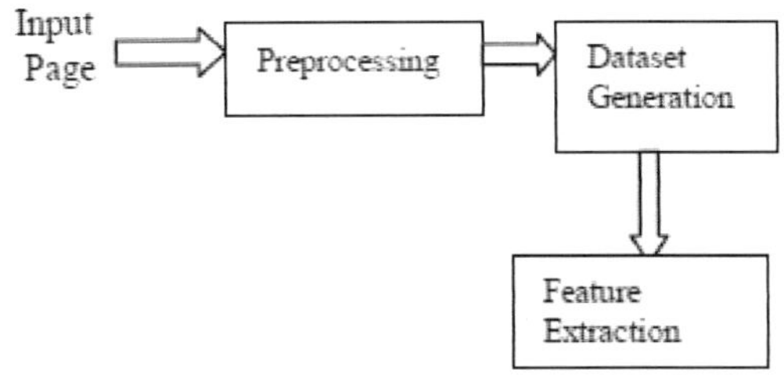

Figure 1. Phases of the OCR system image

In our research work, we had focused on character image classification, segmentation, and recognition. We had implemented the proposed model based on improved deep learning approaches for Telugu Handwritten Character (scanned Handwritten image) segmentation, classification, and detection. A comparison of the proposed model along with other models demonstrates greater performance against similar. The main contribution of this research work is summarized as follows:

1. The proposed model is applied to classify Handwritten Characters.
2. The proposed Telugu Character Recognition transfer learning model "*TCR_new_model.h5*" is designed from the VGGNET model.
3. For the models, from encoding to decoding unit concatenation is applied to the feature maps.
4. The proposed model provides a good outcome compared to other models.

The paper is organized as follows: Section 2 discusses related work. Section 3 describes the methodology and proposed architecture, section 4 explains the dataset and experimental results. Section 5 describes the conclusion and future work.

2. Related Work

A global survey has done on Optical Character Recognition, Segmentation, and Deep Learning approach where various learning techniques, models, and recently proposed training approaches are discussed.

Atul Negi et al. [1], implemented a model using zoning and structural feature vector to recognize the isolated text and they define cavities as a new set of structural features for recognition. In the segmentation process, they gave input as a document image, and that image is analyzed in 2 phases- top-down analysis and bottom-up analysis. The proposed method achieved an accuracy of 97%-98%.

Muni Sekhar Velpuru et, al.[4],they surveyed Optical Character Recognition, Segmentation, and Deep learning. Optical Character Recognition (OCR) is used to recognize text in the images. Images mean scanned documents or pictures. By using OCR, there is no need to digitize the image text. Another advanced technology that is extensively used in OCR is deep earning. Segmentation is used to segment the words into characters.

K. Mohana Lakshmi et al. [10], developed an efficient algorithm to identify individual handwritten Telugu characters based on HOG (histogram of oriented gradient) features and handwritten Telugu text efficiently. Experimental results are analyzed using the Bayesian network algorithm. Depending upon the true positives and true negatives accuracy has been calculated. They achieved the recognition rate for the Telugu script is 87.5%.

C. Vikram et al. [6], proposed a model for the recognition of Telugu handwritten characters, they approach the pre-processing and recognition. The preprocessing is based on ANN (Artificial Neural Network) and MLP (multi-layer perceptron). The recognition is based on hybrid optical HMM/ANN (Hidden Markov model/Artificial Neural Network) and the Multi-Layer Perceptron model applies to handwritten Telugu scripts. For recognition of unconstrained offline handwritten text lines, they used a hybrid HMM/ANN (Hidden Markov Model/Artificial Neural Network) system that is proposed by

Salvador Espanaet [1]. The Proposed scheme has a recognition accuracy of 85%.

Panyam Narahari Sastry et al. [8], In this paper, zonal based feature extraction is used for feature extraction, the character image is divided into a predefined number of zones and a statistical feature is computed from each of these zones. The features of all the zones in the image from a feature vector which is used for handwritten character recognition. In this, they have used NNC (nearest neighbor classifier) i.e., Euclidean distance for identifying and classification for Telugu text. Using the zoning method the recognition accuracy found to be 78%.

K.Vijay Kumar et al. [11] proposed online handwritten character recognition for the Telugu language using support vector machines. To interpret characters automatically automated character recognition is used and it is useful while processing a large number of handwritten characters. To overcome the variations in font and size of Handwritten Character Recognition they used the following techniques. 1) to review the issues and techniques for HCR which are: i) Preprocessing ii) Step stroke Pre-classification iii) Feature Extraction, and 2) to review the support vector machine (SVM) as the classification method. The proposed model achieved an accuracy of around 82.96%.

Konkimalla Chandra Prakash et al [9]; developed the dataset and classifier also introduces pre-processing, segmentation (word segmentation, character segmentation), recognition, and convolution neural network (CNN). The performance of the OCR system depends on its classifier. They have used SVM based classifiers for character classification. They have used 2CNN architecture for classifying the characters. They have developed an android app for this project, which converts an online image to text with industry-standard; it works on any android device. It uses a camera or gallery for images and it is publicly available. The experimental results show a better accuracy of about 98.60%.

SD Prasad et al [15]; to improve the accuracy in Telugu Character recognition they used Zonal-based Feature Extraction. In Zonal based Feature Extraction, they presented two methods: 1) based on Genetic Algorithm and uses Adaptive Zoning topology with extracted geometric features. 2) Zoning is done in a static way and uses distance, density-based features. For feature extraction, they use a zone-based approach. And also, they use 3 types of features. They are distance, density, and geometric based features.

C.Vasantha Laxmi et al [2]; proposed A high accuracy OCR System for Printed Telugu Text. In Pre-processing technique grayscale image is converted to a binary image, neural recognizers are used for identification of basic symbols which founds the fundamental units for segmentation purpose. The main feature of the proposed model is that it can handle multiple sizes and multiple fonts. That outcome of it can be directly implemented to other Indian languages software which supports translation facility into Telugu.

A Ashiqzzaman et al [12]; the author has proposed deep learning neural networks using appropriate activation function and regularization layer. Developed a method that can increase the accuracy of this process by using a convolution neural network (CNN) in place of MLP. CMATERDB Arabic handwritten digit dataset is used to evaluate experimental results. This proposed model will give 97.4% accuracy.

R Vaidya et al [13]; Author has proposed an innovative method for offline handwritten character detection using deep neural networks. They have designed a segmentation-based handwritten character recognition system for image segmentation. They had used Open CV for performing Image Processing and for training a neural network they used Tensor Flow. To train the neural network they comprised a stack of Convolution layer and MAX-pooling layer followed by a fully connected layer.

Meiyin Wu et al [14]; In this, they applied deep learning to real-world handwritten character recognition, and explored the two mainstream algorithms of deep learning: the Convolutional Neural Network (CNN) and the Deep Belief Network (DBN). They conduct the performance evaluation for CNN and DBN on the MNIST database and the real-world handwritten character database. And they achieved the classification accuracy rate of CNN and DBN on the MNIST database is 99.28% and 98.12%. And on the real-world handwritten character database is 92.91% and 91.66%. They analyzed differences between CNN and DBN by comparing the experimental results.

Table 1. Summary of Handwritten character recognition using different algorithm & methods

S.NO	Author Name	Methods Used	Dataset	Validation accuracy
1.	Atul Negiet al. [1]	Zoning and structural Feature Vector	Telugu character dataset	92%
2.	K. MohanaLakshmi et al. [10]	Histogram of Oriented gradient Features	Telugu word images	87.5%.
3.	C. Vikram et al. [6]	Multi Layer Perceptron And Hidden Markov Model	Telugu Scripts	85%
4.	Panyam Narahari Sastry et al. [8]	Zonal Based feature Extraction	Telugu handwritten characters	78%
5.	K. Vijay Kumar et al. [11]	Support Vector Machine	Own dataset	82.96%
6.	Konkimalla Chandra Prakash et al [9]	SVM based Classifier	Own dataset	98.60%,
7.	SD Prasad et al [15]	Zonal based Feature Extraction	Telugu characters	
8.	C. Vasantha laxmi et al [2]	A High accuracy OCR system	Basic symbols	
9.	A Ashiqzzaman et al[12]	Deep Learning Neural Network	CMATERDB Arabic Hand written digit dataset	97.4%
10.	R Vaidya et,al[13]	Deep Neural Network	MNIST	
11.	Meiyin Wuet,al[15]	Convolutional Neural Network	MNIST and Real World hand Written Database	91.66%

3. Methodology

According to the literature, survey researchers had developed many applications in the field of handwritten character recognition for the automatic computer-aided detection system. The main function of handwritten character recognition is classifying and recognizing the handwritten characters. The image of the handwritten text may be sensed "offline" from a piece of paper by optical scanning (Optical Character Recognition) or Intelligent word Recognition. A handwriting recognition system handles formatting, performs correct segmentation into characters, and finds the most plausible words.

The three objectives of the work are:

- Assess the current level of technology in Telugu OCR
- Identify opportunities and challenges facing by Telugu OCR
- Converting ancient Telugu documents into digital records (Converting Telugu handwritten literature into digital records.)

Image Acquisition: Handwritten TELUGU Character Dataset includes 1600 characters. It has been divided up into training and testing sets of 70% and 30%. The images are of grayscale. The size of each image is 52x52. In the preprocessing stage image is resized into the following sizes: 224x224. 128x128,

64x64 and 52x52 from the dimensions 52x52. Each image is normalized as shown in figure 3. Shows one scanned character sample image on the left side and its corresponding recognition map with the ground truth on the right side.

A. Data Collection: This dataset consists of 1400 unique characters, 52 sets of each character handwritten images written by 52 distinct writers. In addition, 12 people under 18 years, 7 people above 45, and the remaining are engineering graduates to enable different handwriting styles. It has the following advantages compared to the offline handwritten Telugu characters.

- Segmentation of characters not required. Because each character scanned as a different image with a good resolution.
- No salt-and-pepper noise, no chance for eroded margins, no skewed while scanning
- No need to spend time on character extraction, only focus on character recognition.

This dataset covers most of all Telugu characters in the literature approximately 1600 characters. It is the largest dataset available for Telugu characters as of now. Moreover, each character image is multiplicities with 12 using Image Data Argumentation (shifting, rotation, and sheering effect).

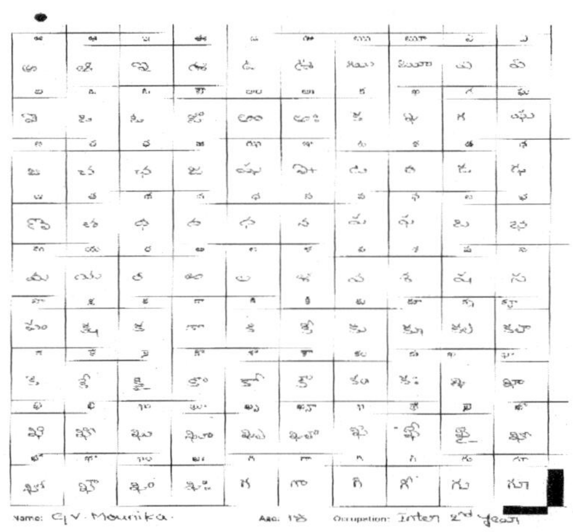

Fig.2 Dataset Collection

B. Data Analysis: in this, we created a dataset from collected data. And we collected data from various people. We divided the data as a testing dataset and training dataset. We train the dataset using VGGNET-16.

C. Validation: we cross-check the results from the test dataset.

Architecture: by using the transfer learning algorithm we had proposed a Visual Graphics Group-16 model The VGGNET can be pre-trained so that it works as a deep feature extractor. CNN's are data-driven learning, helps in higher feature extractors, highly representative but they need sufficient training data. The proposed model has three models. These three models are considered as they had performed well on Telugu Handwritten Character recognition tasks. The proposed Telugu Character Recognition transfer learning model "*TCR_new_model.h5*" is designed from the VGGNET model with different input sizes.

Table 2. Models and Input Sizes

Model No.	Input Sizes
Model 1	(224, 224,3)
Model 2	(128,128, 3)
Model 3	(64, 64, 3)
Model 4	(52, 52, 3)

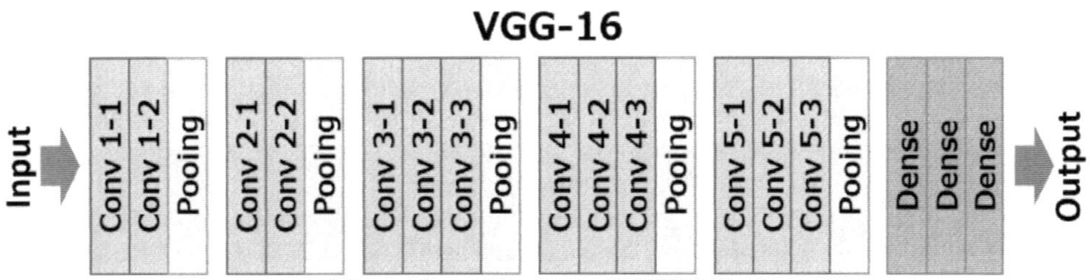

Figure 3. Architecture of VGGNET-16

Table 3. input size and parameters of CNN Model

Model No.	Input Size	Output Total Parameter	Trainable Parameter	Non Trainable Parameters
Model 1	(224, 224,3)	15,116,112	401,424	14,714,688
Model 2	(128,128, 3)	14,845,776	131,088	14,714,688
Model 3	(64, 64, 3)	14,734,586	30,456	14,714,688
Model 4	(52, 52, 3)	14,722,896	8,208	14,714,688

Configuration of VGG-16:

ConvNet Configuration					
A	A-LRN	B	C	D	E
11 weight layers	11 weight layers	13 weight layers	16 weight layers	16 weight layers	19 weight layers
input (224 × 224 RGB image)					
conv3-64	conv3-64 **LRN**	conv3-64 **conv3-64**	conv3-64 conv3-64	conv3-64 conv3-64	conv3-64 conv3-64
maxpool					
conv3-128	conv3-128	conv3-128 **conv3-128**	conv3-128 conv3-128	conv3-128 conv3-128	conv3-128 conv3-128
maxpool					
conv3-256 conv3-256	conv3-256 conv3-256	conv3-256 conv3-256	conv3-256 conv3-256 **conv1-256**	conv3-256 conv3-256 **conv3-256**	conv3-256 conv3-256 conv3-256 **conv3-256**
maxpool					
conv3-512 conv3-512	conv3-512 conv3-512	conv3-512 conv3-512	conv3-512 conv3-512 **conv1-512**	conv3-512 conv3-512 **conv3-512**	conv3-512 conv3-512 conv3-512 **conv3-512**
maxpool					
conv3-512 conv3-512	conv3-512 conv3-512	conv3-512 conv3-512	conv3-512 conv3-512 **conv1-512**	conv3-512 conv3-512 **conv3-512**	conv3-512 conv3-512 conv3-512 **conv3-512**
maxpool					
FC-4096					
FC-4096					
FC-1000					
soft-max					

Figure 4. configuration of VGGNET-16

VGGNET-16 architecture contains six layers and one input layer and one output layer as show in figure (3). The first layer of architecture contains two convolutional layers and one max pooling layer. The second layer has two convolutional layers and one max pooling layer. The third layer contains three convolutional layers and one max pooling. The fourth layer has three convolutional layers and one max pooling. The fifth layer contains three convolutional layers and one max pooling layer and the last layer contains three dense layers.

Let's consider input to the network an image of dimensions(224,224,3) as shown in table (2). And the first 2 layers have 64 channels of 3*3 filter size and the same padding. After that, a max pool layer of stride(2,2), the 2 layers which have convolution layers of 256 filter size and the filter size is (3*3). This process is followed by a max pool layer and a max-pooling layer is the same as the previous layer. There are two sets of three convolutional layers and a max pool layer. Each has 512 filters of (3*3) instead of 11*11 in AlexNet and 7*7 in ZF-Net. But in some of the layers, also uses 1*1 pixel which is used to manipulate the number of input channels. And also there is a padding of 1-pixel done after each convolution layer to prevent the spatial feature of the image.

The proposed VGGNET-16 model is the building block of the convolutional units and architecture of the VGGNET-16 shown in Figures 4 (b) and the configuration of VGGNET-16 is shown in figure (2).

4. Result Analysis

The dataset covers most of all Telugu characters in the literature approximately 1600 characters as shown in figure (2). It is the largest dataset available for Telugu characters as of now. Moreover, each character image is multiplicities with 12 using Image Data Argumentation (shifting, rotation, and sheering effect).

The experimental analysis is evaluated using Google Colab. VGGNET-16 algorithm is applied to the HandWritten Telugu Character dataset. This dataset covers most of all Telugu characters in the literature approximately 1600 characters.

For quantitative analysis of the experimental results, the following performance metrics are considered, including accuracy (AC), Dice coefficient (DC). To do this we also use the variables True Positive (TP), True Negative (TN), False Positive (FP), and False Negative (FN).

Confusion Matrix: it is a particular table format that allows the performance evaluation of an algorithm The overall accuracy is calculated using an Equation.

$$AC = \frac{TP + TN}{TP + TN + FP + FN} \tag{1}$$

Table 4. Accuracy and loss

Model #	Accuracy	Loss
Model 1	92%	0.5
Model 2	80.5%	0.6
Model 3	90%	0.2
Model 4	50%	1.5

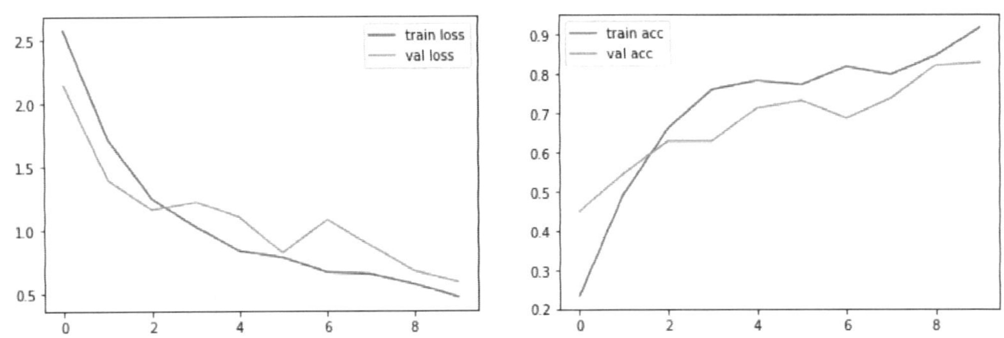

(a) (b)

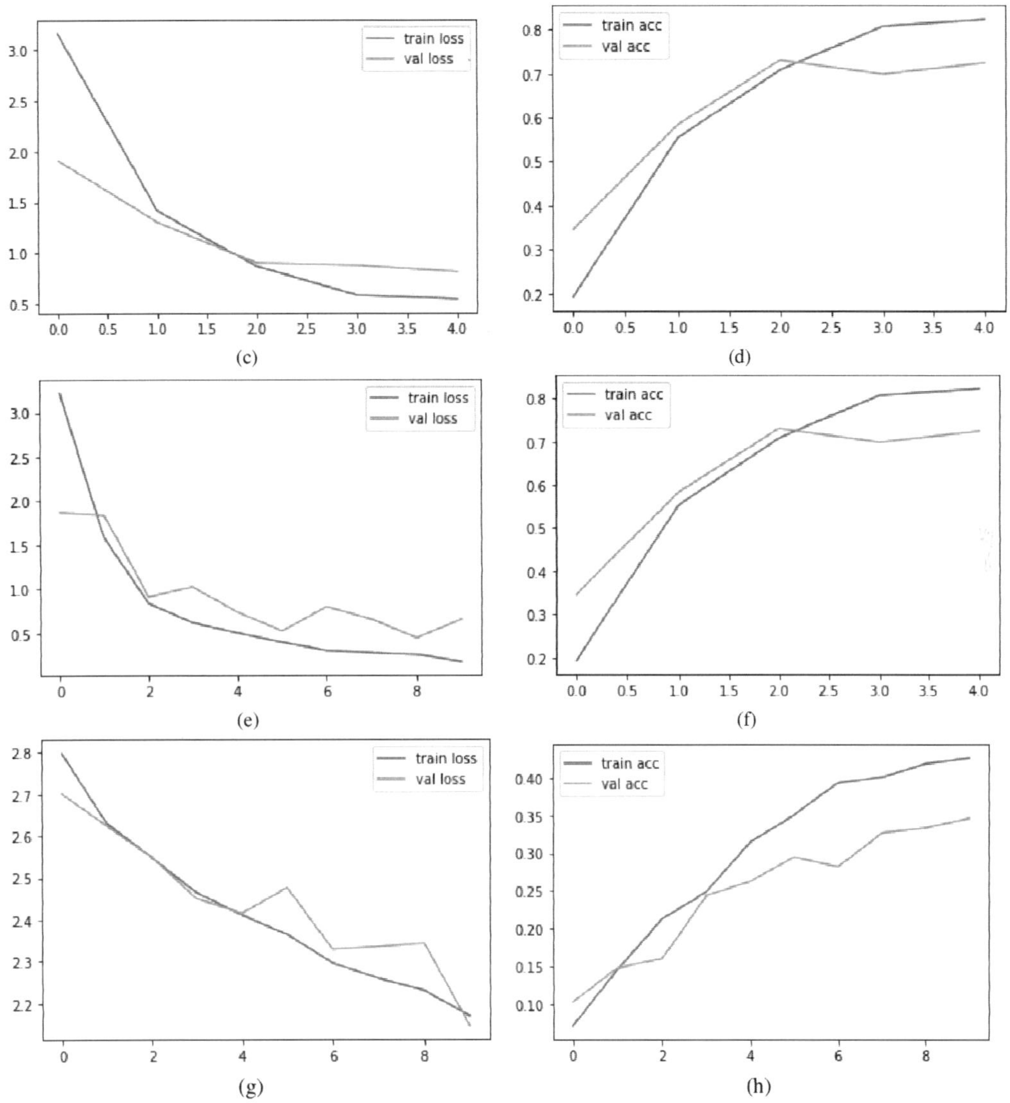

Graph 1. (a,b) represents Model 1, (c,d) represents Model 2,(e,f) represents Model 3, (g,h) represents Model 4

In the above graph 1.(a), (c),(e), and (g) represents training loss and validation loss of model 1,2,3,4 respectively. and graph 1.(b),(d),(f), and (h) represents training accuracy and validation accuracy of Model 1,2,3,4 respectively, where x-axis reprsents number of epochs and y-axis represents acuuracy and loss percentages. Model 1 achived 92 percent accuracy and 0.5 loss. Model 2 achived 80.5 percent of accuracy and 0.6 loss. 90 percent accuracy and 0.2 loss achieved by Model 3 where as Model 4 achieved 50 percent accuarcy and loss of about 1.5.

4. Conclusion

In this paper, we are implemented in our own dataset and used a model called VGG-16. By using the VGG-16 we are divided the data into testing and training set. We trained and tested the data using VGGNET-16 and achieved good accuracy. By using this method we can directly recognize the Characters with no need to do segmentation and all because we scanned each and every character and created our own dataset and by using that dataset only we trained the network and we tested some documents and we can easily digitize the Telugu documents. This will help everyone who wants to store their Telugu documents into digital documents.

Acknowledgment:
The work is carried out Under the Collaborative Research project of TEQIP-III/JNTUH file no: JNTUH/TEQIP-III/CRS/2019/CSE/12 and Titled " **Deep Learning Aided-OCR for Handwritten Telugu Character".**

References

[1] Atul Negi, Chakravarthi Bhagavathi, B. Krishna "An OCR System for Telugu". IEEE, 2001, 0-7695-1263.
[2] C. Vasantha Lakshmi, C Patvardhan "A high accuracy OCR System for Printed Telugu Text", IEEE,2003, 0-7803-7651.
[3] Atul Negi, Nikhil Kasinadhuni", Localization and Extraction of Text in Telugu Document Images", IEEE,2003, 0-7803-7651.
[4] Muni Sekhar Velpuru, Priyadarshini Chatterjee, G Tejasree, M Ravi Kumar, S Nageswara Rao "Comprehensive study of Deep learning-based Telugu OCR", IEEE ICSSIT 2020, 22 August 2020.
[5] Neerugatti Varipally Vishwanath, Manjunathachari and Satyaprasad, "Handwritten Telugu Composite Character
Recognition Using Morphological Analysis", International Journal of Pure and Applied Mathematics, 2018,119,18 2018, 667-676.
[6] Chakravarthy Bhagvati, Tanuku Ravi, S. Mahesh Kumar, Atul Negi," On Developing High Accuracy OCR Systems for Telugu and Other Indian Scripts", Proceedings of the Language Engineering Conference, IEEE,2002,0-7695-1885-0/02.
[7] Atul Negi, Nikhil Kasinadhuni," Localization and Extraction of Text in Telugu Document Images".
[8] Panyam Narahari Sastry, T.R.Vijaya Lakshmi, N.V.Koteswara Rao, T.V.Rajinikanth, Abdul Wahab, Telugu Handwritten Character Recognition using Zoning Features", IEEE,2014, 978-1-4799-6541-0/14.
[9] Konkimalla Chandra Prakash, Y. M. Srikar, Gayam Trishal, Suraj Mandal, Sumohana S. Channappayya, "optical Character Recognition(OCR) for Telugu: Database, Algorithm, and Application", IEEE, 2018, 978-1-4799-7061-2/18.
[10] K. Mohana Lakshmi, T Ranga Babu," A New Hybrid Algorithm for Telugu Word Retrieval and Recognition" .nternational Journal of Intelligent Engineering and Systems, Vol.11, No.4, 2018.
[11] K. Vijay Kumar, R. Rajeshwara Rao, "Online Handwritten Character Recognition for Telugu Language Using Support Vector Machines", International Journal of Engineering and Advanced Technology, ISSN: 2249 – 8958, Volume-3, Issue-2, December 2013.
[12] Rakesh Achanta, Trevor Hastie, "Telugu OCR Framework Using Deep Learning",
[13] C.V.Chakradhar, B.Rajesh, M. Raghavendra Reddy, "A Study on online Handwritten Telugu Character Recognition", International Journal of Science, Engineering and Technology Research (IJSETR) Volume 5, Issue 7, July 2016.
[14] N. Shobha Rani, T. Vasudev," Post-Processing Methodology for Word Level Telugu Character Recognition Systems using Unicode Approximation Models",IEEE, 2015, 978-1-4673-6667-0/15.
[15] C. Vasantha Lakshmi, Ritu Jain, C. Patvardhan, "OCR of Printed Telugu Text with High Recognition Accuracies",Springer,2006,786-795.
[16] Manpreet Kaur, Ashok Bathla, "A Review on Different Techniques for Character Segmentation for Hindi Handwritten Words", International Journal Peer Reviewed Journal Refereed Journal Indexed Journal, 2454-6615.
[17] K.Karthick, K.B.Ravindrakumar, R.Francis, S.Ilankannan, "Steps Involved in Text Recognition and Recent Research in OCR; A Study" International Journal of Recent Technology and Engineering, 2277-3878, Volume-8, Issue-1, May 2019.
[18] Amit Choudhary, "A Review of Various Character Segmentation Techniques for Cursive Handwritten Words Recognition", International Journal of Information & Computation Technology. 0974-2239 Volume 4, Number 6, 2014.

[19] Amjad Rehman, "Cursive Overlapped Character Segmentation: An Enhanced Approach",
[20] Rajiv Kumar and Amardeep Singh, "Character Segmentation in Gurumukhi Handwritten Text using Hybrid Approach", International Journal of Computer Theory and Engineering, Vol. 3, No. 4, August 201.
[21] A.Venkata Srinivasa Rao, Madasu Subbarao, Nenkkanti Venkata Rao, A.S.C.S.Sastry, L.Pratap Reddy, "Segmentation of Touching Handwritten Numerals and Alphabets", 2009 Second International Conference on Computer and Electrical Engineering.
[22] S.E. Benita Galaxy, S. Selvin Ebenezer, "Enhancement of segmentation and zoning to improve the accuracy of handwritten character recognition", International Conference on Electrical, Electronics, and Optimization Techniques (ICEEOT) – 2016.
[23] Aravinda C.V, Dr.Prakash H.N, "A Review On Various HandWritten Cursive Characters Segmentation Techniques", IEEE, 2016, 978-1-5090-2399-8/16.
[24] Li Chen, Song Wang, Wei Fan, Jun Sun, Satoshi Naoi, "Cascading Training for Relaxation CNN on Handwritten Character Recognition", 2016 15th International Conference on Frontiers in Handwriting Recognition.
[25] Akm Ashiquzzaman and Abdul Kawsar Tushar, "Handwritten Arabic Numeral Recognition using Deep Learning Neural Networks", IEEE, 2017, 978-1-5090-6004-7/17.
[26] Kian Peymani, Mohsen Soryani, "From Machine Generated To Handwritten Character Recognition; A Deep Learning Approach", 3rd International Conference on Pattern Recognition and Image Analysis (IPRIA 2017) April 19-20, 2017.
[27] Martin Rajnoha, Radim Burget, Malay Kishore Dutta, "Handwriting Comenia Script Recognition with Convolutional Neural Network", IEEE,2017, 978-1-5090-3982-1/17.
[28] Ruyu Zhang, Qingqing Wang, Yue Lu, "Combination of ResNet and Center Loss Based Metric Learning for Handwritten Chinese Character Recognition", 2017 14th IAPR International Conference on Document Analysis and Recognition.
[29] Bishwajit Purkaystha, Tapos Datta, Md Saiful Islam," Bengali Handwritten Character Recognition Using Deep Convolutional Neural Network", International Conference of Computer and Information Technology (ICCIT), 22-24 December 2017.
[30] Mohammad Agung Wibowo, Muhamad Soleh, Winangsari Pradani, Achmad Nizar, " Handwritten Javanese Character Recognition using Discriminative Deep Learning Technique", International Conferences on Information Technology, Information Systems and Electrical Engineering.

Real Time Emotion Detection of Humans Using Mini-Xception Algorithm

Syed Aley Fatima, Ashwani Kumar*, Syed Saba Raoof
Research Scholar, Department of Computer Science and Engineering
Vardhaman College of Engineering, Hyderabad, India.

*Department of Computer Science and Engineering
Vardhaman College of Engineering, Hyderabad, India.

*ashwani.kumarcse@gmail.com

Abstract. In the past few decades there has been operating analysis on emotion expression recognition due to the great intra-class deviation also it is still challenging. Maximum number of research work performs the best on controlled datasets (i.e., small datasets with less features), whereas it fails to operate well and it's still challenging on datasets varies variations in images and even in partial faces. In modern years, many works have introduced an end-to-end plan for emotion expression recognition, utilizing deep learning models. Although emotion recognition is a great task, there still seems to be a huge area for development. In this paper, we developed a mini-Xception based on Xception and Convolution Neural Network (CNN), which is easy to concentrate on great parts like the face, and conclude important improvements to earlier works. We validated our model by creating a real-time vision system which accomplishes the tasks of face detection, and emotion classification simultaneously in one blended step using our proposed mini-Xception architecture. We still utilize a visualization technique that is ready to detect important face sectors because recognizing various emotions, based on the classifier's output. For experimental analysis we had used FER-2013 dataset and results manifest that the proposed method can efficiently perform all the tasks like detection and classification with seven different emotions using with Mini-Xception algorithm and achieved accuracy around 95.60%.

Keywords: *Mini-Xception, Emotion expression, Convolution neural network (CNN), Face detection.*

1. Introduction

"Emotion" is described by many research studies but there is no particular definition in all the literatures about the emotion. Reflection or actualization of feeling can be defined as "Emotion". It can be either sham or real not only feeling [1]. For instance, feeling of pain straight forwardly describe feeling, where as emotions cannot be felt completely and exactly. An emotion describes the inner esoteric internal state of affairs. In various fields of research like, psychology, health, bio-medical engineering, and even in neuroscience emotion play a vital role and it has became an immense research subject. Emotion detection is an important research area in biomedical engineering [2]. Recent research studies in this field focus one motion detection and diagnosing psychological disorders by automated computer-aided systems. Various methods were implemented by researchers for emotion detection like multimodal, Galvanic Skin Response (GSR), Electro Encephalography (EEG), Facial Expression; visual scanning behavior etc. [3], over past few years, familiarization of deep learning has been a great progress in the field of image classification. Convolutional neural networks (CNNs) are one of the mostly used and popular deep learning algorithms used for implementation of image segmentation, recognition and classification.

Deep learning based algorithms are one of the mostly used methods to detect the state of emotion of human beings (e.g., anger, disgust, fear, happy, sad, surprise, and neutral) [4]. We had developed deep learning based algorithm Mini-Xception. The main aim of the proposed model is to automatically detect emotions and to predict emotional condition with high accuracy. In this method, for analyzing experimental results we had used labeled facial expression images from FER dataset. This mages are given as input to the model created and it is trained by these images. Then, the proposed model makes a determination which facial expression is performed.

This research article is organized as follows: Section 1 indicates the introduction of emotion detection with deep learning algorithm. Section 2 discusses related work. Section 3 describes methodology and proposed architecture. Section 4 explains about dataset, and experimental results. Section 5 describes conclusion and future work.

2. Related Work

Global survey has done on DL approaches where various learning techniques, models, and recently proposed training approaches are discussed. Shervin Minaee [5] implemented a model on Facial expression recognition using Attention Convolution Network. FER-2013, CK+, FERG JAFFE dataset are used for training models and 93.3% accuracy is achieved. D Y Lilianasuntha et al. [6] developed a model on Emotion recognition from facial expression using deep Convolutional neural network. CK+ dataset are used for training models and 92.81% accuracy is achieved. Jyostna Devi Bodapati, N. Veeranjaneyulu. [7], proposed a Facial Emotion Recognition Using Deep CNN Based Features. For feature extraction pre-trained convolution neural network model (VGG16) is used.CK+ dataset are used for training model and 86.04%accuracy is achieved. Nithya Roopa.s [8], in this paper they proposed an Emotion Recognition from Facial Expression using deep learning. The proposed model is Inception Net v3 is applied and Kaggle's Facial Expression Recognition Challenge and Karolinska Directed Emotional Faces (KDEF) datasets are used for training model and 39% accuracy is achieved. Panagiotis Giannopoulos [9], in this paper they proposed" Deep Learning approaches for Facial Emotion Recognition". In this paper they proposed, the performance of two known deep learning approaches (Google Net and Alex Net) on facial expression recognition. FER-2013 dataset are used for training models and 83% accuracy is achieved.

Junnan Li and Edmund Y. Lam. [10] in this paper they proposed a "Facial Expression Recognition using Deep Neural Networks". CK+ dataset are used for training and testing model and91.9% accuracy is achieved. Arushi Raghuvanshi [11] "Facial Expression Recognition with Convolutional Neural Networks". VGG16, the model from Caffe's Model Zoo which was trained on Image Net, and VGG Face, a network trained on a facial recognition data set. There trained and tested there models on the data set from the Kaggle Facial Expression Recognition Challenge, and FER dataset are used for training and testing model and 0.60% accuracy is achieved. K. Sravanthi, G.Jaya suma [12] developed a method "A Prediction of Emotions for Recognition of Facial Expressions Using Deep Learning, CK+, JAFE and BU-3DFE dataset are used for training and testing model and 100% accuracy is achieved. VICTOR-EMIL NEAGOE [13] developed "A Deep Learning Approach for Subject Independent Emotion Recognition from Facial Expressions". There have focused on two "deep" neural models: Convolutional Neural Networks (CNN) and Deep Belief Networks (DBN). One has selected the Support Vector Machine (SVM) as a benchmark algorithm. JAFE dataset are used for training and testing model and 95.71% accuracy is achieved. Fatima Zahra Salman [14] proposed an "Emotion Recognition from Facial Expression Based on Fiducial Points Detection and using Neural Network". There detect and track 49 Fiducial points using a powerful and recent Supervised Decent Method (SDM). The proposed approach has achieved an emotion recognition accuracy of 99% on the CK+ database, 84.7% on the Oulu-CASIA VIS database, and 93.8% on the JAFFE database. The following table shows the comparative study of related work.

Table I: Summary of related work using different algorithm & methods.

S.NO.	Authors	Methods Used	Dataset Used	Accuracy Measures
1.	Shervin Minaee	Attention Convolution Network	FER-2013, CK+, FERG & JAFFE	93.3%
2.	D Y Lilianasuntha et al.	Deep Convolutional Neural Network.	CK+	92.81%
3.	Jyostna Devi Bodapati, N. Veeranjaneyulu	Deep CNN Based Features, (VGG16).	CK+	86.04%
4.	Nithya Roopa.s	Deep Learning	Karolinska Directed Emotional Faces (KDEF)	39%
5.	Panagiotis Giannopoulos	Google Net and Alex Net.	FER-2013.	83%
6.	Junnan Li and Edmund Y. Lam	Deep Neural Networks.	CK+.	91.9%
7.	Arushi Raghuvanshi	Convolutional Neural Networks.	Kaggle Facial Expression Recognition and FER.	0.60%
8.	K.Sravanthi, G.Jaya suma	Deep Learning.	CK+, JAFE and BU-3DFE.	100%
9.	Victor-Emil Neagoe	Support Vector Machine (SVM).	JAFE.	95.71%
10.	Fatima Zahra Salman	Supervised Decent Method (SDM).	CK+, Oulu-CASIA VIS, JAFFE.	99%

3. Proposed Method

We had implemented the proposed model based on improved deep learning approaches for emotion recognition, classification, and detection. An enhanced recognition model has been developed utilizing residual convolution networks for recognizing the emotions. Recognition is done through Mini-Xception based models. Comparison of the proposed model along with other models demonstrates greater performance against other algorithms. By using Xception model we had proposed a mini-Xception for emotion detection. Our proposed model is inspired by the Xception algorithm. This architecture combines the use of residual modules [6] and depth-wise separable convolutions. Residual modules modify the desired mapping between two subsequent layers, so that the learned features become the difference of the original feature map and the desired features. Consequently, the desired features H(x) are modified in order to solve an easier learning problem F(X) such that:

$$H(x) = F(x) + x \tag{1}$$

In proposed model parameters are reduced from the Convolutional layers, which was done by using depth-wise separable convolutions. Depth-wise separable convolutions are composed of two different layers: depth-wise convolutions and point wise convolutions.

The main purpose of these layers is to separate the spatial cross-correlations from the channel cross correlations [1]. They do this by first applying a D×D filter on every M input channels and then applying N 1×1×M convolution filters to combine the M input channels into N output channels. Applying 1×1×M convolutions combines each value in the feature map without considering their spatial relation within the channel. Depth-wise separable convolutions reduces the computation with respect to the standard convolutions by a factor of 1 N + 1 D2 [2]. A visualization of the difference between a normal Convolution layer and a depth-wise separable convolution. The proposed architecture contains 4 residual depth-wise separable convolutions where each convolution is followed by a batch normalization operation and a ReLUs activation function. The last layer applies a global average pooling and a soft-max activation function to produce a prediction. This architecture has approximately 60,000 parameters. Figure 1(a) (b) displays our complete architecture which is referred as mini-Xception.

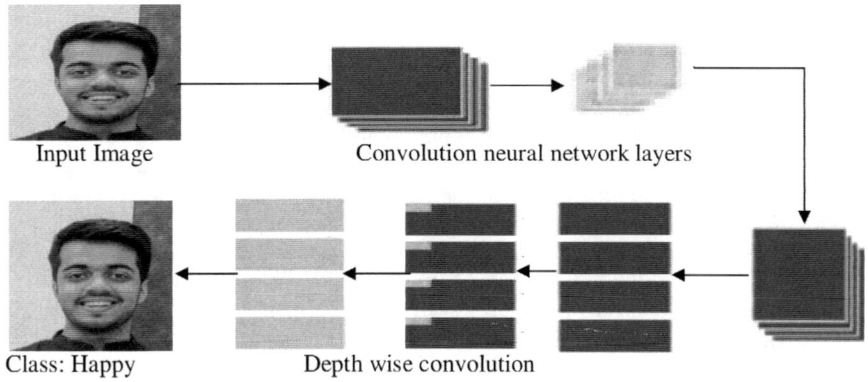

Figure 1(a). Architecture of Proposed Model

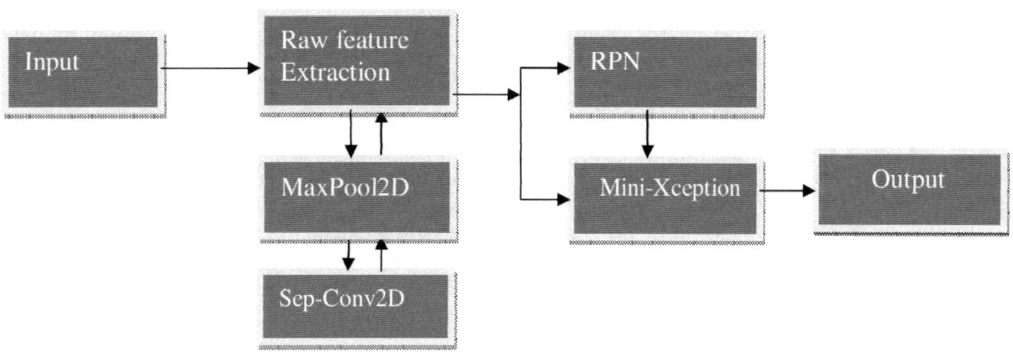

Figure 1(b). Architecture of Proposed Model

We tested this architecture in the FER-2013 dataset and obtained the accuracy of 89.96% for the emotion detection task. Real-time image and video analysis is also done based on back-propagation. Given a mini-Xception with ReLU function for the intermediate layers, back propagation takes the derivative of every element (x, y) of the input image I with respect to an element (i, j) of the feature map f^L in layer L. The reconstructed image R filters all the negative gradients; consequently, the remaining gradients are chosen such that they only increase the value of the chosen element of the feature map. Following a fully ReLU mini-Xception reconstructed image in layer l is given by:

$$Rli,j \;=\; (Rl+1\,i,j \;>\; 0)*Rl+1\,i,j \qquad (2)$$

Algorithm

Description of our proposed model is mentioned below:

Method: Emotion Expression and Recognition using Mini-Xception.

Input: FER-2013 Dataset.

Output: Classification of Emotions.

Step 1: Load the FER-2013 dataset.

Step 2: Partition the dataset into training and testing.

Step 3: Apply the preprocessing techniques.

Step4: Build the model using Mini-Xception algorithm.

Step5: Test data is given to Mini-Xception for classification.

Step6: Calculate the Accuracy.

4. Methodology

Initially the face region is extracted from the given face images using proposed face detection algorithm. Then from the cropped face regions, deep features are extracted from the last fully connected layer of mini-Xception. Based on the experimental results we observed that the proposed method is simple and gives better performance compared to the other methods. Mini-Xception, a pre-trained convolution model, [21] is known for its state of the art performance in various applications [21] like image classification, object detection etc. It is trained on Image Net dataset. After training of proposed architecture, the trained model was tested in real time. First of all, human faces were detected. After that, the detected images were sent to the model and the classes they belong to were queried. This process was performed on every 30 frames that occurred every second of the camera image obtained in real time.

5. Experimental Analysis

The experimental analysis is evaluated using Python IDLE tool. Mini-Xception algorithm is applied on emotion detection dataset. Dataset contains 35,887 images. Accuracy is calculated from the confusion matrix.
Dataset: To train our models we used dataset: FER-2013. The FER-2013 dataset had been created by acquiring and combining the result of Google image search for every particular emotion [22]. Each and every image in FER dataset is labeled image and it consists of seven emotion images i.e., anger, disgust, fear, happy, sad, surprise, and neutral. This dataset consists of 35,887images in the training set, 3,500 labeled images in the test set, and 3,500 images in the development set. It consists of pair of posed and un-posed identification images, these images are in grayscale and pixel value is 48x48. Even Though this dataset consists of labeled emotions, it was discovered that the proposed model achieved better accuracy. Here are some examples of the FER 2013 dataset recognized using proposed algorithm i.e., Mini-Xception. Figure 2 represents emotions like anger, disgust, fear, happy, sad, surprise, and neutral respectively.

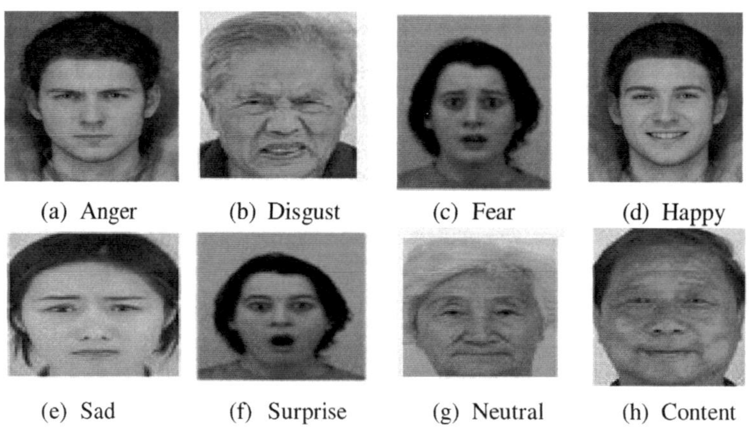

Figure 2. Emotions recognized using Mini-Xception proposed algorithm.

Accuracy is calculated from the confusion matrix. For quantitative analysis of the experimental results, the following performance metrics are considered, including accuracy (AC). To do this we also use the variables True Positive (TP), True Negative (TN), False Positive (FP), and False Negative (FN). Confusion Matrix: It is a particular table format that allows the performance evaluation of an algorithm [23-25].

Table 2. The following table shows confusion matrix of proposed algorithm.

Expression	Anger	Disgust	Fear	Happy	Sad	Surprise	Neutral
Anger	60.0	0.1	0.0	0.0	25.0	0.0	10
Disgust	0.0	55.0	0.5	0.7	11.9	25.0	0.0
Fear	12.9	0.01	65.78	0.5	0.12	17.8	0.0
Happy	0.02	0.5	0.02	95.55	0.08	14.5	0.78
Sad	25.0	11.9	0.12	0.08	85.0	0.05	7.50
Surprise	0.0	25.0	17.8	14.5	0.05	73.68	11.9
Neutral	10.0	0.0	0.0	0.78	7.50	11.9	89.66
			Average Accuracy =95.60				

The overall accuracy is calculated using.

$$AC = \frac{TP+TN}{TP+TN+FP+FN} \tag{3}$$

Table 3. The following table demonstrates the performance measures calculated by using accuracy, precision, and recall.

Test size	Accuracy	Precision	Recall
90%-20%	95.60	93	90

6. Conclusion

In this paper, we implemented Mini-Xception model which is an enhanced model of Xception architecture using residual networks for Emotion expression and Recognition. Using the FER-2013 database gives better performance than the existing method. There are seven types of emotion expressions such as (e.g., anger, disgust, fear, happy, sad, surprise, and neutral.) we tried to recognize. In modern years, many works have introduced an end-to-end plan for emotion expression recognition, utilizing deep learning models. Although emotion recognition is a great task, it still seems emotion huge area for development. The accuracy obtained for Emotion expression and recognition using Mini-Xception is 95.60% and precision and recall rate is 93% and 90% respectively. Further the accuracy can be increased by training the Mini-Xception algorithm using the original image dimensions of 48x48. The number of Convolutional layers and the size of filter can also be increased to improve accuracy.

References

[1] S. Minaee and A. Abdolrashidi. Deep-emotion: Facial expression recognition using Attentional Convolutional network. arXiv preprint arXiv:1902.01019, 2019. 1, 2.1, 5.1, 5.2.
[2] D Y Liliana [2019] "Emotion recognition from facial expression using deep Convolutional neural network". ". 2018 International Conference of Computer and Informatics Engineering (IC2IE)
[3] Jyostna Devi Bodapati, N. Veeranjaneyulu "Facial Emotion Recognition Using Deep Cnn Based Features", International Journal of Innovative Technology and Exploring Engineering (IJITEE) ISSN: 2278-3075, Volume-8Issue-7, May 2019.
[4] Nithya Roopa. S " Emotion Recognition from Facial Expression using Deep Learning", International Journal of Engineering and Advanced Technology (IJEAT) ISSN: 2249 – 8958, Volume-8 Issue-6S, August 2019.
[5] Panagiotis Giannopoulos, Isidoros Perikos and Ioannis Hatzilygeroudis,"Deep Learning Approaches for Facial Emotion Recognition", Springer International Publishing AG 2018.

[6] Junnan Li and Edmund Y. Lam, "Facial Expression Recognition using Deep Neural Networks. 978-1-4799-8633-0/15/$31.00 ©2015 IEEE.

[7] Arushi Raghuvanshi, Vivek Choksi, (2017) "Facial Expression Recognition with Convolutional Neural Networks". 978-1-7281-3783-4/20/$31.00 ©2020 IEEE.

[8] K.Sravanthi, G.Jaya Suma,"A Prediction of Emotions for Recognition of Facial Expressions Using Deep Learning", International Journal of Recent Technology and Engineering (IJRTE) ISSN: 2277-3878, Volume-8, Issue-2S11, September 2019.

[9] A VICTOR-EMIL NEAGOE, ANDREI-PETRU BĂRAR, NICU SEBE, PAUL ROBITU," A Deep Learning Approach for Subject Independent Emotion Recognition from Facial Expressions".978-1-7281-3783-4/20/$31.00 ©2020 IEEE.

[10] Fatima Zahra Salmam, Abdellah Madani, and Mohamed Kissi. "Emotion Recognition from Facial Expression Based on Fiducial Points Detection and using Neural Network", International Journal of Electrical and Computer Engineering (IJECE) Vol. 8, No. 1, February 2018, pp. 52 – 59.

[11] Syed Aley Fatima, Ashwani Kumar, Ajay Pratap, Syed Saba Raoof, " Object recognition and detection in remote sensing images". 2020 International Conference on Artificial Intelligence and Signal Processing (AISP) 978-1-7281-4458-0/20/$31.00 ©2020 IEEE.

[12] Jiaxing Lia, Dexiang Zhanga, Jingjing Zhanga, Jun Zhanga, Teng Lia, Yi Xiaa, Qing Yana, and Lina Xuna. "Facial Expression Recognition with Faster R-CNN", International Congress of Information and Communication Technology (ICICT 2017).

[13] Najmeh Samadiani, , Guangyan Huang , Borui Cai , Wei Luo , Chi-Hung Chi , Yong Xiang and Jing He "A Review on Automatic Facial Expression Recognition Systems Assisted by Multimodal Sensor Data". International Congress of Information and Communication Technology (ICICT 2017).

[14] Mengyi Liu, Ruiping Wang, Shaoxin Li, Shiguang Shan, Zhiwu Huang, Xilin Chen "Combining Multiple Kernel Methods on Riemannian Manifold for Emotion Recognition in the Wild". Sensors 2019, 19, 1863; 10.3390/s19081863 www.mdpi.com/journal/sensors.

[15] Asim Munira, Ayyaz Hussaina,d, Sajid Ali Khan, Muhammad Nadeema, Sadia Arshid "Illumination invariant facial expression recognition using selected merged binary patterns for real world images". **http://dx.doi.org/10.1145/2663204.2666274.**

[16] Junkai Chen, Zenghai Chen, Zheru Chi, and Hong Fu "Facial Expression Recognition in Video with Multiple Feature Fusion", IEEE Transactions on Affective Computing • Optik 158 (2018) 1016–1025.

[17] Gil Levi, Tal Hassner "Emotion Recognition in the Wild via Convolutional Neural Networks and Mapped Binary Patterns", November 9, 2015.

[18] P. Shen, ShangfeiWang, and Zhilei Liu "Facial Expression Recognition from Infrared Thermal Videos", Intelligent Autonomous Systems 12, AISC 194, pp. 323–333.

[19] André TeixeiraLopes, EdilsondeAguiar, AlbertoF.DeSouza, ThiagoOliveira-Santos "Facial expression recognition with Convolutional Neural Networks: Coping with few data and the training sample order", 19July2016.

[20] Kaihao Zhang, Yongzhen Huang, Yong Du, and Liang Wang, "Facial Expression Recognition Based on Deep Evolutional Spatial-Temporal Networks", IEEE TRANSACTIONS ON IMAGE PROCESSING, VOL. 26, NO. 9, SEPTEMBER 2017.

[21] Shangfei Wang • Chongliang Wu • Menghua He • Jun Wang • Qiang Ji "Posed and spontaneous expression recognition through modeling their spatial patterns", Machine Vision and Applications (2015) 26:219–231. ", IEEE TRANSACTIONS ON IMAGE PROCESSING, VOL. 26, NO. 9, SEPTEMBER 2017.

[22] Alex Krizhevsky, Ilya Sutskever, Geoffrey E. Hinton. Image net classification with deep Convolutional neural networks. In Advances in neural information processing systems, pages 1097–1105, 2012.

[23] Andrej Karpathy. Cs231n Convolutional neural networks for visual recognition. http://cs231n.github.io/convolutional-networks/, 2015. [Online; Accessed 24 May 2016].

[24] Andrej Karpathy, George Toderici, Sanketh Shetty, Thomas Leung, Rahul Sukthankar and Li Fei-Fei. Large-scale video classication with Convolutional neural networks. In Proceedings of International Computer Vision and Pattern Recognition (CVPR 2014), 2014.

[25] Challenges in representation learning: Facial expression recognition challenge: http://www.kaggle.com/c/challengesin-representation-learning-facial-expression recognition

challenge. ", IEEE TRANSACTIONS ON IMAGE PROCESSING, VOL. 26, NO. 9, SEPTEMBER 2017.
[26] Liu, Ping, Shizhong Han, Zibo Meng, and Yan Tong. "Facial expression recognition via a boosted deep belief network." In Proceedings of the IEEE Conference on Computer Vision and Pattern Recognition, pp. 1805-1812. 2014.

AutoAssociative Memory for Face Recognition using High-Dimensional Octonion Data

Shoba Rani Salvadi [1], Dr.D.Nagendra Rao[2], Dr.S Vatsal[3]
[1] Associate Professor, Department of Computer Science and Engineering, Vardhaman College of Engineering, Shamshabad, Telangana, India.
reddymallashobarani2@gmail.com
[2] Principal & Professor, Electronics and Communication Engineering, JBREC, Moinabad, Telangana, India.
[3] Professor, Electrical and Electronics Engineering, Institute of Aeronautical Engineering, Telangana, India

Abstract:

Uniqueness or individuality of an individual face is the representation of one's identity. Recent research in color image processing focusing on octonion algebra, which is the generalization of complex and quaternion algebra. In this paper, the face of an individual is used for identification making automatically. For this, an autoassociative memory is used for the recognition of images or videos with 8-dimensions. Octonion algebra is not commutative and not associative but it is an alternate algebra for representing the phase and angle of images effectively. The prospective experiments are conducted on color images for image reconstruction, denoising and efficiently retrieving blurred images in true color images.

Keywords: Associative memory, autoassociative memory, octonion-valued, quaternion-valued networks, face recognition.

1. Introduction

The mainstream of earlier research efforts has focused on unique fingerprint-based biometric systems. Different applications using various forms of technologies for access control and authentication purpose such as banking systems with username and password, employee authentication in the organization with the biometric system, etc. Due to the rapid growth of computer technology all these access control mechanisms facing a lot of problems like losing fingerprint, unauthorized person using username, and passwords(stolen). It is not reliable even though it is effective for purpose of detecting a person. To overcome fit falls of fingerprint-based biometric systems a pioneering identification and authentication system is required. It is not reliable even though it is effective for the person

Compared with other biometric features, such as palms, ears, irises, gait, and voice, the human face has many distinct advantages. So face recognition (FR) based identification and authentication system taken into consideration in research because it is natural, easy to use, and nonintrusivity. Automatic face recognition is not only alternate to fingerprint biometric system has many uses such as crowd surveillance and border control.

In recent years, a digital image plays a unique role in the advent of technological advancements by extracting useful information from images. It focuses on two tasks; the first one is an improvement of pictorial information for human interpretation and the second one is the processing of image data for storage, transmission, and representation for autonomous machine perception.

Face recognition is an application of associative memory pattern recognition problems. Face recognition is an emerging technology in the era of computer vision (CV) from the long term onwards because of the machine learning algorithms and CVNN usage in the data processing. Neural networks Associative memory is taken initiation in the research from human being's memory because of artificial neural networks inspired by biological neural networks with a mathematical representation of the model. Human memory depends totally on association with objects frequently seen together to become linked in our mind; once we attempt to retrieve information one thing reminds us of another that reminds us of yet another, and so on. Recalling and remembering things by associating with objects is called associative memory. Simply associative memory means storing the patterns in memory and recalls the stored pattern from memory bypassing the input pattern. Face recognition is a type of associative memory because images are going to be stored in the database first (training) and then recalling the image by giving a query image as input. Face recognition using CVNNs obtained enormous results because it can process the two-dimensional data promptly and treats as a single entity.

Complex number with four-dimensional data (4D) is quaternion algebra with one real and three imaginary units appropriate for the representation of three color channels (R, G, and B) in image processing [6]. Quaternion-valued neural network in color image processing showed an amazing result in terms of image reconstruction, face recognition with color, and denoising [7]. A quaternion-based optimization problem is quite difficult to solve because of its non-commutative property. Extended complex and quaternion numbers are required.

Inspired by the extraordinary results of quaternion algebra in color image processing, this paper addresses an associative memory model for color face recognition using complex number octonion, which is an 8-dimensional hypercomplex non-associative algebra. Due to the non-associative property of the octonion algebra applications are limited compared to quaternion algebra to but it can solve physical problems.

Therefore, this paper aims to design and investigate OVNNs for associative memory-based face recognition by mathematical examination and procedures. This technique makes it conceivable to represent the data in an octonion-valued entity. Still octonion number system facing a computer simulation problem but few authors representing theoretically.

The rest of this paper's aspects are composed as follows. Section II, provides some basic knowledge on octonions, the persistent time OVNNs model, and the discrete-times partners are presented. In Section III, autoassociative memory with storage and recall phases of images along with Hebbian learning is discussed, which can understand affiliated recollections. In Section IV, some adequate experiments with results are discussed. And finally, in Section V some conclusions are drawn and guidelines for future research.

2. Basic concepts of octonion Algebra

Octonions, which can articulate as doubled quaternion numbers [34], have been used in signal and image processing, and we believe that they can also be used effectively in many image recognition applications. Octonions are an extension of quaternion and complex algebra and it is a type of Cayley-Dickson algebra with dimension 2^n. The octonion is the 8-dimensional nonassociative algebra over the real numbers, which consists of one real part and seven imaginary parts with basis elements $\{1, e_1, e_2, e_3, e_4, e_5, e_6, e_7\}$ can be written in the form

$$\mathbb{O} := a_0 + \{ \sum_{q=1}^{7} [a]_q \, e_q \, | [a]_0, [a]_1 \ldots.., [a]_7 \in R \} \quad (1)$$

where a_0 and a_q are real numbers, and e_q is imaginary units.

Table 1: Octonion basis elements multiplication rules.

×	e_0	e_1	e_2	e_3	e_4	e_5	e_6	e_7
e_0	e_0	e_1	e_2	e_3	e_4	e_5	e_6	e_7
e_1	e_1	$-e_0$	e_3	$-e_2$	e_5	$-e_4$	$-e_7$	e_6
e_2	e_2	$-e_3$	$-e_0$	e_1	e_6	e_7	$-e_4$	$-e_5$
e_3	e_3	e_2	$-e_1$	$-e_0$	e_7	$-e_6$	e_5	$-e_4$
e_4	e_4	$-e_5$	$-e_6$	$-e_7$	$-e_0$	e_1	e_2	e_3
e_5	e_5	e_4	$-e_7$	e_6	$-e_1$	$-e_0$	$-e_3$	e_2
e_6	e_6	e_7	e_4	$-e_5$	$-e_2$	e_3	$-e_0$	$-e_1$
e_7	e_7	$-e_6$	e_5	e_4	$-e_3$	$-e_2$	e_1	$-e_0$

where the imaginary units satisfy $e_i^2 = -1$ and multiplication rules between them are given in Table 1.

An octonion can be represented by a vector as $a = a_0 + \vec{a}$, where a_0 and $\vec{a} = a_1 e_1 + a_2 e_2 + a_3 e_3 + a_4 e_4 + a_5 e_5 + a_6 e_6 + a_7 e_7$ are called the real and vector part of $a(\mathbb{O})$.

The imaginary units $e_0, e_1, \ldots.., and\ e_7$ abide by the following rules:

$$e_i e_j = - e_j e_i \neq e_j e_i \ \forall i \neq j, 0 < i, j \leq 7, \text{ and} \quad (2)$$

$$(e_i e_{j)}) e_k \neq e_i(e_j e_k) = -e_i(e_j e_k) \ \text{ for } i \neq j \neq k \quad (3)$$

From the above rules and multiplication table, it is clear that the multiplication of octonions imaginary units is not commutative and nor associative. The algebra properties of the octonion are addition between two octonion numbers are component-wise, which is the same as complex numbers.

If the octonion $a = a_0 + a_1 e_1 + a_2 e_2 + a_3 e_3 + a_4 e_4 + a_5 e_5 + a_6 e_6 + a_7 e_7$, the conjugate of a is defined as $\bar{a} = a_0 - \sum_{q=1}^{7} a_q e_q$ and $\overline{a1\, a2} = \overline{a2}\ \overline{a1}$.

The norm of octonion a is represented as

$$|a| = \sqrt{a\,\bar{a}} = \sqrt{\bar{a}\,a} = \sqrt{a_0^2 + a_1^2 + a_2^2 + a_3^2 + a_4^2 + a_5^2 + a_6^2 + a_7^2}\,. \quad (4)$$

and norms are real numbers. From this definition, we can find the inverse of a as $a^{-1} = \frac{a^-}{N(a)}$

Features of the image are mapped with an input layer with octonion representation, which means that N features are divided into one real N/8 and seven imaginary features from the remaining N/8. Because every octonion disintegrate into a real part and imaginary part as $Re\,(a) := \frac{1}{2}(a + \bar{a})$ and $Im(a) := \frac{1}{2}(a -$

$\bar{a}$). The octonions are alternative algebra, by the definition of the multiplication table of octonion we have

$$(e_i\, e_i)\, e_j = -\, e_j$$
$$= -\, e_k\, e_i$$
$$= e_i\, e_k$$
$$= e_i\,(e_i\, e_j)$$

and in the same way

$$(e_j\, e_j)\, e_j = e_k\, e_j$$
$$= -\, e_j\, e_k$$
$$= -\, e_i$$
$$= e_i\,(e_j\, e_j)$$

Multiplication of octonions is not associative from table1.
According to the Cayley-Dickson construction of octonion $\mathbb{O}$ can be represented in quaternion algebra as :

$$\mathbb{O} = a' + a''e, \quad a', a'' \in \mathbb{Q}; \mathbb{Q} = \{a_0 + a_1\, i + a_2\, j + a_3\, k \mid a_0, a_1, a_2, a_3 \in \mathbb{R}\} \qquad (5)$$

Here, $\mathbb{Q}$ is the quaternion algebra with {1, i, j, k}, $i^2 = j^2 = k^2 = -1$, $ijk = -1$ and e is imaginary unit [9].

3. AutoAssociative Memory

To store and recall the face images by using linear autoassociative memory was first proposed by Kohonen in the year 1999 [9] [10]. He used two autoassociative memories on the symmetry property of an image to quickly locate all the candidates' human faces with all possible sizes and regions for storing and validation (recall) purpose. The fundamental function of an associative memory through self-organizing is to store a set of patterns X^p, p=1.....n pairs in memory and retrieves the corresponding output pattern Y^q for a given input pattern. The input stimulus and the response of the output pattern represent feature vectors of images.

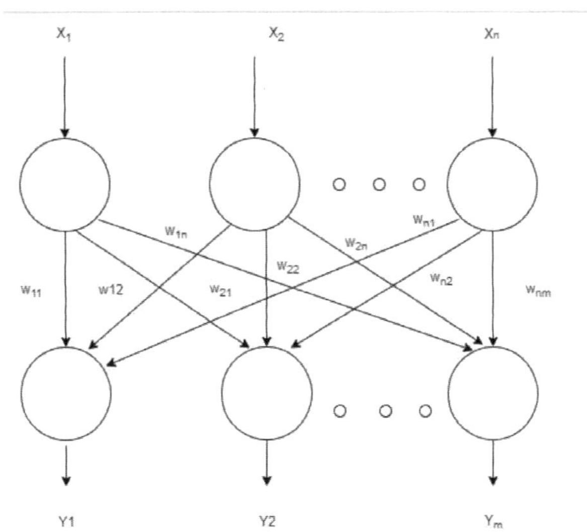

Fig. 1.Associate Memory model

Associative memory main function is to recall the perfect pattern from a stored pattern based on a given input pattern; this can be done in two phases:

A. *Storage Phase*

Construction of weight matrix **w** (associative memory) with paired input and output patterns (X_p, Y_q) is called encoding, during the storage phase constructed weight matrixes are stored in the memory. Features of the images are stored in the memory in the form of a weight matrix, which is computed as

$$W_{pq} = (x_p)^T (y_q) \quad (6)$$

Where $(x_p)^T$ and y_q represents the transpose of input pattern and output patterns of the associative memory.

B. *Recall Phase*

The process of retrieving a stored pattern from memory for a given input pattern is called decoding, during this phase the output pattern y_q is computed as

$$y_q = \sum_{p=1}^{n} w_{pq} x_p \quad (7)$$

Output units y_q dependent on activation function which is represented in bipolar from

$$Y_q = \begin{cases} +1 & if\ x_p > 1 \\ -1 & if\ x_p = 0 \end{cases} \quad (8)$$

If the input patterns are mutually orthogonal then perfect retrieval is possible, which means input pattern and output patterns are equal or identical $x_p = y_q$, this type of memory is called autoassociative memory.

$$x_p x_q = \begin{cases} 1 & if\ p = q \\ 0 & if\ p \neq q \end{cases} \quad (9)$$

Otherwise, if input patterns are not orthogonal, the retrieved pattern is different from the given input pattern because of crosstalk between the input patterns .i.e noise, this type of memory is called as heteroassociative memory. In this paper, autoassociative memory model is used to stores the images in the memory in the form of features, and recognition is done based on the input image features.

C. *Learning Rule*

To reduce error or noise present in the recalled image is achieved by updating the weight matrixes which are computed during the storage phase via the Hebbian learning rule until reaching the stable state

$$W_{new} = W_{old} + \eta \, \Delta \, W_i \qquad (10)$$

where $Wnew$, W_{old} are new and old weight matrices, η constant learning rate and $\eta \, \Delta \, W_i$ is change in the connection weight matrix.

4. Experimental Results and Discussion

We demonstrate our method by applying the multiple channel technique for image recalling and denoising. Quaternions are triplets and are perfectly mapping with RGB as a one to one. The proposed octonion auto-associative memory is an extension of quaternion associative memory with increased channels that is from 3 imaginary units to 7 imaginary units. Figure 2 shows the color image denoising with four images. Compared to complex, quaternions based denoising of images octonions yielded better results. We have taken 30 images with 25 patterns as a training data set which means training of associative memories or storing and testing data set with 40 images with 43 patterns for recall purpose used. This method gives better results when compared to quaternion-valued associative memory when learning $\eta = 15$ and 20. Figure 2 shows the original color images in (a), blurred color images with Gaussian noise 50 in (b), and in (c) retrieved images are present.

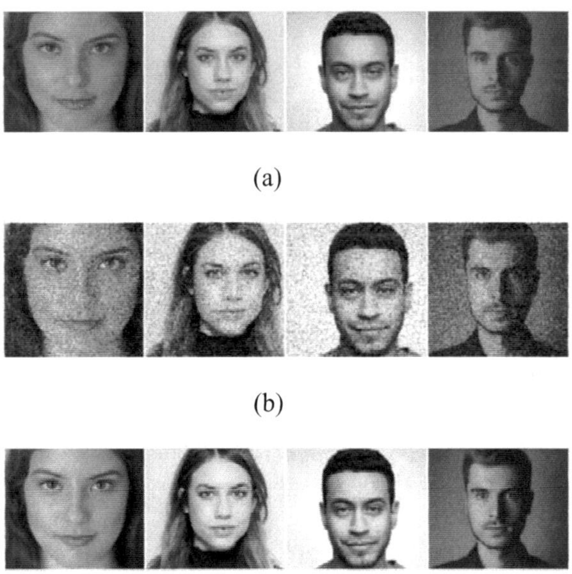

(a)

(b)

(c)

Figure 2.Autoassociative memory based (a) original images (b) Blurred images (c) Recalled images.

For experiments, all images with a size of 512 × 512 are considered patches of size 8 × 8 but we use dictionary size for image denoising with 64 × 256. Results are compared with QVNN and our model

IOP Conf. Series: Materials Science and Engineering 1042 (2021) 012028 doi:10.1088/1757-899X/1042/1/012028

OVNN gives better results at smaller noise levels when compared to higher noise levels. At higher noise levels OVNN produces the same results as QVNN.

(a)

(b)

(c)

Figure 3 large images (a) stored image or original color image. (b) Blurred image. (c) Final retrieved image from associative memory.

And figure 3 shows an experiment on a large image, in this, also retrieved image the same as an original image with 8-dimensions. From Figures 2 and 3 the retrieved images are perfect because we stored images in the form of weight matrixes or vectors in associative memory. From the flexibility of autoassociative memory, we retrieved perfect images the same as original ones, even though passing noise images as input patterns in the recalling process.

5. Conclusion

In this paper, we proposed octonion autoassociative memory for face recognition, which is an extension of quaternion, complex, and real numbers. The important points of octonion algebra, such as norm and conjugate properties are discussed, and octonion is not commutative and not associative stressed. Associative memory applications for image recognition are also discussed here.In the future, we plan to examine the performance of this model specifically in terms of associative memory storage capacity and noise tolerance.

References

[1] Y. Xu, L. Yu, H. Xu, H. Zhang, and T. Nguyen, "Vector sparse representation of color image using quaternion matrix analysis," IEEE Transactions on image processing, vol. 24, no. 4, pp. 1315–1329, 2015.
[2] Xiaofeng Chen, Qiankun Song, and Zhongshan Li, "Design and Analysis of Quaternion-Valued Neural Networks for Associative Memories", IEEE TRANSACTIONS ON SYSTEMS, MAN, AND CYBERNETICS: SYSTEMS, 2168-2216 _ 2017 IEEE.
[3] Qingxiang Feng, Yicong Zhou "Color Face Recognition using High-Dimension Quaternion-based Adaptive Representation" 19 Nov 2017.
[4] Emad S. Jaha1 and Lahouari Ghouti2 ," Color Face Recognition Using Quaternion PCA" Nov 2011.
[5] Mu-Chun Su and Chien-Hsing Chou "Application of Associative Memory in Human Face Detection" 1999.
[6]. T. A. Ell, N. Le Bihan, and S. J. Sangwine, Quaternion Fourier transforms for signal and image processing. John Wiley & Sons, 2014.

[7]. C. Zou, K. I. Kou, and Y. Wang, "Quaternion collaborative and sparse representation with application to color face recognition," IEEE Transactions on image processing, vol. 25, no. 7, pp. 3287–3302, 2016.

[8] Srd¯an Lazendi´c_, Hendrik De Bie_ and Aleksandra Piˇzuricay "Octonion Sparse Representation for Color and Multispectral Image Processing", 26th European Signal Processing Conference (EUSIPCO), ISBN 978-90-827970-1-5 2018
[9] T. K. Kohonen, Associative Memory: A System Theoretic Approaches, Springer, Berlin, 1977.
[10] T. Kohonen, E. Oju, and P. Lehtio, "Storage and processing of information in distributed associative memory systems," Parallel Models of Associative Memory, G. E. Hinton and J. A. Anderson, editors, pp. 49-81. Erlbaum, Hillsdale, 1981.
[11] Călin-Adrian Popa, Octonion-valued neural networks, Artificial Neural Networks and Machine Learning-ICANN, 2016.
[12] Y. Tian, "Matrix representations of octonions and their applications," Advances in Applied Clifford Algebras, vol. 10, no. 1, pp. 61–90, 2000.
[13] J. Mairal, M. Elad, and G. Sapiro, "Sparse representation for color image restoration," IEEE Transactions on image processing, vol. 17, no. 1, pp. 53–69, 2008.
[14] Mingming Li, Shuzhi Sam Ge, Fellow, IEEE, and Tong Heng Lee, "Content-Driven Associative Memories for Color Image Patterns "IEEE TRANSACTIONS ON CYBERNETICS 2168-2267.
[15] T. Dray and C. A. Manogue, The geometry of the octonions. World Scientific, 2015.
[16] Sushil Kumar, Bipin Kumar Tripathi, " On the High Dimentional Information Processing in Quaternionic Domain and its Applications", IJAAS Vol. 7 No. 2, pp. 177-190 , June 2018.
[17] Jiasong Wu, Ling Xu, Youyong Kong, Lotfi Senhadji, Huazhong Shu," Deep Octonion Networks ", Computer Vision and Pattern Recognition, March 2020.
[18] Călin-Adrian Popa, Global Exponential Stability of Octonion-Valued Neural Networks with leakage delay and mixed delays. Neural Networks, 105 (2018) 277-293.
[19] E. Hitzer, T. Nitta, Y. Kuroe, Applications of clifford's geometric algebra, Advances in Applied Clifford Algebras 23 (2013) 377–404.
[20] J. Wright, Y. Ma, J. Mairal, G. Sapiro, T. S. Huang, and S. Yan, "Sparse representation for computer vision and pattern recognition," Proceedings of the IEEE, vol. 98, no. 6, pp. 1031–1044, 2010.

Detecting Sybil Attack In Wireless Sensor Networks Using Machine Learning Algorithms

Mandala Mounica[1], Mrs.R.Vijayasaraswathi[2] , Mrs. R.Vasavi[3]

[1]M.Tech Student of Software Engineering, Dept. of Computer Science & Engineering, VNR VignanaJyothi Institute of Engineering and Technology, Hyderabad, Telangana, India
[2]AssistantProfessor,Dept.ofComputerScience&Engineering,VNRVignanaJyothiInstitute Of Engineering and Technology, Hyderabad, Telangana, India
[2]AssistantProfessor,Dept.ofComputerScience&Engineering,VNRVignanaJyothiInstitute Of Engineering and Technology, Hyderabad, Telangana, India

monicagoud9@gmail.com
vijayasaraswathi_r@vnrvjiet.in
vasavi_r@vnrvjiet.in

Abstract. In sensitive areas such as battlefields, a Wireless Sensor Network (WSN) is especially in military and civilian applications and it is of utmost importance to develop security in these networks. In various respects, this can improve the quality of life. But to be used for protection reasons in multiple situations such as implementation.There is a high risk of being exposed to multiple viruses and hacking attacks. Unauthorized APs for information protection needs to be detected. Any malicious attacks against these networks, Like Sybil Attack leads to breach of security by enacting as a node that illegitimately declares several false identities at the same time.This misleads valid nodes, and they presume each of those identifiers as actual independent nodes by accident. Thus we proposed a machine learning model to detect Sybil attack in network where rawtraffic data has been collected and is used identify authorized and unauthorized APs in an integrated wired/wireless environment.
Keywords: Wireless Sensor Network, Machine Learning, Sybil attack, NSL-KDD Dataset.

1. Introduction

Since Smart gadgets are constantly utilizing wireless sensor networks, it is difficult without Wi-Fi to find positions in our lives.Wi-Fi is used in offices, bars, military stations, schools and public institutions. The creation of low-power, low-cost, small multifunctional sensor nodes has been enabled by numerous advances made in the field of wireless communication and electronics research. A wireless sensor network consists hundreds of thousands lightweight, inexpensive sensor nodes operate together to provide the capacity for data collection and environmental monitoring as shown in Fig 1. In other words, sensor nodes transfer them, step by step, to sink node(s), after sensing information from the environment[27].

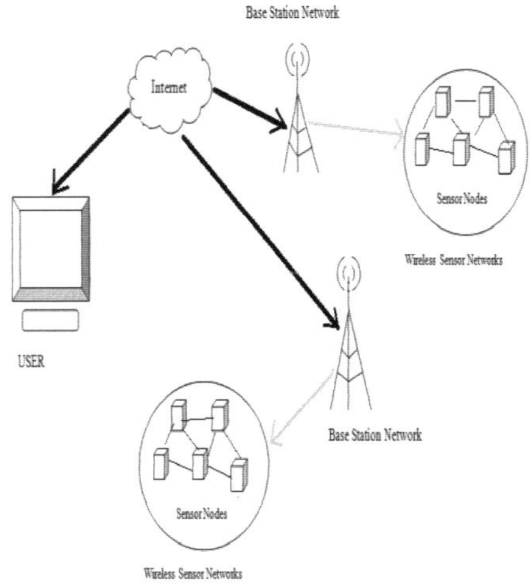

Figure 1 Wireless Sensor Network System

There is a risk of hooligans, interference, and modification of data packets due to the existence of transmitted communication. It is also important to include protection in this area to protect the network from these threats in order to safely route the data in the network[28].

The most harmful attack for wireless sensor network is Sybil attack which oughts to cause security breach.Sybil Attack is type of attack in peer-to - peer networks where several identities are successfully run at the same time by a node in the network and compromises the authority / power in credibility schemes as shown in Fig 2. The key goal of the assault is to obtain the majority of the network in order to carry out unlawful acts on the infrastructure (in compliance with laws and regulations laid down in the network). A single entity (computer) can establish and maintain several identities (user accounts, accounts on an IP address). These numerous false identities tend to be valid, identical identities for outside observers.

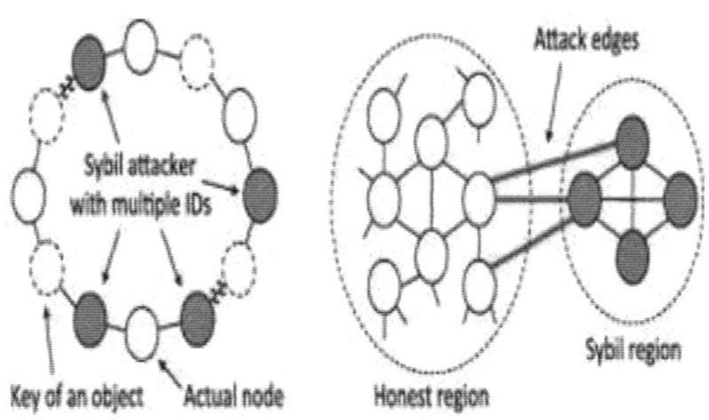

Figure 2 Sybil Attack Occurrence Wireless Sensor Network Systems

1.1. Types of Sybil attack

- In direct attack, truthful nodes directly affected by node(s) of Sybil.

- The truthful node(s) is targeted in an indirect assault by node that connects directly with Sybil node(s). As it under the destructive control of Sybil node(s), this middle node corrupted.

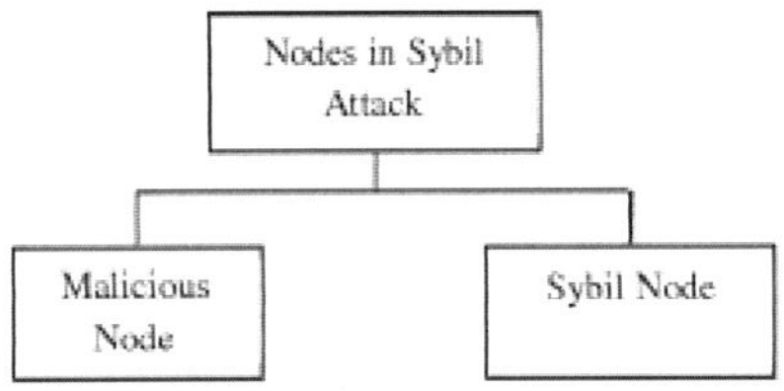

Figure 3 Sybil Attack Nodes

Figure 3 depicts types of nodes in Sybil attack

- Malicious hub/Sybil Attacker: hub that parodies characters of different hubs.
- Sybil hubs: Additional characters are made pernicious hubs known Sybil hubs.

2. Related Work

Wireless sensor networks are distinct from conventional networks that need creative energy-sensitive and real-time routing solutions, reliability, fault identification, data integrity. Machine Learning aims to solve functional problems as well as non-functional requirements.[1] Mohammad Abu Alsheikh et al. information about WSN 's functional problems such as routing, location, data aggregation, data aggregation Attacks are rapidly rising and become lethal threats to Networks. [2] Recommended monitoring method that assists in the identification of Attack traffic. Attacks such as Botnet Assault can assault IoT machines and can allow them to be more easily hacked than desktop computers. Network-based IoTBotnet Attacks Identification was introduced in [3] YairMeidan, Michael Bohadana et al. They classified advantages by using their method to detecting infected IoT such as resistance to heterogeneity, open world,performance.Sybil attack is extremely dangerous against sensor and impromptu structures where multiple characters are claimed illegitimately by a node. R's suggested a Random Password Evaluation system [RPC]. Amuthavalli et al. [4] which encourage the deployment and control of a node's location, thus preventing the attack on Sybil. In defining the Sybil threat, the RPC approach is complex and precise. This approach increases the transfer of data in the network and can increase the throughput as well. Sybil Attack's Methods and Materials are provided at [5] by Udaya Surya et al. To configure, energy-efficient, promising nodes in the network, they graded the Network Model for Sybil attack. A WSN subsystem is separated by means of services that would satisfy WSN specifications. [6] Palak et al. introduced the Sybil Attack. They divided Sybil attacks into multiple types of fake or compromised images, direct or indirect communications, concurrent or non-simultaneous communications. Various methods of Sybil attack have been reviewed. [7] The key management protocol is intended to have a secure communication framework that uses computationally cheaper symmetric cryptography than asymmetric cryptography, which allows the energy-efficient protocol. A decentralized technique is proposed in [8] by EhsanUllahWarriach et al to find errors in WSNs. Secret Markov Models (HMMs) are used to base this methodology on a statistical approach. The collected data was broken down into two forms as a supervised machine learning proposal: a training set and a test set. In actual cases where faults are basically offset faults, stuck-at fault and benefit fault, the suggested approach has worked well.. [9]The authors suggested a classification strategy by SVM. It splits down the data into two classes: one for normal measurement and one for inaccurate measurement. To identify the data, the decision function is then implemented in the cluster head.

In [10], the authors separated the faults into two categories: (1) Persistent and (2) intermittent faults. Persistent errors are persistent errors while intermittent mistakes are temporary mistakes The paper [11] contains a comprehensive bibliography of and implementations of computer algorithms. Supervised and unsupervised learning frameworks were discussed in the report..Sunil Ghildiyal et al [12] addressed small processing capacities and fewer resources to render WSN nodes very vulnerable to a range of attacks. [13]Nodes conserve scarce resources and, like any strong system inside the network like BS, they have to be covered by any outside support. Only sophisticated security processing and security algorithms for the whole network can be executed by BS. [14]The proposed solution to Sybil's attack based redistributed sensor node keys, embedded at the time of development.ImranMakhdoom et al [15] identifies various defences proposed against Sybil Attack andProposed epic One Way Code Attestation Protocol (OWCAP) for remote sensor organizations, a financially savvy and secure code certification scheme that protects against Sybil Attack [16][17][18]. The statistical test and the neural network are merged by Manikopoulos and Papavassiliou. To get similarity measures, first run the Klomogrov-Smirnov test and classify network packets using 5 neural networks. [19]. By identifying and defining consistencies and trends in training data[20], the adoption of statistical methods to boost computer performance.[21] Suggested a stable neighbour authentication protocol on HB adaptation, validation convention for RFID gadgets. The utility of HB shared-key, unidirectional verification conventions permits them hypothetically ideal for asset confined gadgets as RFID labels.Suggested approach in [22] defined Sybil attacks via the positioning of anchor nodes. The main idea behind this method was that two nodes cannot be present at same location at same time and [23] uses received signal strength indication (RSSI) based methodologies. They all used the nodes signal strength to judge whether a node is Sybil in nature or not. C. Komar et al[24] developed the notion of preferred routes for Trespassers (TFP) and given method can estimate possibility of detecting surveillance network. Both computational approaches and simulation findings confirm the precision of the suggested consistency measure. S. A variety of current methodologies for detecting Sybil attacks have been reviewed by Sharmila et al[25] and algorithm suggested detecting Sybil assaults in remote Sensor Network.It is observed after identification, the ratio of throughput and packet transmission has increased. SunilGhildiyal et al[26] addressed reduced capability and less control WSN nodes to render vulnerable to a range of attacks.

3. Proposed System

This research is aimed at creating system that detects Sybil and other threats, using a machine learning model trained for efficiency and accuracy in detection by machine learning algorithms varies details in system Architecture as shown in Fig 4.It's hard to spot an intrusion because it appears close to the victim-end's legitimate network traffic. In the meantime, Sybil and other attacks on victim systems (ddos, u2r, and probe) must be produced over time. Otherwise, the network / system services would not be malicious.Therefore, the designed Machine learning approach used to identify Sybil attack based on the raw traffic internet data. We practise our machine learning model with a broad dataset, NSLKDD 2017 Dataset (We use NSLKDD short in rest paper). Our monitoring strategy uses a series continuous network packets and capable of learning subtle distinction between threats and legal ones. Historical information fed the machine learning model classify Sybil and other attacks (ddos, probe, u2r). It helps to identify recurring patterns of Sybil, DDoS, U2R, Probe attacks and locate in long-term traffic chain.

Machine learning is typically defined by sensor network programmers set of techniques and algorithms used to construct prediction models. Machine learning professionals consider it rich area with very wide themes and trends. Among those want to add machine learning WSNs, knowing certain themes would be helpful. Machine learning algorithms have enormous versatility advantages applicable to various WSNs implementations. An access point in WSN base station which transmits and receives data (sometimes called a transceiver). An access point links users inside the network to other users and can also act as an interconnection point between the WSN fixed wire network.

IOP Conf. Series: Materials Science and Engineering 1042 (2021) 012029 doi:10.1088/1757-899X/1042/1/012029

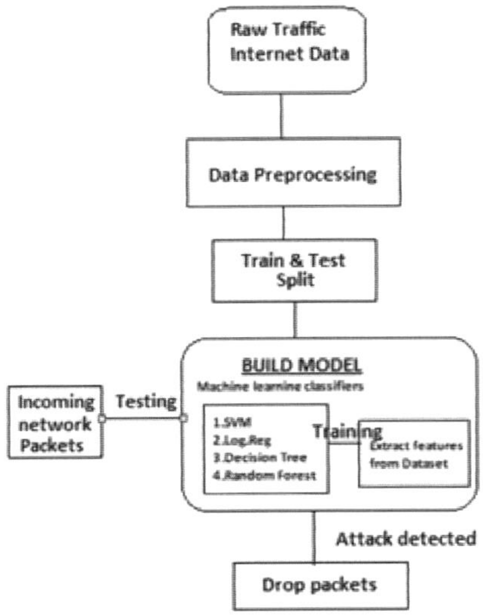

Figure 4: System Architecture

Gathering data

Raw network traffic data involves the analysis of many attributes that emphasize for prediction of target variable. This module involves gathering data from mass amount of internet traffic records that collected and bundled into dataset for real-time analysis. The study involves the detection of attacks in dataset records.

Pre-processing data

Data pre-processing is a vigorous technique in the machine learning process. The categorical data is converted to binary data by clearing the missing values, null values, and inaccurate values, data polishing. The dummy values are added for different attributes to form a dataset and are sent to the training and testing phase.

Feature Extraction and Transformation

First concentrate network traffic fields from NSLKDD informational index. Table I a few models and kinds of fields. We use preparing highlights. Unique in relation to other AI strategies applied on Sybil, Probe, U2RDDoS location, we don't have to choose measurable highlights in model.

Table I: Examples and Types of organization traffic fields

Field	Field Example	Field Type
duration	5607	Continuous
src_bytes	232	Continuous
Protocol_type	tcp	Nominal
Num_failed_logins	5	Continuous
Num_file_creations	1	Continuous
Num_shells	0	Continuous
Su_attempted	0	Binary
Count	1	Discrete
srv_count	1	Discrete
Serror_Rate	0.5	Discrete
dst_host_serror_rate	0.01	Discrete

srv_serror_rate	0.67	Discrete
dst_host_srv_serror_ra te	0.03	Discrete

Table I classifies three types of fields: Binary fields, nominal field,continous fields. For substance of most text and trivial areas, we turn the representation. We convert them to binary Boolean fields like TCP / UDP port number.After element change, we get $m \times n'$ network, where m shows number of parcels and n' demonstrates number of new highlights For e.g., we encode TCP and UDP port number to 16-digit paired rundown.

3.1.Machine Learning based Approach

Algorithms are crucial in the analysis of hidden data knowledge. Because of their precision and speed, machine learning is commonly used in attack detection. We have suggested a framework where machine learning algorithms are applied to effectively detect the attacks with performance, the data collected is that there is no raw traffic internet data in the necessary format to collect the data and bring it into the required format using different data processing techniques to train the model, so the user only has to upload data after developing the system.There are four different machine algorithms in this project it is a comparison study of the algorithms and comparison study of the heart disease data set and real-time dataset. The machine learning algorithm incorporates three different types of learning.

Supervised Learning

Here the algorithm uses the set of variables to generate a method were it maps inserted query to the desired outputs. It consists of the target which is projected from the set of independent variables; the training come to an end till the desired accuracy is attained on the trainingdataset shown in Fig 5. Various examples of Supervised Learning are Strategic Regression, Decision Tree, Support vector machine and Random Forest, and so on

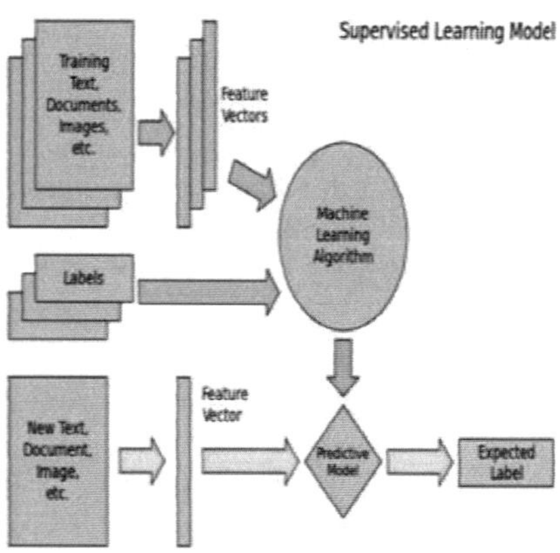

Figure 5: Supervised learning model

Unsupervised Learning

Here thetechnique makes use of clustering the population in distinct clusters. Itis used for segmenting the all given population into the desired groups that arespecified as showninFig 6. Some of the unsupervised learning isApriori algorithm, k-meansclustering, etc.

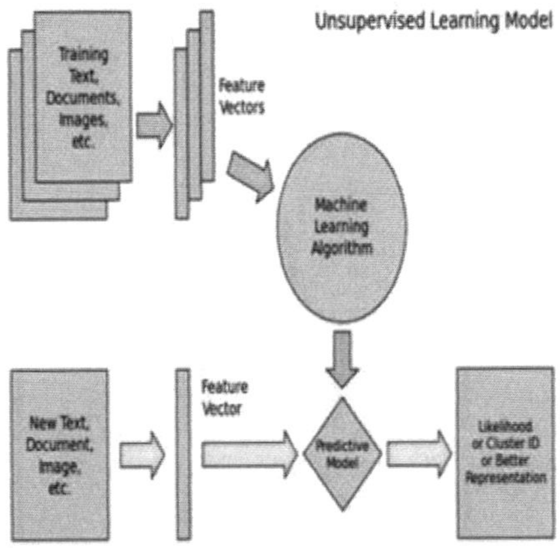

Fig 6: Unsupervised learning model

Reinforcement Learning

Using this technique was the mechanism is trained to make choices by itself.The machine is put in an environment where it itself trains using the trial and errormethod where it learns from the previously experienced encounters to gather the bestknowledge and to make appropriate decisions. Example is Markov decision process.

A. Logistic Regression

Strategic relapse is a calculation utilized for arrangement. It is utilizedto estimate discrete values such as yes, no, 0, 1, true and false .It studies the relationship between the binary response variable from the set of the predictors. This method is a statistical technique for evaluating a dataset where there are one or more independent attributes that define the effect; the outcome is measured using discrete values. The objective of the logistic regression is to obtain the optimal fitting model which determines the relation between the explanatory and the response variable. It generates the coefficient of the formula to predict logit transformation of the presences of the probability of the characteristics of the interest.

$$S(z) = \frac{1}{1 + e^{-z}}$$

The values predicted are mapped to probabilities; the above sigmoid function is used. Using the sigmoid function and decision boundaries the prediction function is made.

$$z = \beta 0 + \beta 1x1 + \beta 2x2 + \cdots$$

The above expression is of the multiple explanatory variables.

B. .Random forest (RF)

This approach forms a combination of the decision trees with every feature provided into every decision tree. It is very simple and flexible to use algorithm avoiding tuning hyper parameter it always gives the best results. It provides both classification and regression tasks. It uses the supervised learning technique. It creates decision trees by randomly choosing the different attributes present in the dataset available and then gets the predictions from the trees and chooses the best solution. It also indicates the attribute importance. It is considered as the highly accurate and strong method it does not have an over fitting problem because it takes the average of all the predictions made. It handles missing values very efficiently by median replacement and proximity weighted average. It provides with importance of the relative feature which helps in picking the feature contributing most in the classification.

C. Support vector machine

This regulated calculation for AI utilized for issues like arrangement or relapse. To change over the information, it utilizes method called portion stunt and afterward finds suitable limit between the potential yields based changes [29]. Predictive analysis data classification algorithm that assigns one of the labeled categories to new data components. In certain instances, a binary classifier implies that two potential goal values are present in the data in question. In specific, it is widely used in the classification of support vector machines. On two samples x and x, 'the RBF kernel, expressed feature vectors in some input space specified:

$$K(x, x') = \exp\left(-\frac{||x - x'||^2}{2\sigma^2}\right)$$

$||x - x'^2||$perceived as squared Euclidean separation between two component vectors between the two featurevectorsσ. An equivalent definition involves parameter.

$$\gamma = \frac{1}{2\sigma^2}$$

$$K(x, x') = \exp(\gamma||x - x'||^2)$$

Since estimation of RBF bit diminishes with separation and reaches between zero (in the breaking point) and one (when x = x'), prepared understanding as similitude measure.

D. Decision Tree

Paradigm in which each option is centred on the constant time analysis of two numbers is called the Decision Tree Paradigm. The technical difficulty of sorting and searching has been added.The decision tree paradigm is computing model in terms of computational complexity in which algorithm is assumed effectively decision tree, i.e. series branching operations centred similarities such quantities, assigning unit computational expense to the similarities.

The "tests" or "queries" are considered branching activities. The algorithm in question can see Boolean function computation in this context.

$$f: \{0,1\}^2 \rightarrow \{0,1\}$$

Where the input is a set of questions and the final judgement is the result. Each question can rely on past queries Differing versions of decision tree models is applied on basis of complexity of operations allowed in determiningsingle contrast and way of branching.

Decision tree models help establish lower limits for the theory of complexity in certain forms of computational problems and algorithms. Decision tree complexity or query complexity of this issue or calculation communicated as far as choice tree model known as being computationally intricate.

3.2. Analysis of Algorithms

In the process of anomaly detection, machine learning algorithms are applied to the DDoS, Sybil, u2r, investigate.

- A distributed denial of service (DDoS) attack is deliberate effort to interrupt traffic of targeted website, or network by cascade of Internet traffic flooding the object or the surrounding networks.
- The Sybil assault is a major disruptive assault on the sensor network in which many genuine identities and forged identities are used to access a network illegally.
- User to Root (u2r): Intruder has local access to the victim computer and is attempting to obtain rights for super users.
- Probe: Intruder seeks to obtain details regarding the target host.

IOP Conf. Series: Materials Science and Engineering 1042 (2021) 012029 doi:10.1088/1757-899X/1042/1/012029

The most frequently used algorithms include Backing Vector Machine, Decision Tree, irregular timberland, calculated relapse are executed on dataset for identification of assaulted bundle.

3.3.Analysis of Performance

Eventually, to label each instance as benign or anomalous, we add the customised model to features derived from continuously observed packets. Marked instances are then used to assess if the whole accompanying stream is benign or anomalous. Consequently, where an anomalous stream is observed, a warning should be released, since it may suggest fraudulent behaviour on the provided dataset. Thus notifies each model accuracy shows how effectively detecting the attacks in dataset.

4. Results

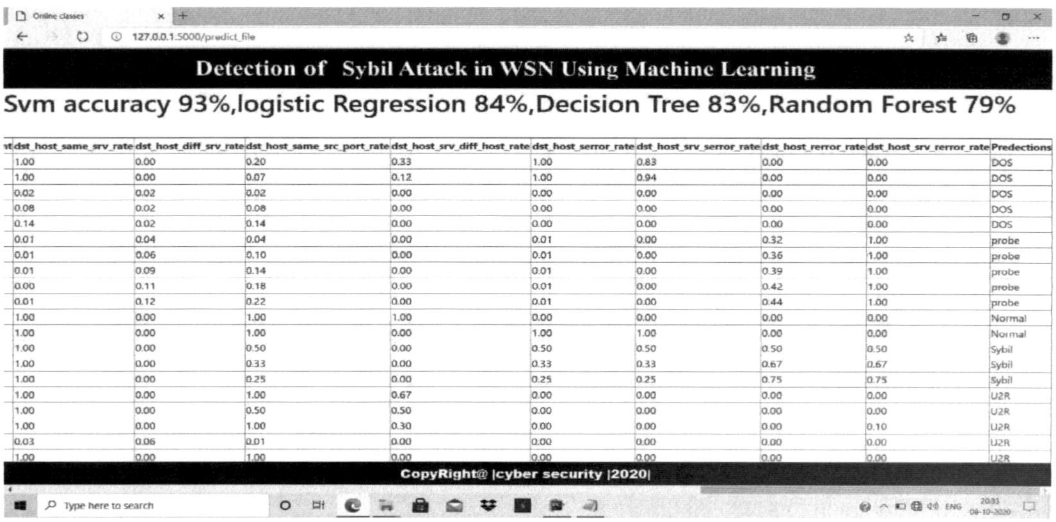

dst_host_same_srv_rate	dst_host_diff_srv_rate	dst_host_same_src_port_rate	dst_host_srv_diff_host_rate	dst_host_serror_rate	dst_host_srv_serror_rate	dst_host_rerror_rate	dst_host_srv_rerror_rate	Predections
1.00	0.00	0.20	0.33	1.00	0.83	0.00	0.00	DOS
1.00	0.00	0.07	0.12	1.00	0.94	0.00	0.00	DOS
0.02	0.02	0.02	0.00	0.00	0.00	0.00	0.00	DOS
0.08	0.02	0.08	0.00	0.00	0.00	0.00	0.00	DOS
0.14	0.02	0.14	0.00	0.00	0.00	0.00	0.00	DOS
0.01	0.04	0.04	0.00	0.01	0.00	0.32	1.00	probe
0.01	0.06	0.10	0.00	0.01	0.00	0.36	1.00	probe
0.01	0.09	0.14	0.00	0.01	0.00	0.39	1.00	probe
0.00	0.11	0.18	0.00	0.01	0.00	0.42	1.00	probe
0.01	0.12	0.22	0.00	0.01	0.00	0.44	1.00	probe
1.00	0.00	1.00	1.00	0.00	0.00	0.00	0.00	Normal
1.00	0.00	1.00	0.00	1.00	1.00	0.00	0.00	Normal
1.00	0.00	0.50	0.00	0.50	0.50	0.50	0.50	Sybil
1.00	0.00	0.33	0.00	0.33	0.33	0.67	0.67	Sybil
1.00	0.00	0.25	0.00	0.25	0.25	0.75	0.75	Sybil
1.00	0.00	1.00	0.67	0.00	0.00	0.00	0.00	U2R
1.00	0.00	0.50	0.50	0.00	0.00	0.00	0.00	U2R
1.00	0.00	1.00	0.30	0.00	0.00	0.00	0.10	U2R
0.03	0.06	0.01	0.00	0.00	0.00	0.00	0.00	U2R
1.00	0.00	1.00	0.00	0.00	0.00	0.00	0.00	U2R

Figure 7: Result of Detecting Attacks in Uploaded Dataset

Dataset uploaded underwent through model for Detection processing each record indicates in predictions whether they are Normal,Ddos,Sybil or U2R attacks by classifiers as showninFig 7 by introduced machine learning model.

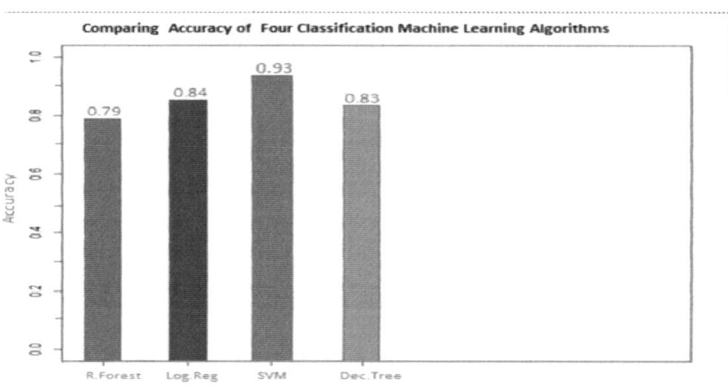

Figure 8: Accuracies in Algorithms

In proposed Machine Learning model, as mentioned in section-III four classifier algorithms ie.,SVM, Random forest, Decision tree and logistic Regression algorithms as showninFig 8 used to determine how effectively does each classifier detect attacks in network data by accuracy performance.

5. Conclusion

It is apparent from doing the aforementioned analysis that the accuracy of the support vector machine algorithm is of the highest precision. The variables of Predictor are used from the dataset that was generated for the prediction of attacks (DDOs, R2L, Probe, Sybil, and Normal) from real-time data obtained from web sources. For the estimation of the outbreak, there were four main models used. The first model used was Random Forest, which gave 79 percent accuracy. Logistic regression was the second model developed, which gave 84 percent precision. The support vector machine's third model gave 93 percent precision and the fourth Decision Tree model gave 83 percent precision. It is therefore assumed that the Support vector machine model with the highest precision obtained can be used effectively by 93 percent and also the best predictors of the Attack presence (DDOs, R2L, Probe, Sybil, and Normal) in the uploaded dataset from the variable significance graph.

Future studies will require device extension since data set attacks can be closely studied under networks that currently use construction models that have the potential to foresee future demands, preferences, requirements, and attack detection needs. For different attacks under consideration, there is an ability to achieve prediction method, since it is easy to terminate anything from happening until it alters the current state of nodes in networks by creating a real-time website where nodes go through detection and drop clean data packets.

REFERENCES:

1. Mohammad Abu Alsheikh, Shaowei Lin, DusitNiyato, and Hwee-PinkTan, "AI in Wireless Sensor Networks: Algorithms, Strategies, and Applications", in IEEE Communication Surveys and Tutorials, Vol. 16, No. 4, Fourth Quarter, 2014.
2. Xiaoyong Yuan, Chuanhuang Li, Xiaolin Li, "Profound Defense: Identifying DDoS Attack through Deep Learning", in IEEE International Conference, 2017.
3. YairMeidan, Michael Bohadana, Yael Mathov, YisroelMirsky, "N-BaIoT:Network-based Detection of IoTBotnet Attacks Using Deep Auto encoders", Published by the IEEE Computer Society, 2018.
4. R. Amuthavalli, Dr. R. S. Bhuvaneswaran, "Recognition and Prevention of Sybil Attack in Wireless Sensor Network Employing Random Password Comparison Method", in Journal of Theoreticaland Applied Information Technology, 2014.
5. UdayaSuriya Raj Kumar Dhamodharan and RajamaniVayanaperumal, "Distinguishing and Preventing Sybil Attacks in Wireless Sensor Networks Using Message Authentication and Passing Method", in Hindawi Publishing Corporation e Scientific World Journal, 2015.
6. Palak, "Survey on the different Sybil Attack Detection Techniques in Wireless Sensor Network", in International Journal of Computer Applications, Volume 164 – No 1, April 2017.
7. SonaMalhotra, PankajRathee, "Avoidance of Sybil Attack utilizing Cryptography in Wireless Sensor Networks", IJIRST – International Journal for Innovative Research in Science& Technology, 2015.
8. EhsanUllahWarriach, Kenji Tei, "Flaw Detection in Wireless Sensor Networks: A Machine Learning Approach", 2013, distributed in 2013 IEEE sixteenth International Conference on Computational Science and Engineering.
9. SalahZidi, TarekMoulahi, and BechirAlaya, "Flaw Detection in Wireless Sensor Networks through SVM Classifier", distributed in IEEE Sensors Journal, Vol.18, No. 1, January 1, 2018.
10. ThahaMuhammed, Riaz Ahmed Shaikh, "An examination of deficiency recognition procedures in remote sensor organizations" distributed in the Journal of Network and Computer Applications.

11. Uday Shankar Shanthamallu, Andreas Spanias, CihanTepedelenlioglu, and Mike Stanley, "A Brief Survey of Machine Learning Methods and their Sensor and IoTApplications", distributed in eighth International Conference on Information, Intelligence,Systems and Applications, 2017.
12. Sunil Ghildiyal1, Ashish Gupta2, Nitesh Tomar3, AnupamSemwal, "Examination of Sybil Attack in Wireless Sensor Networks", International Journal of Engineering Research& Technology (IJERT), Vol. 3 Issue 5,pp.845-848. May–2014.
13. V.Sujatha, E.A.MaryAnita, "FEM-Hybrid Machine Learning Approach for the Detection of Sybil assaults in remote sensor organizations" distributed in International Journal of Innovative Technology and Exploring Engineering (IJITEE), Vol.8, May 7, 2019.
14. P. RaghuVamsi and Krishna Kant, "Sybil Attack Detection utilizing Sequential Hypothesis Testing in Wireless Sensor Networks," in IEEE,2014.
15. ImranMakhdoom, MehreenAfzal, ImranRashid, "A Novel Code Attestation Scheme Against Sybil Attack in Wireless Sensor Networks" in National Software Engineering Conference,2014.
16. C. Manikopoulos and S. Papavassiliou, "Organization interruption and deficiency location: a measurable irregularity approach," IEEE Communications Magazine, vol. 40, no. 10, pp. 76–82, 2002.
17. S. Seufert and D. O'Brien, "AI for programmed safeguard against disseminated disavowal of administration assaults," in 2007 IEEE International Conference on Communications. IEEE, 2007, pp. 1217–1222.
18. P. A. R. Kumar and S. Selvakumar, "Circulated refusal of administration assault recognition utilizing a gathering of neural classifier," Computer Communications, vol. 34, no. 11, pp. 1328–1341, 2011.
19. A. H. Duffy, "The 'what' and 'how' of learning in plan," IEEE Expert,vol. 12, no. 3, pp. 71–76, May/Jun. 1997.
20. P. Langley and H. A. Simon, "Uses of AI and rule enlistment,"Commun. ACM, vol. 38, no. 11, pp. 54–64, Nov. 1995.
21. Jasmine Norman, Paulraj Joseph "Secure Neighbor Authentication in Wireless Sensor Networks," in second International Conference on Wireless Communication, Vehicular Technology, Information Theory and Aerospace &Electronics Systems Technology, Chennai, Feb. 28, 2011-March. 3, 2011.
22. Bin TIAN, Yizhan YAO, Lei SHI, Shuai SHAO, Zhaohui LIU and Changxing XU "A Novel Sybil Attack Detection Scheme for Wireless Sensor Network," in fifth IEEE International Conference on Broadband Network and Multimedia Technology, Guilin, Nov. 17-19, 2013.
23. Shanshan Chen, Geng Yang, Shengshou Chen "A Security Routing Mechanism against Sybil Attack for Wireless Sensor Networks," in IEEE International Conference on Communications and Mobile Computing, Shenzhen, April. 12-14, 2010.
24. C. Komar, M. Y. Donmez, and C. Ersoy, "Discovery nature of outskirt observation remote sensor networks in the presence of trespassers'favourite ways," Computer Communications, vol. 35, no. 10, pp. 1185–1199, 2012.
25. S. Sharmila and G. Umamaheswari, "Discovery of Sybil assault in portable remote sensor organizations," International Journal of Engineering Science and Advanced Technology, vol. 2, pp. 256–262, 2012.
26. Sunil Ghildiyal1, Ashish Gupta2, Nitesh Tomar3, AnupamSemwal, "Investigation of Sybil Attack in Wireless Sensor Networks", International Journal of Engineering Research and Technology (IJERT), Vol. 3 Issue 5,pp.845-848. May – 2014.
27. R. V. Saraswathi, L. P. Sree and K. Anuradha, "Dynamic and probabilistic key administration for disseminated remote sensor organizations," 2016 IEEE International Conference on Computational Intelligence and Computing Research (ICCIC), Chennai, 2016, pp. 1-6, doi: 10.1109/ICCIC.2016.7919666.

28. R.V.Saraswathi, L. P. Sree, and K. Anuradha, "Multi-stage Key Management Scheme for Cluster based WSN,", International Journal of Communication Networks and Information Security, Vol. 10, No. 3, pp. 552, 2018.
29. R. VijayaSaraswathi, L. PadmaSree and K. Anuradha"Backing Vector Based Regression Model to Detect Sybil Attacks in WSN" distributed in International Journal of Advanced Trends in Computer Science and Engineering, Volume 9,no.3,May-June 2020.

IOP Conf. Series: Materials Science and Engineering 1042 (2021) 012030 doi:10.1088/1757-899X/1042/1/012030

A literature surveyonrecurrent attentionlearningfortext classification

Ayesha Mariyam[1], SK Althaf Hussain Basha[2],S Viswanadha Raju[3]

[1]Department of Computer Science and Engineering, Jawaharlal Nehru Technological University,Hyderabad, India
[2]Department of Computer Science and Engineering, A1 Global Institute of Engineering and Technology, Markapur, India
[3]Department of Computer Science and Engineering, JNTUH College of Engineering Jagtial, Nachupally,Jagtial, India

[1]ayesha.mariyam84@gmail.com
[2]althafbashacse@gmail.com
[3]svraju.jntu@gmail.com

Abstract: Witha rapid rise of complex data every year needs more enrichment in machine learning methods to provide vigorous and accurate data classification. Deep learning models such as Convolutional Neural Networks (CNN), Recurrent Neural Networks (RNN), Long Short-Term Memory(LSTM) have accomplished to obtain better results in the domain of computer vision, object recognition, speech recognition and natural language processing compared to traditional machine learning algorithms. This paper mainly discusses about the blending of attention mechanism with various deep learning models for text classification which improves the performance of text classification task.

1. Introduction

Text classification is a task of natural language processing which assigns predefined labels to text according to its contents. Text classification is an essential component in many applications such as sentiment analysis, document classification, email filtering etc. Traditional methods such as bag of words (BOW), n-grams, Term Frequency – Inverse Document Frequency (TF-IDF) use statistical methods to represent the text as feature vectors and then classify these features using Naïve Bayesian, SVM, KNN [2]. These methods are straightforward and robust but they often neglect the context-dependent data or word sequence in passsage which influences the classification accuracy and also have the problem of sparsity [5].
However, in parallel with the emergence of deep learning, NLP has procured new facets which allows machine to analyze, realize and extract meaning out of texts. Deep learning models such as Convolutional Neural Network(CNN) and Recurrent Neural Network(RNN) based on Long Short- Term Memory(LSTM) outperforms standard methods of text classification[3].
RNN mainly handles sequence learning problems. One of the key ideas of RNN is to connect previous information with the present processing unit, that is to have a store for the previous

information. Conceptually, RNN can record any long-term information but practically it is not possible due to gradient exploding and vanishing problems. The solution of this problem is to alter the framework of the RNN, by including gated activation function to obtain the trade-off between the old and the new time information. Both Long Short-Term Memory(LSTM) and Gated Recurrent Unit(GRU) designs include gated activation functions which makes the network to learn long-term dependency data and reduce the gradient exploding and vanishing problems[1].

Attention mechanism was initially introduced for Machine Translation.But, now it has become a main concept in neural network studies. Attention is used extensively within the Artificial Intelligence (AI) community as an essential unit of neural frameworks for huge number of purposes such as Speech and Computer Vision, Natural Language Processing, Statistical Learning etc. With the help of human biological systems, the instinct behind attention can be well explained. For instance, our visual processing system focus its attention on particular segments of the image, while neglecting other irrelevant information in a way that helps in perceptron. Similarly, in many tasks such as speech, vision or language, some portion can be more relevant than other [6].
Attention mechanism improves the capacity of RNNs. Attention mechanism makes the network to seek information from inputs and proceed to suceeding state on the accordance of it. Word vector representation is another factor which influences the working capacity of NLP tasks in RNN [3]. In Natural language Processing, the attention mechanism is normally used with the Encoder-Decoder Model [6].
The rapid growth in casting attention in neural network is essentially due to three reasons: Firstly, these models are now revolutionary for different problems such as Sentiment Analysis, POS tagging, Question Answering, Sentiment Analysis etc. Secondly, theyhave been abundantly used for improving understanding of neural networks, which are in other respects reviewed as black-box models. Third, they surpass some issues with Recurrent Neural Networks(RNNs) such as performance degradation as input size increases and inefficient computation evolved from sequential processing of input[7]. Therefore, in this paper we intent to impart a brief, yet comprehensive study on attention modeling for text classification.
The paper is organized as, section II discusses about the literature survey of different deep neural networks with attention mechanism for text classification. Section III concludes with a future scope.

2. Literature Survey

The following is the literature survey conducted on various attention models for text classification:

Guogiang Zhong, et al [1],proposed a RNN model named as Recurrent Attention Unit(RAU),which absolutely combines the attention mechanism in the center of Gated Recurrent Unit (GRU) by adding an attention gate. This upgrades the GRU's capacity to remember long-term memory and quickly get rid of unimportant matter.RAU has an ability of obtaining useful information from sequential data by extensively determining a sequence of locations and concentrate on those regions during learning.
The experiments performed on Recurrent Attention Unit shows that it is appropriate for language modeling, sentiment analysis, image classification.RAU performance is outstanding compared to both LSTM and GRU.

Jin Zheng et al [2], proposed a hybrid bidirectional recurrent convolution neural network attention-based model, named as BRCAN to address the challenges with the multi-class text classification and the fine-grained sentiment analysis. This model integrates the Bi-LSTM and CNN constructively with the benefit of the word2vec model and attention method for attenuated text classification.
In BRCAN, a Bi-directional recurrent construct acquires contextual information and vulnerability of sentences and a maximum pool layer of CNN examines which words are essential for text classification.Four text classification datasets, which are separated into topic classification and sentiment classification are used to assess the proposed model. The topic classification dataset has Yahoo! Answers and Sogou News. The sentiment classification dataset consists of Douban movie top250 short reviews full and Douban movie top250 short reviews polarity, Yelp Reviews full and Yelp Reviews polarity.The BRCAN model achieves outstanding performance in comparison with other alternative models on all tasks.
The proposed framework is usually not restricted to text classification problem but can also be implemented for other approaches [8] which is significant for future research.

Ankit Kumar et al [3], proposed a new mechanism of attention based recurrent neural network for sentence classification. This model is designed on attentional implementation of Bidirectional and Vanilla Recurrent Neural Network framework on top of two kinds of recurrent cells, Long Short-Term Memory (LSTM) and Gated Recurrent Unit (GRU) for sentence classification.
These are called as Attentional LSTM(ALSTM) and Attentional GRU(AGRU), and Attentional Bi-LSTM and Attentional Bi-GRU(ABGRU).
Word2Vec model is used and the resulting feature vectors are fed as input to this model. The words are initialized randomly which are not present in word2vec model.
To check the efficiency of this proposed model, its performance is analogized with the other methods on six datasets namely CR, MPQA, MR, Subj, SST-1 and SST-2which confirms that this model has a cabaility to store long-term dependencies in the problem of sentence classification.

Jun He et al [4], proposed a hierarchical attention networks for document classification which has two distinguishing properties. Firstly, it has a hierarchical design which reflects the hierarchical structure of documents. Secondly, it has two stages where attention mechanisms are applied that is at the word-level and sentence-level.
This model gradually builds a document vector by merging important words into sentence vectors and then merging important sentences vectors into document vectors.Experiments done with this model on six enlarged datasets outperforms the previous models by considerable amount.

Zichao Yang et al [5], has proposed a new document classification model. which is based on recurrent attention learning which needs to glimpse only a few groups of words. This is preferable for long document because it does not make use of all detailed information.With respect to the coarse document text, only a few glimpses are utilized to make the selective feature for classification. This problem is addressed as a reinforcement learning task in which the actions are motivated for right classification and the actions are demotivated for wrong classification.
This method intent to prepare an agent which handles how to thereafter look at the next position in a given time steps to get correct classification.
This model is evaluated with the arXiv papers data set which excels in performance in comparison with CNN model which employs different length convolution kernels.
Motivated by [4], in future, there is a scope of improving the controller network by inspecting the hierarchical attention mechanism.

3. Conclusion

The main intention behind this study is to give brief idea on text classification using Attention Mechanism. The study shows that all of the models can be applied to increase the accuracy of text classification over different datasets but with a constraint of tuning hyper parameters of the model. The aim behind this survey is to develop an appropriate and effective model for text classification.

References

[1] Guogiang Zhong, Guohua Yue and Xiao Ling,2018, Recurrent Attention Unit,*arXiv: 1810.12754.* [Online].Available: https://arxiv.org/abs/1810.12754.
[2] Jin Zheng, Limin Zheng,2019,A Hybrid Bidirectional Recurrent Convolutional Neural Network Attention-Based Model for Text Classification,*IEEE Access*,7,106673- 106685.
[3] Ankit Kumar and Reshma Rastogi (nee Khemchandani),2018,Attentional Recurrent Neural Networks for Sentence Classification, *Innovations in Infrastructure, Advances in Intelligent Systems and Computing 757*,549-559.
[4] Jun He,Liqun Wang, Liu Liu, Jiao Feng,Hao Wu,2019, Long Document Classification from Local Word Glimpses via Recurrent Attention Learning," *IEEE Access*,7,40707-40718.
[5] ZichaoYang,Diyi Yang,Chris Dyer,Xiaodong He,Alex Smola,Eduard Hovy,2016, Hierarchical Attention Networks for Document Classification, *Proceedings of NAACL- HLT*,1480–1489.
[6] Sneha Chaudhari,GungorPolatkan, Rohan Ramanath, Varun Mithal,An Attentive Survey of AttentionModels,*arXiv:1904.02874v1*.[Online].Available: https://arxiv.org/abs/1904.02874.
[7] DzmitryBahdanau,Kyung Hyun Cho, Yoshua Bengio,2016, Neural Machine Translation by Jointly Learning to Align and Translate, *arXiv:1409.0473v7,"*,1-15
[8] Xiaoyu Guo, Hui Zhang, Haijun Yang, Lianyuan Xu, Zhiwen Ye,A single attention-based combination of CNN and RNN for relation classification, *IEEE Access*,4.

IOP Conf. Series: Materials Science and Engineering 1042 (2021) 012031 doi:10.1088/1757-899X/1042/1/012031

Human Action Recognition Using A Novel Deep Learning Approach

Badhagouni Suresh Kumar[1], Dr S.Viswanadha Raju[2], Dr H.Venkateswara Reddy[3]

Department of CSE,Vardhaman College of Engineering.Hyderabad, India[1,3]
Deparment of CSE, JNTUH College of Engineering ,Nachupally[2]

Sureshkumargoud2006@gmail.com[1],viswanadha_raju2004@yahoo.co.in[2],h.venkateswarareddy@vardhaman.org[3]

Abstract. Behavior human analysis is always a significant aspect in societal communication. The human behavior analysis is developed based on few factors like human activity and action recognition. Human action recognition is an significant feature in different safety fields. The assessment of the action recognition algorithm depends on the appropriate extraction and the learning data. In the human action recognition, classification plays the major role so in order to this effectively Gated Recurrent Neural Network is used with an increased computation level. Feature extraction is one of the essential factor in human action recognition it will influence the performance and computation time of the algorithm. This paper presented an approach for human action recognition based on new mixture deep learning model. The proposed method is evaluated on the different data sets like UCF Sports, KTH and UCF101.On UCF Sports data set the proposed method has given an average of 96.8%.

1.Introduction:

In the present world , there are lot of domains showing interest in this field for the real world applications[1] and concentration in human recognition mounting in last few years. There are lot application domains which are showing more interest in this field[2]. Vera et al in [3] has developed an application that directly interact with humans based on their head movements and from the past few years there are lot of application are proposed based on human action recognition mostly in surveillance system. The main challenges in human action recognition systems are extracting feature from the source images and classification is another challenge.

Gaussian mixture model (GMM) and Kalman filters are the best suitable methods for detect and extracting features from the images. Gated Recurrent Neural Networks(GNN) are used to collect these features of each and every frame and predicts the action movement of human. The proposed method is developed based on method proposed by Neziha Jaouedi etc in[24] new hybrid deep learning model for human action recognition, in this paper they have proposed method using GMM to detect the human motion detection , Kalman filter for Noise reduction and GNN for Predicting human action but the proposed method in this paper mainly deals in Noise reduction that is Kalman filter. The Kalman filter is mostly used method that shows the best performance in working with conditions that a noise infect input video frames. Kalman filter is time dependent recursive mechanism, but Kalman filter has two major drawbacks in the moving object tracking.

First, in order to get current position estimation it takes only one earlier position of the target. secondly it cannot take into account the stored trajectories of the same environment. So to overcome these two problems Probabilistic Kalman filter has been developed[25]. The proposed method in this paper is, to find human motion GMM is used and Probabilistic method is used in order filter in the noise part in the videos frames and finally to predict the motion action GNN is used. This paper organized in the following way in the next section2 is discussed about Related work and section3 deals with proposed method , section3 is discussed about results and conclusion

2. Related work:

There is a lot of work is going on in human action recognition . Human Detection and motion tracking are the two important factors to recognize the actions of the human. Many methods are used background subtraction for human detection. Gussian Distribution is one of best method for human detection in video background[4].Sigma-Delta is also one of human detection method developed based on background subtraction[5]. Kim et al. proposed a method to identify foreground and background[6].According to method , to identify background in the frame a codebook is developed with lot of words. Gussian Mixture Model(GMM) is successful model for noisy backgrounds[7].All the methods are useful for the human detection but in action recognition motion tracking is one of essential task. Optical Flow and Histogram of Optical Flow and motion boundary histograms(MBH) are proposed and those are used to detect and track the moving object[8].

Among all these techniques optical flow did not perform well in tracking the object when there is lighting changes[9]. In order to overcome these type of problems Kalman filter is proposed[14].This method is very successful because of its linear quadratic nature. Kalman filter is used is used to track single moving object in different environments[10]. Recently a new approach called Deep learning is becomes the most successful method. This method is used to learn the feature from the frames, and frames may be different types like videos ,speech etc..[22][11].These deep learning methods have been trained by a large set of labeled data and a neural network with many hidden layers.3D Convolution Neural Network(CNN) is proposed with multiple 3D deep leading to identify human action recognition. This method given good results in different data sets [12].Several applications used the Recurrent Neural Network(RNN)[13].Long and short term memory unit(LSTM) is used in RNN for successful prediction speech recognition, video capturing [15,16,17].Gated Recurrent Neural Network(GRU) is simplified version of LSTM[18].GRU is deployed on large data like network traffic data and computer vision application[19].Recently [20] a method proposed with the combination of GRU and Support Vector Machine. Kalman filter is one of the successful method deployed on human track system, but it suffers with two major drawbacks[21].Firstly , to predict the current position of the object it will consider only one previous position of the target object. Second issues is it cannot consider the stored data of the same environment. In order to overcome this problem a novel method Probabilistic Kalman Filter(PKF) is proposed[23].

3. Proposed method:

Human Motion tracking: To Improve the performance of the GRU is better to have good classification methods and Feature Extraction is one best way to video classification. Extraction of features will reduce computation time. To provide best features, GMM and Probabilistic Kalman Filter is applied on video frames. Background modeling is essential task in finding moving person detection in dynamic scene. Spotting the human movement in the video frame GMM is the most effective model. In Gaussian parameters like variance, amplitude and mean are optimized based on the Expectation maximization(EM) algorithm.

3.1 Probabilistic Kalman Filter:

In the process of PKF, development of probabilistic Graph will accomplish objectives. To fulfill the objectives, PKF utilized some videos information taken from the source of internet that consisting stand-camera which is fixed with a minimum-scaling and consisting of normal and abnormal human movements. Consider there are N number of videos and each video is having m frames with size of (v,h).Using Practicle tracker a trajectory matrix is developed with size of $2\times m\times N$. In ordered to store these trajectories in to a proper format Fuzzy C-means is used. Now consider bj, L is the probability of Sj, where Sj is beginning point of the frame. Transition from node Si to node Sj is aij so in each step the next position can be calculated based on the following:

$$\text{next position}=\max\left(P(\text{Current position})\times P(\text{Transition})\times P(\text{Observation})\right)$$

so finally for this sequence of frame the target is obtained

$$M^*=\text{Max}(P(S/Y1:L) \qquad (1)$$

In the equation M* is the set that consist of gained observation and S is node and Y is number of nodes.

The probabilistic graph development comprises of making the weights by nodes and edges. The working nature of probabilistic graph is represented in Figure1.

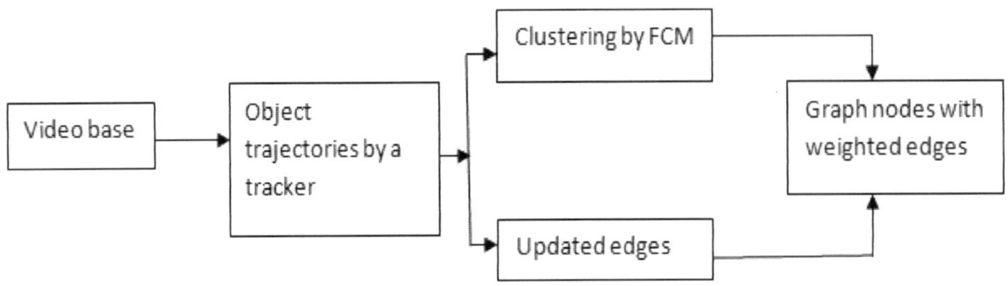

Figure 1: *This Diagram represents the procedure of the probabilistic graph. Primarily, videos are tracked and then it is classified by FCM method.*

The acquired core points for FCM are gained from the nodes of the graph. The weights of two similar nodes are increased by one for each consecutive trajectory points.
According the working sketch of PKF, initially the videos are send to the precise tracker, and the outcomes from this tracker considered as ground truth.

Trackers can be of different types, instance we may have particle ,KLT trackers etc.. In this PKF we are utilizing the particle tracker. All the selected trackers acquire directions in all objectives in the videos; at that point all gained trajectories are kept on grid with size of $2 \times m \times n$ and it is called as a trajectory matrix for the given N videos. Currently, it requires an suitable representation for stored direction, it can be done by appropriate grouping strategy. for instance Fuzzy C-means(FCM) is the appropriate strategy for keeping trajectories.
The nodes of the graph can be characterized as Nodes = $\{X_i|i = 1, 2...c\}$

In this c represents the quantity of clusters, X_i presents focal point of the ith cluster acquired from the FCM. As it were, focal point of each cluster is taken as the node of the ideal graph. To get the different results for the distinct values in FCM it is essential for user to decide the number of cluster. The decision of this hyper boundary relies upon the empty area of the environment free space in environment and the quantity of items. There are 50 clusters are chosen and selected clusters are pre-owned videos, because the selected videos are not packed. In the packed domain the movement of the objects is very slow and the movements of the objects in continuous two progressive frames are become difficulty to recognize.

At this situation, nodes in the graph have to be very near to one another, making the quantity of node significantly more to cover all the moderate movements, depending on the quantity of items and unreserved area in the recorded area, the quantity of 300 node clusters should be adequate for swarmed conditions, by taking 30-60 groups might convey satisfactory outcomes to the isolation situations. The quantity of clusters is absolutely relies upon the selected videos. So in the direction of complete the graph, at first, every nodes are associated with one another to a weighted value as zero. In this situation, for every progressive focuses on all gathered trajectories. Pair of nodes have been selected from the graph, and these selected nodes are having the base separation to the sequential focuses and also the weights of edges is expanded by one.
This strategy has to apply for all the pair of nodes in every direction. These weighted edges represents the crowed area in the environment. Development of the graph is finished by finding the weights between the pair of nodes. In the remaining areas in the graph, this chart will be used for PKF forecast. This graph is utilized in tracking process of human in videos. If an user select a point in the initial frame it will be tracked by PKF.
Figure2 describes various required steps for the PKF development. In the figure2 it is clearly shows that ,when the target object is found then two filters related to Kalman are applied. This is known as PKF prediction part.

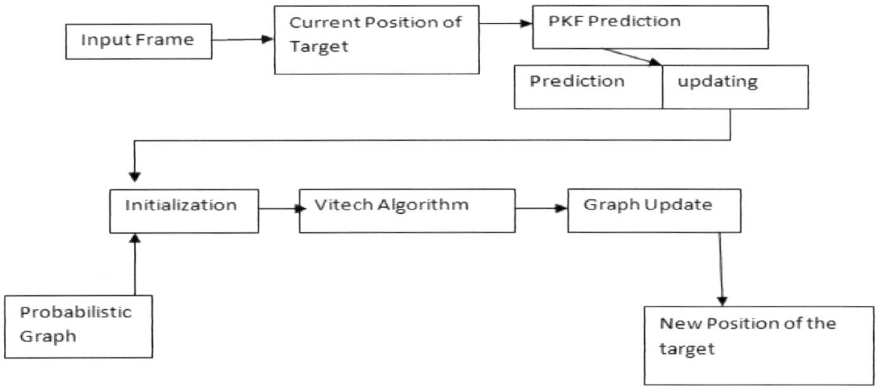

Figure 2: *The diagram represents the PKF structure in one frame. Initially, the filter is applied to calculate and update the next position.*

The inputs to the PKF are taken from the Kalman filter estimated points. These estimated points are having the newly refined.The estimated point by the Kalman filter is the input to PKF, which obtains a newly refined estimation with the help of an estimation of the next moves from the graph.

The assessed point by the Kalman channel is the contribution to PKF, which acquires a recently refined assessment with the assistance of an assessment of the following moves from the chart. The three periods of PKF could be depicted as the PKF's introduction, locate the most plausible way by the Viterbi calculation, and the chart refreshing. In the accompanying, these means are fastidiously portrayed.Since each target has an uncommon direction, directions have the non-fixed nature. This possessions is the purpose behind adding the refreshing stage to the probabilistic chart after the PKF assessment. The diagram refreshing intends to limit the separation between the current estimation yobs and the successful node Vi (a node which is like the current area of the objective) in the time k, subject to Vi,k and Vi,k−1 have the base minimum

Algorithm2:The Graph Updating stage
Step 1: Set the term α, y_{obs} as a learning terms
Step 2: Update the successful node Vi
Step 3:Calculate Vi,k=Vi,k-1+$\frac{1}{1+\alpha}(y_{obs} - V_{i,k-1})$
Step 4: Return Probabilistic graph

Algorithms 2 shows the general steps for the Graph refreshing. It is wanted to acquaint a calculation with present all the past advances.

3.2 GRNN model for action classification: For Human action recognition GRNN provides the two gates , reset gate and update gate. The task of the reset gate is to determine how add the new input frame with previous memory and update gate controls what to keep from the previous memory.GMM and Probability Kalman Filter is applied on the received input and it provides the output as moving person only. The output is given as input to GRNN . The hidden layers in the network will extract the features and prediction module is used to predict the action. The prediction can done in training and testing phases. Training phase in Gated Recurrent Neural Networks:

- Input the video frames with the moving person
- Provide the learning parameters weights and biases with arbitrary values.
- calculate the cell states of GRU via the input frames and its learning parameters values.
- Regulate the weights by an optimization function. The weights are customized by the Back propagation through the time and reduce error and loss function.

The modification is carry out by the subsequent equations.

$$W^{t+1} = W^{t} - \gamma \frac{\partial E}{\partial W^{t}} \quad \text{------------(1)}$$

$$U^{t+1} = U^{t} - \gamma \frac{\partial E}{\partial U^{t}} \quad \text{-------------(2)}$$

where γ represent the learning rate, $\frac{\partial E}{\partial W^t}$ and $\frac{\partial E}{\partial U^t}$ are the values of gradient, and E is represents Error and Loss Function. After completion of training phases, Save that trained GRNN model. In Testing phase accumulate the model using a different video. The outputs of the proposed method are the division of a probability of every class, where each class represents one person action. lastly, the test video fit in on class which has a maximum of probability.

4. Experimental Analysis

4.1 Dataset and experimental setup

The proposed method is applied on UCF Sports action data set and UCF101 and KTH datasets. UCF Sports dataset is consisting of 150 videos, each video is with resolution of 720 × 480, consisting of 10 human actions. UCF1012 dataset contains 13,320 video, with 150 frames for each video with the resolution of 320 × 240, for 101 human actions. The KTH is the mainly used public dataset for human action. It holds 6 styles of video actions and resolution of 160 × 120 (walking, jogging, running, boxing, hand-waving and handclapping). The frames number used for video classification is around 100 frames per video.

4.2 Results and discussion:

All the experiments are done on laptop with core i5 processor and code developed on python programming. The given datasets are divided as training data set and testing data set for video classification. To proposed model actually done in four steps. Initially, GMM and Probabilistic Kalman filter are applied on the given data sets to extract human motion in each frame of KTH dataset. Secondly,KNN is applied with GMM and Probabilistic Kalman Filter for person action classification.Thirdly,The GRU model is applied. For the training, in the experiment of proposed the method used 100 frame in the inputs, GRU hidden nodes are 32,batch size as 128, epochs as 30, and finally optimizer is Stochastic Gradient Descent ,loss function as categorical cross-entropy is used. lastly,150,150,1700 videos from KTH,UCF sports and UCF data sets are used for better classification.

in fact, proposed method used a mixture of deep learning methods with motion tracking. Tracking of Human in the every video frame are extracted by GMM and PKF, and the results of these methods are used in GRNN model. GRNN will extract the feature and predict human action. In the initial stage videos are divided into training and testing purposes. So to do action classification, provide the test videos as input to GRNN model.

Table 1 represents the comparison of the performance with other methods on the date sets KTH, UCF Sport and UCF 101 Human actions dataset. Certainly proposed approach, accomplished an 96. 72% accuracy rate in KTH data set, 89.98% under dataset UCF Sports and 90.31% accuracy in UCF101. On the whole, proposed method provides similar results with the best related work on datasets. Fig3 show the comparison graph between proposed method and all existing methods on KTH data set.

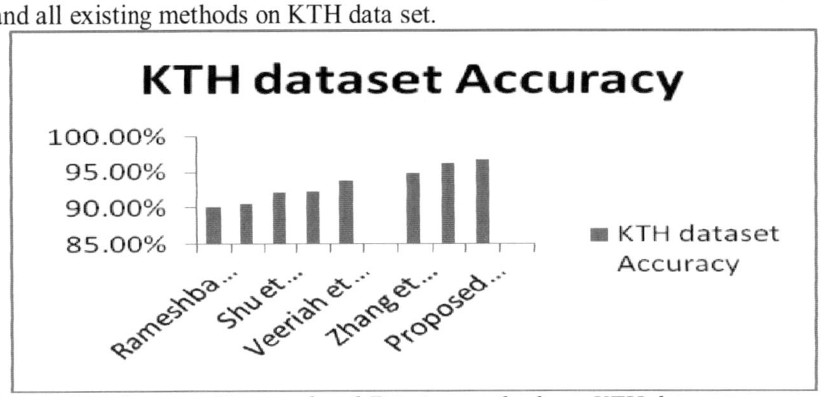

Fig3: Accuracy Comparison between Proposed and Existing methods on KTH data set

Fig4 show the comparison graph between proposed method and all existing methods on UCF data set.

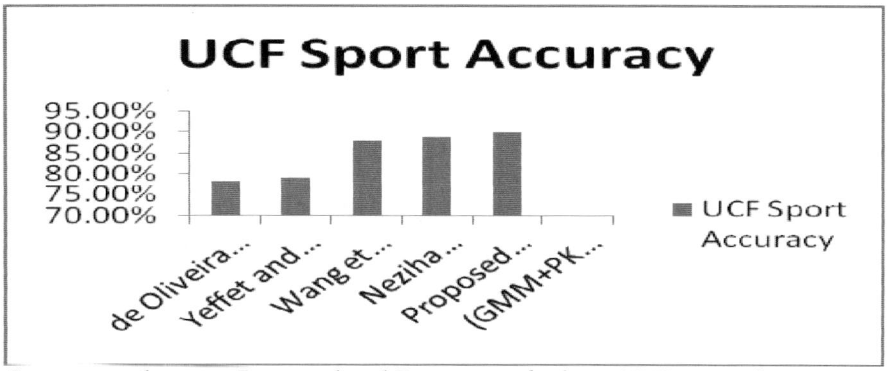

Fig4: Accuracy Comparison between Proposed and Existing methods on UCF sport data set

Table1: comparison table with all other methods on the date sets KTH, UCF Sport and UCF 101dataset

KTH dataset		UCF Sport		UCF 101	
Method	Accuracy	Method	Accuracy	Method	Accuracy
Rameshbabu et al., 2012	90.2%	de Oliveira Silva et al., 2017	78.46%	Soomro et al., 2012	44.5%
Grushin et al., 2013	90.70%	Yeffet and Wolf, 2009	79.2%	Karpathy et al., 2014	65.4%
Shu et al., 2014	92.30%	Wang et al., 2011	88%	Idrees et al., 2016	78%
Geng and Song, 2015	92.49%	Neziha Jaouedi et al.,2019	89.01%	Neziha Jaouedi et al.,2019	89.30%
Veeriah et al., 2015	93.96%	Proposed method (GMM+PKF+GRNN)	89.98%	Proposed method (GMM+PKF+GRNN)	90.31%
Zhang et al., 2017	95%				
Neziha Jaouedi et al.,2019	96.30%				
Proposed method (GMM+PKF+GRNN)	96.72%				

5. Conclusion:

The proposed work presents a new method in the field of Human action recognition. The proposed method worked on the analysis of video frame content and feature extraction.GMM and PKF are used for tracking the motion features of a human in videos frames.

Using Gated Recurrent with Recurrent neural network, other features like characteristic of the frames in videos series will be extracted. The key benefit of the proposed method is analysis and extraction of features are done in best way in the each and every frame. In Human action recognition also proposed method played the important role. The experimental results of the proposed method demonstrate the better performance of the proposed approach to gain the good classification rate. In fact proposed technique can be applied in a different fields like human activity analysis, medical fields depends on body posture analysis, intellectual interface using hand gestures and like other applications. For upcoming work, the objective is to assess proposed approach to shrink the time of video classification and to improve the classification rate of UCF Sport, UCF101 and more challenging datasets

6.References

1.Baccouche, Moez, Mamalet, Franck, Wolf, Christian, Garcia, Christophe, Baskurt, Atilla, 2011. Sequential deep learning for human action recognition. In: International Workshop on Human Behavior Understanding 2011. https://doi. org/10.1007/978-3-642-25446-8_4.

2.Cho, K., van Merienboer, B., Gulcehre, C., Bahdanau, D., Bougares, F., et al., 2014. Learning phrase representations using RNN encoder–decoder for statistica machine translation. In: Proceedings of the Empirical Methods in Natural Language Processing, Oct. 25–29, Association for Computational Linguistics,

3. Doha, Qatar, pp. 1724–1735. https://doi.org/10.3115/v1/D14-1179. Choi, Sungjoon, Kim, Eunwoo, Songhwai, 2013. Oh human behavior prediction for smart homes using deep learning. In: IEEE RO-MAN: The 22nd IEEE International Symposium on Robot and Human Interactive Communication Gyeongju, Korea, August 26–29, 2013. https://doi.org/10.1109/roman.2013. 6628440.

4.Chorowski, J.K., Bahdanau, D., Serdyuk, D., Cho, K., Bengio, Y., 2015. Attention-based models for speech recognition. Adv. Neural Inf. Process. Syst.

5.de Oliveira Silva, Vinicius, de Barros Vidal, Flavio, Romariz, Alexandre Ricardo Soares, 2017. Human action recognition based on a two-stream convolutional network classifier. In: International Conference on Machine Learning and Applications. https://doi.org/10.1109/icmla.2017.00-64.

6.Deldjoo, Yashar, Quadrana, Massimo, Elahi, Mehdi, Cremonesi, Paolo, 2017. Using mise –en – scene visual features based On MPEG7 and Deep learning for movie recommendation, ArXiv, 2017, arXiv: 1704.06109.

7.Dey, N., Dey, N., Ashour, A., Acharjee, S., 2016. Applied Video Processing in Surveillance and onitoring Systems. IGI Global.

8.Eyben, F., Wollmer, M., Schuller, B., Graves, A., 2009. From speech to letters – using a novel neural network architecture for grapheme based ASR. In: Proceedings of the IEEE Workshop on Automatic Speech Recognition and Understanding, Nov. 13-Dec. 17. IEEE Xplore Press, Merano, Italy, pp. 376–380. https://doi.org/ 10.1109/ASRU.2009.5373257.

9.Geng, Chi, Song, JianXin, 2015. Human action recognition based on convolutional neural networks with a convolutional auto-encoder. In: 5th International Conference on Computer Sciences and Automation Engineering (ICCSAE 2015) https://doi.org/10.2991/iccsae-15.2016.173.

10.Geng, C., Song, J.X., 2015b. Human action recognition based on convolutional neural networks with a convolutional auto-encoder. Proceedings of the 5th International Conference on Computer Sciences and Automation Engineering, (SAE' 15). Atlantis Press. https://doi.org/10.2991/iccsae-15.2016.173.

11.Gowsikhaa, D., Abirami, S., Baskaran, R., 2012. Automated human behavior analysis from surveillance videos: a survey. Artif. Intell. Rev. https://doi.org/10.1007/ s10462-012-9341-3.

12.Graves, A., Jaitly, N., 2014. Towards end-to-end speech recognition with recurrent neural networks. In: Proceedings of the 31st International Conference on International Conference on Machine Learning, Jun. 21–26, JMLR.org, pp. 1764– 1772.
13.Graves, A., Liwicki, M., Fernández, S., Bertolami, R., Bunke, H., et al., 2009. A novel connectionist system for unconstrained handwriting recognition. IEEE Trans. Patt. Anal. Mach. Intell. 31, 855–868. https://doi.org/10.1109/TPAMI.2008.137.
14.Grushin, A., Derek, D.M., James, A.R., Ajay Mishra, 2013. Robust human action recognition via long short-term memory. In: The 2013 International Joint Conference on Neural Networks (IJCNN), 2013, pp. 1–8.
15.Hochreiter, S., Schmidhuber, J., 1997. Long short-term memory. Neural Comput. 9, 1735–1780. https://doi.org/10.1162/neco.1997.9.8.1735.
16.Idrees, Haroon, Zamir, Amir R., Jiang, Yu-Gang, Gorbane, Alex, Laptev, Ivan, Sukthankar, Rahul, Shaha, Mubarak, 2016. The THUMOS challenge on action recognition for videos ‘‘in the wild”. Comput. Vis. Image Underst. https://doi. org/10.1016/j.cviu.2016.10.018.
17.Jang, D.S., Jang, S.W., Choi, H.I., 2002. 2D human body tracking with structural Kalman filter. Patt. Recognit. 35, 2041–2049. https://doi.org/10.1016/S0031- 3203(01)00201-1.
18.Jaouedi, Neziha, Zaghbani, Soumaya, Boujnah, Noureddine, Bouhlel, Mohamed Salim, 2016. Human motion detection and tracking. In: The 9th International Conference on Machine Vision 2016. https://doi.org/10.1117/12.2268539.
19.Jaouedi, Neziha, Boujnah, Noureddine, Htiwich, Oumayma, Bouhlel, Mohamed Salim, 2016. Human action recognition to human behavior analysis. In: The International Conference on Sciences of Electronic, Technologies of Information and Telecommunications (SETIT 2016). https://doi.org/10.1109/SETIT.2016.7939877.
20.Yang, Jian Bo, Nguyen, Minh Nhut, San, Phyo Phyo, Li, Xiao Li, Krishnaswamy, Shonali, 2015. Deep convolutional neural networks on multichannel time series for human activity recognition. In: Proceedings of the Twenty-Fourth International Joint Conference on Artificial Intelligence (IJCAI 2015), pp. 3995–4001.
21.Kale, G.V., Patil, V.H., 2016. A study of vision based human motion recognition and analysis. Int. J. Ambient Comp. Intell. (IJACI) 7 (2), 75–92.
22.Karpathy, A., Toderici, G., Shetty, S., Leung, T., Sukthankar, R., Fei-Fei, L., 2014. Largescale video classification with convolutional neural networks. IEEE Conf. Computer Vision Pattern Recog. https://doi.org/10.1109/cvpr.2014.223.
23.Kim, H., Chalidabhongse, T.H., Harwood, D., Davis, L., 2005. Real-time foregroundbackground segmentation using codebook model. Real-Time Imaging 11, 172– 185. https://doi.org/10.1016/j.rti.2004.12.004.
24. Neziha Jaouedi et.al. A new hybrid deep learning model for human action recognition, Journal of King Saud University – Computer and Information Sciences. https://doi.org/10.1016/j.jksuci.2019.09.004
25. F. Farahi and H.S. Yazdi, Probabilistic Kalman filter for moving object tracking, Signal Processing: Image Communication (2019), doi: https://doi.org/10.1016/j.image.2019.115751.

An Approach to detect fault text in articles

Pakala Prahasit Reddy, Yempati Prasheela, Avula Uday Kumar Reddy, Rajanikanth Aluvalu

Computer Science and Engineering, Vardhaman College of Engineering, Hyderabad, Telangana, India

rajanikanth.aluvalu@gmail.com

Abstract. The problem of Fault text (that contains fake news) has spread enormously in recent years. Social media is the major source for this type of news because it is easily and freely accessible to everyone and this type of news can be shared easily and rapidly, this enabled the widespread of news, so there is a need to detect this type of news. This news will mislead the users creating chaos and unrest. So the detection has become mandatory to avoid chaos and to maintain the pleasant news. We use Machine Learning Algorithms and basic Natural Language Process techniques to detect this type of news. In this work, Machine Learning Algorithms are used to parse the article and extract the summary from the article, this summary is used to predict the credibility. A passive-aggressive classifier and multinomial naive Bayes algorithms are used to train the model. A web app is created using the Flask API. One can put the URL in the web page to detect the news as "FAKE" or "REAL".

Keywords: TF-IDF vectorizer, Flask Framework, Passive Aggressive Classifier, Multinomial Naïve Bayes, Fake text, Real text.

1 INTRODUCTION

With the increase in technology, data became easily accessible to everyone free of cost. The Internet is the major source to provide freely accessible data, but the trustworthiness of information provided by them relies on a wide variety of factors. A large amount of data is produced through electronic media, online, and print media, but it is difficult to tell whether the data is genuine or not.

In the last few years, there has been an extensive increase in the use of social media and browsing the internet. Taking this as an advantage some people have started publishing incredible news to the users. In which there is a discrepancy between the title and the text of that article. In many news articles, fake news spread around us has resulted in chaos among people and are unable to predict the credibility of the news. Also, in this hectic life schedule, one does not have time to check the credible nature of the news. This type of news creates chaos and unrest in society [1]. For example, in the 2016 US elections, there are a lot of sources suggest that President Trump had won the election due the fake propaganda and also creating fake accounts on Facebook to spread propaganda to impact the opinion of voters [2].

A study portrayed that diffusion of fault text had a non-ignorable impact on USA twenty-sixteen elections [3]. A few facts on fault news in the United States:

- Around 20 crore US population gets the news updates from social media [4]
- Fault text has a lion's share on Facebook than other external sources. [5].

In recent years the study on this area has increased through Artificial Intelligence and Machine Learning. The real objective of this work is to detect whether the data contain any Faulty news (Fake news) or not. If the data contain fault text then it is termed as "FAKE" and if data doesn't contain fault text then it is termed as "REAL". This project uses efficient data mining algorithms [6][7][8] and basic natural language(text) processing techniques [9]. A subset of AI is called machine learning in which statistical methods are used to give systems the ability to learn the data without using explicit programming. Natural language processing is the study of interaction with computers and natural languages particularly dealing with the analysis of natural text [10].

In this paper data is transformed into a TF-IDF vectorizer, then Machine Learning algorithms like Passive-Aggressive classifier and Multinomial Naive Bayes classifier [7] are used to train the model.

A passive-aggressive classifier is an online learning algorithm that works as an interpreter step by step. This algorithm is best useful for analyzing social media data as it gets updated every day. One more algorithm that is in use is Multinomial Naive Bayes which is good for word counts for classification of text (NLP problems). Naive Bayes works well for detecting tag of text, in our case fake or real.
In this paper, the training process is created by using machine learning algorithms and this process is deployed using flask to further analyze the nature of news of other articles. Flask is an API that is used to create the web application, it was developed by Armin Ronacher and it is easy to learn and it has less base code to implement the simple web application [11]. Our project needs only a simple web application to detect the nature of news, so Flask API is suitable for our project.
Finally, if we enter the URL "localhost: 5000" a web page will be displayed where the URL of news is entered to detect the credibility of news.

2. RELATED STUDY

This section consists of previous studies that are already performed to which the present proposed study is related or have similar features. Numerous studies have been done that have a focus on detecting fake news. They have applied different data mining techniques for detection and achieved different probabilities for different methods.

Aswini Thota et all have proposed a process to detect the fake news using a transformed Tf-IDF combined with DNN(stands for Dense neural networks), we're capable to surpass previous process architectures by 2.25% and attained an accuracy of 93.21% on the validation set(test data) [12].

In a dissertation, the author built a model that used features related to frequency-based to train the processes including some classifiers like stochastic gradient classifier, Naïve Bayes classifier, support vector machine classifier, and an online supervised Passive Aggressive Classifier. SVM(support vector machine)and Stochastic gradient classifier algorithm combined with the Tf-Idf vector attains good accuracy. Performance metrics used for this work are classification accuracy, confusion matrix, the area under the ROC curve, and classification report. The train: the test is 68:32 is used for this work [13].

Victoria L. Rubin et all described that there exist many kinds of fault news, each with distinguished potential textual indicators. Thus, existing works plan hand-wrought features which are not only hard but strongly depends on the specific dataset and the accessibility of domain knowledge to plan the apt features [14].

3 PROPOSED MODEL

Proposed model combines Passive Aggressive Classifier (PAC) and Naïve Bayes processes to improve the accuracy of fault text detection. A feature extraction method called TF–IDF vectorizer is used in this training process. The ML, Passive Aggressive classifier is performed using the TF-IDF feature extraction approach [15]. This model is trained to classify fault news as either fake or real. A web page is created and this model is deployed to that page using Flask web framework. Our proposed model works as:

Step 1: Get the data set
Step 2: Apply data cleaning pre-processing
Step 3: Extracting the training data.
Step 4: Generate TF-IDF vectorizer
Step 5: Split the data for test and train
Step 6: Train model using PAC and Naïve Bayes

Step 7: Calculate the accuracy of the model
Step 8: Deploy the model using the Flask web framework
Trained process accuracy is calculated as
Trained process accuracy = (correctly classified samples) / (total samples)

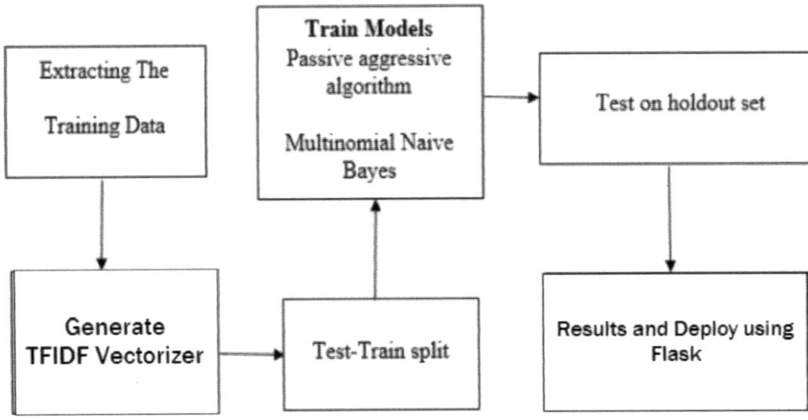

Fig 1: Architectural Model

4 EXPERIMENTAL SETUP

Experimental setup began with importing all the necessary libraries. Then the data is read into a data frame using read_csv() which is a pandas function to read the CSV files. If there is any redundant data in the dataset, the drop duplicates function is used to remove redundant tuples. Then labels are created from the data frame. Secondly, partition the dataset into a training dataset and a test dataset in the ratio 80%,20% respectively. Thirdly, initialize the Tfidfvectorizer with a maximum document frequency of 0.7 using max_df=0.7 where TF (Term Frequency) frequency of a word that appears in a document is coined as Term Frequency [16].
• tf(t,d) = (frequency of term t in a document)/(total frequency of terms in a document).

IDF(Inverse Document Frequency) is the number of times a word appears on the total set of documents [17].

•idf(t) = loge((frequency of documents present) / (frequency of documents with term t present in it)).
tfidf=tf(t,d)*idf(t)

We initialize the Passive Aggressive Classifier

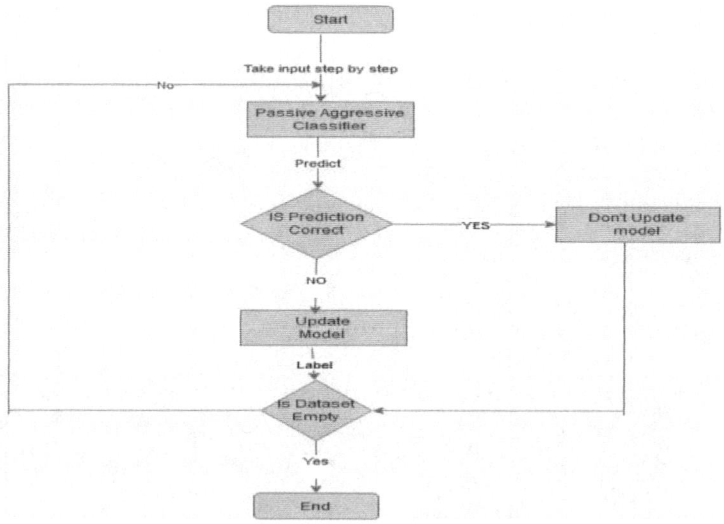

Fig 2: Flowchart of PAC

As shown in figure 2 in the Passive-Aggressive Algorithm, the input data comes in sequential order and the model is updated step-by-step. If the prediction is correct, we keep the model and do not update it. If the prediction is incorrect, we will make changes to the model and update it. We continue this process until the dataset becomes empty.

Then, predict the test set data and calculate the accuracy. Then find the confusion matrix(2*2) to find the percentages of tp,fp, fn, and then plot the confusion matrix[18].

Finally, initialize Multinomial Naïve Bayes, train the model, and dump it into the hard disk.

This Naïve Bayes classifier is derived from the Bayes theorem(A probability theorem).

P(q|a) where q is the total outcomes possible and a is the object which has to be classified, with few features [19].

$P(q|a) = P(a|q) * P(q) / P(a)$

P(q)= priori probability of class)

P(a|q)= posteriori probability of q

P(a) = priori probability of a(predictor)

Our model is trained in the filename model.py

Our flask API is used in the filename app.py

index.html is used to display web page.

We create a normal HTML page used to take a URL from the user by using Flask API. The process is trained by the naïve Bayes so it can give the custom output when custom input. We enter the localhost:5000 to see the webpage. Operate the terminal from your project directory yourself. Produce the training process using machine learning by running the model.py file. This would serialize the trained process as model.pkl or model.pickle Execute app.py using the below command to start Flask API python app.py. By default, the flask will run on port 5000.The user can enter the URL in the webpage, after entering the URL [20], we should press the predict button then we get the nature of the news as "the news is "REAL" or "the news is "FAKE"."

5 RESULT ANALYSIS

In the Existed System, we check the credibility of the news based on the publisher of the news nonetheless of the content of the news. This is never a good approach because a news source could be categorized as mistrustful and still produce a true fact. In our proposed system, We have mainly three files for this work
HTML-index.html, CSS-style.css --frontend.
model.py-training the model.
app.py-for running and initializing the FLASK API.
Our frontend looks like:

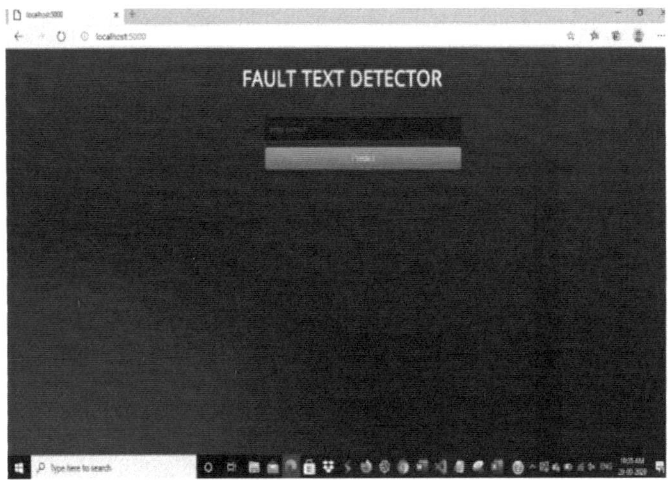

Fig 3: Webpage for user

The dataset news.csv was taken from Kaggle. Dataset was of size 7796 * 4 records. A passive-aggressive classifier is used for calculating the accuracy of the model. We have taken the 80:20 tests: train ratio. Multinomial Naïve Bayes algorithm is used to train the process and then we dump the training process into the disk. The confusion matrix plotting for the model is as shown in figure 4.
After dumping the file into disk, load the model into app.py and run the flask API. After running app.py, the terminal states to visit the browser and open the localost: 5000. The URL gives the webpage in which we can enter the URL to detect the nature of the news.

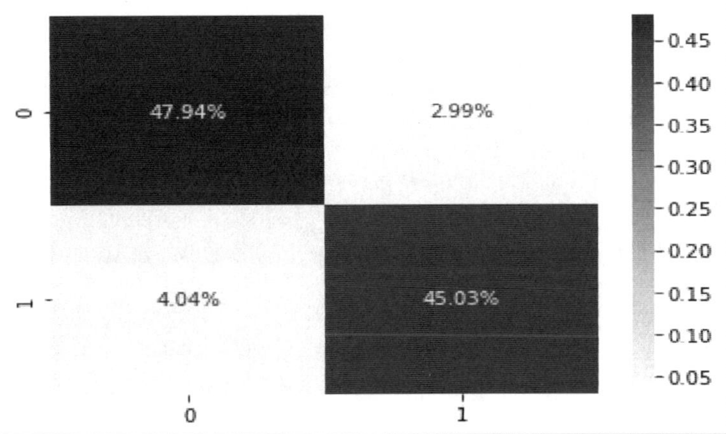

Fig 4: Confusion matrix plotting

6 CONCLUSION

Our model has analyzed the detection of fake news which is now prevalent in social media platforms and websites. We have used the TF-IDF feature extraction technique [21], Passive Aggressive Classifier, and Naïve Bayes Algorithm for training our model. We have achieved an accuracy of 93.94% by our model. Therefore by using Machine Learning techniques we can conclude that any news from a large or small dataset can be classified as fake or not fake with previous data set values in less time which helps the user to believe in particular news that appears on social media or other sources.

7. FUTURE SCOPE

The spread of misinformation has extremely harmful effects on users and the social environment. Fake news is designed to deceive the user which makes it difficult to detect it in the first place. There are many sources from which fake news is spread which causes chaos among the people and society. A future enhancement would be to identify the source of the fake news and to stop the increased spread of fake news on online platforms and in social media platforms. It would also have the capability to track and find the sources of this fake news so that we can stop the people who are trying to address the public with these malicious intentions. They would also identify the social account of the people spreading fake news and rumours so that they can stop them before it's too late. These things can engulf society with positivity and a healthy life. Finally, we want to expand this work by carrying out like analysis on a different dataset such as Twitter, Facebook, and Instagram. By categorizing fake news from social media platforms, we hope to get one step forward towards building a programmed fake news diagnosis platform.

REFERENCES

[1] David O Klein and Joshua R Mueller. Fak news: A legal perspective. 2017

[2] Alessandro Bessi and Emilio Ferrara. Social bots distort the 2016 us presidential election online discussion First Monday, 21(11), 2016.

[3] Hunt Allcott and Matthew Gentzkow. Social media and fake news in the 2016 election. In Journal of Economic Perspective, volume 31, 2017.

[4] Jeffrey Gottfried and Elisa Shearer. News Use Across Social Media Platforms 2016. Pew Research Center, 2016

[5] Craig Silverman and Lawrence Alexander. How teens in the Balkans are duping trump supporters with fake news. Buzzfeed News, 3, 2016.

[6] Rong-En Fan, Kai-Wei Chang, Cho-Jui Hsieh, Xiang-Rui Wang, and Chih-Jen Lin. Liblinear: A library for large linear classification. J. Mach. Learn. Res., 9:1871–1874, 2008.

[7] Harry Zhang. The Optimality of Naive Bayes.

[8] Mykhailo Granik, Volodymyr Mesyura, "Fake news detection using naïve Bayes classifier", Published in 2017 IEEE First Ukraine Conference (UKRON), Kyiv, Ukraine.

[9] James Thorne, Mingjie Chen, Giorgos Myrianthous, Jiashu Pu, Xiaoxuan Wang, and Andreas Vlachos. Fake news stance detection using a stacked ensemble of classifiers. In Proceedings of the 2017 EMNLP Workshop: Natural Language Processing meets Journalism, pages 80–83, 2017.

[10] Steven Bird, Ewan Klein, and Edward Loper. Natural Language Processing with Python. O'Reilly Media, 2009.

[11] Devendra Ghimire, " Comparative study on Python web frameworks: Flask and Django", A Thesis for Bachelor of Engineering, Metropolia University of Applied Sciences, 2020

[12]. Aswini Thota, Priyanka Tilak, Simrat Ahluwalia, Nibrat Lohia, "Fake News Detection: A Deep Learning Approach", A data science review, Southern Methodist University.

[13]. Aayush Ranjan, Associate Prof. Manoj Kumar, "Fake News Detection Using Machine Learning", A Dissertation for Master of Technology in Software Engineering, Delhi Technological University.

[14] Niall J Conroy, Victoria L Rubin, and Yimin Chen.Automatic deception detection: Methods for finding fake news. Proceedings of the Association for Information Science and Technology, 52(1):1–4, 2015.

[15] Vasu Agarwal, H.Parveen Sultana, Srijan Malhotra, Amitrajit Sarkar, "Analysis of Classifiers for Fake News Detection", Volume 165, 2019, Pages 377-383

[16] Simon Lorent, Ashwin Itoo, "Fake news detection using machine learning", A thesis presented for the degree of Master in Data Science, University of Liège.

[17] Stephen Robertson. Understanding inverse document frequency: On theoretical arguments for IDF, 2004

[18] Nicole O'Brien, "Machine Learning for Detection of Fake News", A thesis Master of Engineering in Electrical Engineering and Computer Science, Massachusetts Institute of Technology

[19] Faraz Ahmad and Lokeshkumar R, "A Comparison of Machine Learning Algorithms in Fake News Detection", International Journal on Emerging Technologies, Published by Research Trend, Website: www.researchtrend.net

[20] S. S. Uddin, M. Umar and K. Keerti Chennam, "Dynamic contol structure in gameplay mechanism as a solution to gaming addiction : Tartarus – Conceptual approach using a simple FSM pattern," 2020 International Conference on Computer Communication and Informatics (ICCCI), Coimbatore, India, 2020, pp. 1-8, doi: 10.1109/ICCCI48352.2020.9104164.

[21] M. T. Ahmed, M. M. Hussain and K. K. Chennam, "Designing a consensus ranking algorithm for same domain entities," 2017 2nd International Conference on Communication and Electronics Systems (ICCES), Coimbatore, 2017, pp. 12-16, doi: 10.1109/CESYS.2017.8321253.

IOP Conf. Series: Materials Science and Engineering 1042 (2021) 012033 doi:10.1088/1757-899X/1042/1/012033

5:3 compressor based neural network with LM Algorithm in Multiplier as Application

Lanka Sai Charan*, M Vaishnavi Reddy, N Pooja Reddy and J V R Ravindra

C-ACRL, Department of Electronics and Communications Engineering, Vardhaman College of Engineering, Telangana, India

*lankasaicharan@ieee.org

Abstract. In conventional approach for VLSI, implementing large numbers of operations in parallel is possible with NN. The fundamental task of a neural network hardware is not dependent on the implementation technology and are quite constructive in simulating the digital circuits. These NN representations can be incorporated in various applications where in the behaviour of these circuits are essential to get a solution for discrete problems. Therefore, supervised learning algorithms are used to train the computer software. One such algorithm is the L-M method which is one of the powerful and widely used approaches. This paper accounts to the designing and training of a neural network model as computing technique in 5:3 compressor with Lavenberg-Marquardt Algorithm. The performance parameters are significantly remarkable for real time implementation, are presented in the paper.

1. Introduction

Advancements in the neural network-based applications always need efficient hardware in terms of power, delay, and throughput. Thus, employing such hardware always challenges VLSI research. Introduction of on-chip computation, less data movement - based architectures, developing efficient MAC (Multiply and Accumulate) units and in-memory/near memory concepts [1] are employed in neural network accelerators [2] to perform such powerful computation of neural networks [3]. Each neuron is a combination is multiplying of weights and inputs, and the result is passed to the activation function. Thus, using various algorithms [4], these neurons developed network are trained and utilized to develop various advance technologies like artificial intelligence [5]. Back propagation technique is used in hardware circuits [6] as a learning method for neural network.

A neural network is designed to build a training model that comprises a synthesized model that is competent enough of comprehending the outputs at high accuracy, precision which are extensive to machine learning and artificial intelligence applications that have implemented networks as transitional blocks of functionality. The efficiency of neural networks is depicted by regression, gradient, testing error parameters specifying as the ratio of the number of error outputs by the number of exact outputs such as MSE, SSE MAE, etc. The following paper explicates about the training of a 5:3 compressor-based neural network, along with testing the exemplar, elucidating error parameters, accuracy, and few additional terms such as gradient, adjustment factor (μ) that are measuring factors of a training model.

The rest of the paper is followed by literature survey in section 2, which continuous with proposed work in section 3. Section 4 is inclusive of simulation results, the application of 5:3 compressor in digital circuits in section 5, concluding the paper in section 6.

2. Literature Survey

Research on neural networks and its applications in VLSI domain has commenced 3-4 decades back. The hardware realization of neural networks and its applications were majorly concerned with a decrease in energy, memory, and delay factors. Some of the survey points are as follows-

In [7], the performance evaluation of the on-chip interconnects using the DIL model is prepared using neural network, where performance of neural networks is evaluated at various on-chip interconnect lengths. Conventional methods consume more time in simulating repeatedly at different values of lengths. To reduce this time, neural networks were used where the performance of the system is determined using a regression algorithm. The main purpose of this paper is to provide an automatic performance evaluation method by avoiding traditional time-consuming methods.

Similarly, a high energy efficient MAC-based mixed-signal neuron architecture is presented in [8] by using a small-signal voltage mode multiplication. This design attained high energy efficiency without effecting accuracy parameters in applications of image/speech recognition. This architecture achieved energy efficiency higher than 20 times than that of prior specified neuron designs and around 3 times in case of memristor-based designs.

In [9], neural networks are used to evaluate system parameters, biasing field, and related parameters of a nano device for enhancing better performance devices with low operating power. Parallelism is applied for fast learning thus making an optimized design much faster.

3. 5:3 Compressor Neural Network design

The conception of neural networks is explored by the way biological neurons in human brain function, the way it processes data, links various scenarios. A Neural Network is precisely an interconnected artificial neuron energized 2 by activation functions, weights, training algorithms supporting the network to generate desired outputs. The classification of neural networks relies on considerations of layers, data flow, neurons exercised and their densities, depth activation filters etc [10]. This paper accounts for a training model of a 5:3 compressor that operates with the Levenberg-Marquardt algorithm for high speed and low memory.

A neural network design comprises of input, hidden, and output layers. The peripheral layers are always constant, the hidden layer count keeps altering with the requirement of application. The design evaluates the function with proper selection of weights, bias, training algorithm, and activation function.

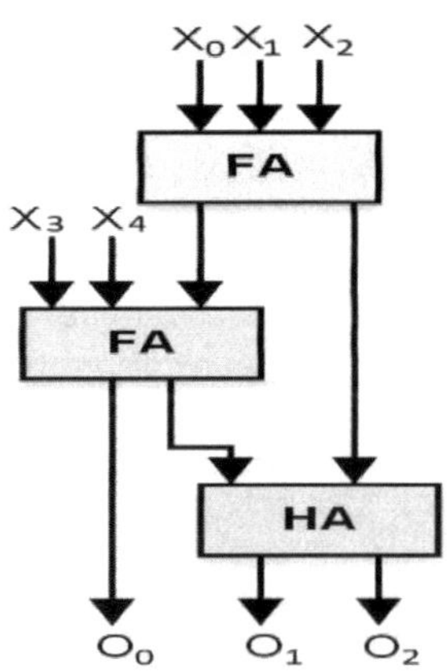

Figure-1: 5:3 Compressor design

The input for 5:3 compressor design as shown in Figure 1, of 5bit size is given as input set with 64 samples consisting of twice 32 input combinations of 5 elements as input layer to Neural network, and the output layer with 64 samples consisting of 3 elements the model is trained thoroughly, rigorously by the Levenberg-Marquardt training (TRAINLM) algorithm. TRAINLM marks its significance by producing a high-speed network that is consuming less time. In order to be an ideal representative of a network in the digital counterpart, each neuron, along with its count is taken into consideration by the algorithm. The network is chosen with a single hidden layer with 18 neurons in it. Thus, a feed-forward back propagation neural network is engaged for the digital circuit. To finalize the training model the parameter considerations are as follows - to provide the network for large training space with 85% of inputs as the training set, along with the maximum limit of epoch value of 1000 for better generalization that terminates the training when the improvement by the mean squared error (MSE) of the validation samples. The training set circuit is shown in Figure 2, the inputs are propagated to the neural network for training, testing, and validation operations. The performance of a network is decided by the error parameters along with neuron count per layer, number of iterations taken by the network out of the epochs provided, the least the value, the efficient is the design.

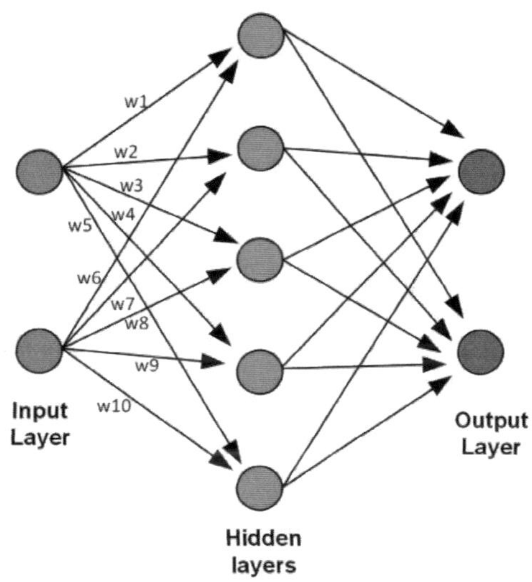

Figure 2: Neural network Model

Compared with the real-valued signals, binary signals have higher possibility rates of performance especially in the case of spiking neurons [11]. The decision making in the selection of neuron count depends on the functional techniques such as underfitting and overfitting. If the test set performance is better than the training set performance, the neuron count must be decreased gradually to equalize the performance rates apart from this observation a typical intuition by many is increasing the accuracy by increasing the neurons.

4. Simulation Results

A neural network is alleged to be completely valid when the set accuracies of train and test are approximately equal. The novel feed-forward back-propagation network for 5:3 compressor is designed in the MATLAB Neural Fitting tool R2017a. The competence of the neural network is illustrated by the parameters in the Table. 1.

While implementing a design it is important to consider the attributes that make the circuit to be impeccable, one such aspect is to find performance of a function in terms of error parameters such as MSE or SSE. The training performance rate is obtained as 6.1425E-23 at epoch 9 as shown in Figure. (3) indicating negligible value. Second, the adjustment factor μ that gets formulated at each iteration with the computation process of the Jacobian Matrix (J) for every iteration within the epoch range. Initially, the network gets started with small (μ) value and gets regulated with each iteration (μ) dec or (μ) inc. The optimization factor that directs the steepest ascent of the loss function is known as Gradient as shown in Figure (4) (5). The overall observation of the network outcomes results in elevated performance of a 5:3 compressor neural network with TRAINLM algorithm. Compressor used as transitional circuitry in electronics proves to be part of deep learning that enhances the circuit functionality.

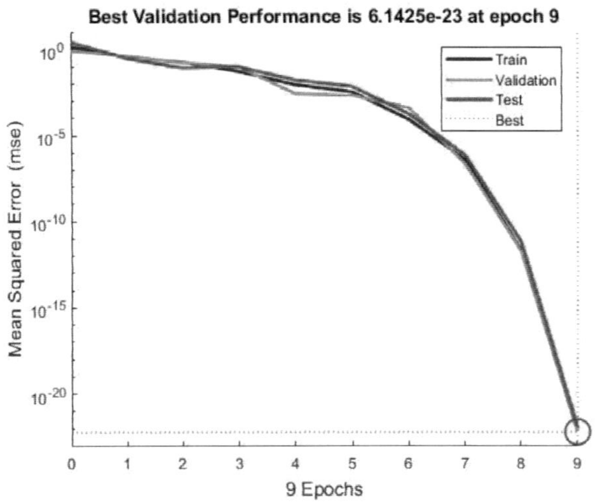

Figure 3: Performance of Neural Network

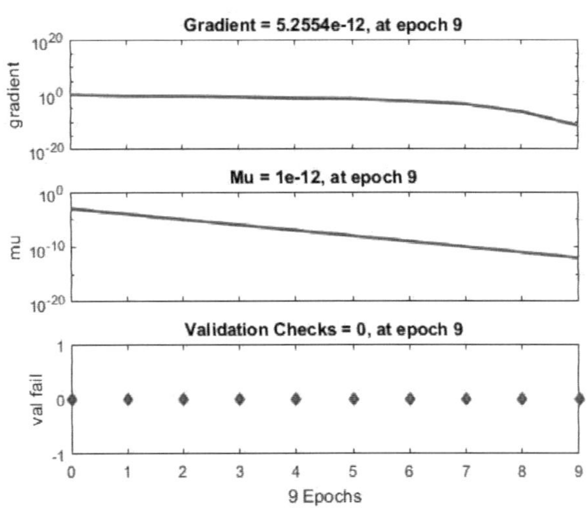

Figure 4: Training State of Neural Network

IOP Conf. Series: Materials Science and Engineering 1042 (2021) 012033 doi:10.1088/1757-899X/1042/1/012033

S. No	Parameter	Value
1	Epoch	9
2	MSE	6.142E-23
3	SSE	1.85E-09
4	(μ)	1.00E-12
5	Gradient	5.26E-12

Table-1: Parameter values for 18 neurons based neural network

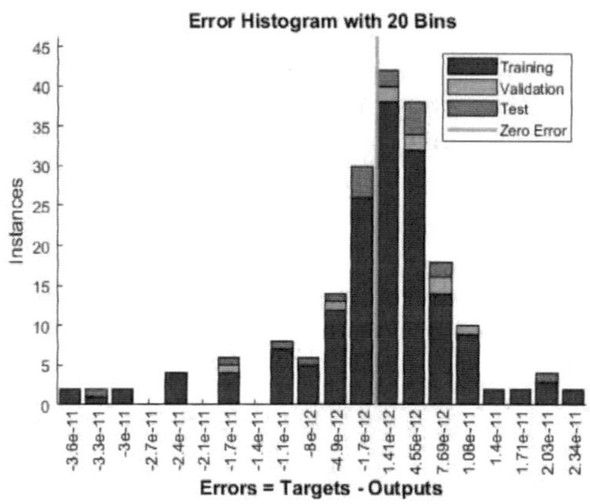

Figure 5: Error Histogram of Neural Network

S. No	Neurons	Training MSE	Testing MSE
1	8	2.87E-10	3.87E-1
2	10	2.02E-10	3E-10
3	15	1.10E-16	5.79E-17
4	18	6.142E-23	1.112E-22
5	20	2.01E-1	3.10E-1
6	22	2.036E-25	2.328E-17
7	25	2.39E-25	3.086E-25
8	30	6.60E-15	8.46E-15

Table-2: Neurons v/s MSE (Mean Squared Error)

Neural network is built by considering various neuron counts as described in the Table. 2. From the table, it is found that 18 neurons gave a better MSE values for both training and testing cases. Training MSE is valued at 6.1425E23 and testing MSE is valued at 1.112E-22. With least count of neurons and better performance, the neural network with 18 neurons in the hidden layer is considered to be a better choice from the Table. 2.

5. Application: 8 × 8 Multiplier

Machine learning (ML) is employed in many applications where large data needs to be compiled and efficient results must be extracted from it. In the field of VLSI, ML is used in finding more accurate output generating circuits. Various ML algorithms are used in finding efficient circuits by considering parameters like error distance, methods of groupings of 1's and 0's in K-maps [12] etc. Designing an efficient 8 x 8 multiplier by considering above mentioned parameters, is the main goal of this paper. Hence by maintaining the accuracy, an algorithm is proposed in building an 8x8 multiplier with the novel 5:3 compressor resulting in least error outputs.

Algorithm:

Step 1: Calculate all partial products and arrange them as shown.

Step 2: If any of the partial products is zero, consider remaining non-zero elements and proceed.

Case 1 – If there are 2 or 3 elements in the column including carry, perform normal addition.

Case 2 – If there are 4 elements in the column including carry, apply them to 4:3 compressor NN model.

Case 3 – If there are 5 elements in the column including carry, apply them to 5:3 compressor NN model. Check for inputsX3 X2 X1at iterations (4,8,15,16,23,27) for output O1 If found, complement the output obtained from NN model Check for inputs X4 X3 X2 X1 at iterations (7, 11) for output O2 If found, complement the output obtained from NN model

Else, continue.

Case 4 – If there are more than 5 elements in the column including carry, apply above cases accordingly to reduce them. Repeat Step 2 until one attains a single row.

6. Conclusion

The concept of 5:3 compressor is implemented using neural networks in MATLAB with the Lavenberg-Marquardt algorithm. Various parameters like MSE, SSE, gradient is considered during analysis and had better results for 18 neuron count in the neural network with the training MSE valued at 6.142E-23 and the testing MSE valued at 1.112E-22. After simulation of a serial-in parallel-out circuit combined with the neural network, it was found that the NN model has worked accurately as that of the digital circuit implementation of 5:3 compressor thus becoming an efficient design of a 5:3 compressor.

References

[1] Junjie Mu and Bongjin Kim. A 65nm logic-compatible embedded and flash memory for in-memory computation of artificial neural networks. In 2020 IEEE International Symposium on Circuits and Systems (ISCAS), pages 1–4. IEEE, 2020.
[2] S Rasoul Faraji, Pierre Abillama, Gaurav Singh, and Kia Bazargan. Hbucnna: Hybrid binary-unary convolutional neural network accelerator. In 2020 IEEE International Symposium on Circuits and Systems (ISCAS), pages 1–5. IEEE, 2020.
[3] Eyal Kolman and Michael Margaliot. Knowledge extraction from neural networks using the all-permutations fuzzy rule base: the led display recognition problem. IEEE transactions on neural networks, 18(3):925–931, 2007.
[4] Laura Wang and Matt Luo. Machine learning applications and opportunities in ic design flow. In 2019 International Symposium on VLSI Design, Automation and Test (VLSI-DAT), pages 1–3. IEEE, 2019.
[5] A Dinu and PL Ogrutan. Opportunities of using artificial intelligence in hardware verification. In 2019 IEEE 25th International Symposium for Design and Technology in Electronic Packaging (SIITME), pages 224–227. IEEE, 2019.
[6] Zheng Tang, OKIHIKO Ishizuka, and HIROKI Matsumoto. Backpropagation learning in analog t-model neural network hardware. In Proceedings of 1993 International Conference on Neural Networks (IJCNN-93-Nagoya, Japan), volume 1, pages 899–902. IEEE, 1993.
[7] Ajita Misra, Diksha Diksha, Yash Agrawal, and Vinay Palaparthy. Performance evaluation of on-chip interconnect system using prospective neural network design. In 2020 IEEE International Students' Conference on Electrical, Electronics and Computer Science (SCEECS), pages 1–4. IEEE, 2020.
[8] Baibhab Chatterjee, Priyadarshini Panda, Shovan Maity, Ayan Biswas, Kaushik Roy, and Shreyas Sen. Exploiting inherent error resiliency of deep neural networks to achieve extreme energy efficiency through mixed-signal neurons. IEEE Transactions on Very Large-Scale Integration (VLSI) Systems, 27(6):1365–1377, 2019.
[9] Subir Kumar Sarkar, Ankush Ghosh, MA Gautham, A Sobha Rani, and Debasis Samanta. An efficient technique of integrating parallel neural networks for faster and power efficient nanodevices for ultradense vlsi cir8 cuits. In 2007 International Workshop on Physics of Semiconductor Devices, pages 232–235. IEEE, 2007.
[10] Ying Ma and Jose C Principe. A taxonomy for neural memory networks. IEEE transactions on neural networks and learning systems, 2019.
[11] Sungho Jo, Jijun Yin, and Zhi-Hong Mao. Random neural networks with state-dependent firing neurons. IEEE transactions on neural networks, 16(4):980–983, 2005.
[12] Lavanya Maddisetti and JVR Ravindra. Machine learning based power efficient approximate 4: 2 compressors for imprecise multipliers. In 2019 32nd International Conference on VLSI Design and 2019 18th Int Conf on Embedded Systems (VLSID), pages 221–226. IEEE, 2019.

AUTHOR INDEX